QUEUEING NETWORKS WITH FINITE CAPACITY

QUEUEING NETWORKS WITH FINITE CAPACITY

Proceedings of the Second International Conference on Queueing Networks with Finite Capacity
May 28-29, 1992 and September 28-30, 1992
Research Triangle Park, NC, USA

edited by

RAIF O. ONVURAL
IBM
Department H83/675
Research Triangle Park, NC, USA

IAN F. AKYILDIZ
College of Computing
Georgia Institute of Technology
Atlanta, GA, USA

1993
NORTH-HOLLAND
Amsterdam – London – New York – Tokyo

ELSEVIER SCIENCE PUBLISHERS
Sara Burgerhartstraat 25
P.O. Box 211, 1000 AE Amsterdam, The Netherlands

ISBN: 0 444 89772 0

This book is printed on acid-free paper.

Printed in The Netherlands.

Dedication

To my wife Nur,
who has always been there
to make sure I am never *blocked*

To Professor Harry G. Perros
who introduced me to
networks with blocking

To Melih and Doruk
who will go much further

---- Raif O. Onvural

FOREWORD

The objective of this workshop was to bring together researchers from many diverse areas, including Computer Science, Computer Engineering, Operations Research, and Industrial Engineering, who are interested in the theory and applications of finite capacity queueing networks and how they relate to their respective fields.

The technical program consisted of twenty-two papers selected by the program committee out of thirty-one papers submitted. There were two keynote speeches given:*G-Networks: Product Form Queueing Networks with (or without) Instantaneous Customer Motion* by Professor Erol Gelenbe and *Discrete Time Queues with Blocking* by Professor Harry G. Perros. We would like to express our thanks to the two speakers for their speeches.

We also would like to thank the members of the program committee and the external reviewers for their meticulous and timely reviews. We are particularly grateful to Mr. Hoyt E. Bowen, IBM, for his support in covering the expenses of the workshop. Special thanks to Dr. Y.C. Liu, Mr. Chien Vu, Mr. Steve Worley, Ms. Betty Lanier, and Ms. Anne Bucki, all of IBM at Research Triangle Park, who helped us at various phases of the organization of the workshop.

We also would like to express our thanks to Dr. Nat Macon at NSF for approving a travel grant enabling us to cover the expenses of some of our speakers, and also would like to acknowledge the support of IBM, College of Computing of the Georgia Institute of Technology, the Center for Communications and Signal Processing of the North Carolina State University, the ORSA/ TIMS Applied Probability Group, and IFIP WG 7.3.

We hope that this workshop have helped towards extending the state of the art in queueing networks with finite capacity, stimulate the exchange of ideas, and create new research initiatives.

Raif O. Onvural, IBM
Ian F. Akyildiz, Georgia Tech.

CONTENTS

PROGRAM COMMITTEE

Workshop Co-Chairmen

Raif O. Onvural, USA
Ian F. Akyildiz, USA

Program Committee

T. Altiok, USA
S. Balsamo, Italy
A. Brandwajn, USA
J. Buzacott, Canada
I. Chlamtac, USA
Y. Dallery, France
A. Ganz, USA
S. Gershwin, USA
L. Gun, USA
D. Kouvatsos, England
J-Y. LeBoudec, Switzerland
H.G. Perros, USA
G. Pujolle, France
R.F. Serfozo, USA
J. MacGregor Smith, USA
H. Takagi, Japan
A. Tantawi, USA
H. Yamashita, Japan

Sponsorship

National Science Foundation
IBM, Networking Systems, Research Triangle Park
College of Computing, Georgia Institute of Technology
Center for Communications and Signal Processing, NC State University
ORSA/TIMS Applied Probability Group
IFIP W.G. 7.3

PARTICIPANTS IN THE REVIEW PROCESS

T. Altiok, Rutgers University, USA
T. Atmaca, Universite P. et M. Curie, USA
S. Balsamo, University of Modeno, Italy
A. Brandwajn, University of California-San Jose, USA
J. Buzacott, York University, Canada
R.F. Chang, IBM, USA
P.F. Chimento, IBM, USA
I. Chlamtac, University of Massachusetts-Amherst, USA
Y. Dallery, Laboratoire MASI, France
L. Donatiello, University of Modeno, Italy
A. Ganz, University of Massachusetts-Amherst, USA
S. Gershwin, Massachusetts Institute of Technology, USA
L. Gun, IBM, USA
D. Kouvatsos, University of Bradford, England
F. Lai, IBM, USA
J-Y. LeBoudec, IBM, Switzerland
H-M. Liang, IBM, USA
Y.C. Liu, IBM, USA
H.G. Perros, North Carolina State University, USA
G. Pujolle, Universite Laboratoire MASI, France
R.F. Serfozo, Georgia Institute of Technology, USA
H. Takagi, IBM, Japan
A. Tantawi, IBM-Research, USA
T.E. Tedijento, IBM, USA
I. Viniotis, North Carolina State University, USA
H. Yamashita, Sophia University, Japan
D. Yao, Columbia University, USA
L. Zhuang, University of Waterloo, Canada

Queueing Networks with Finite Capacity
R.O. Onvural and I.F. Akyildiz (Editors)

TWO QUEUES IN TANDEM WITH DROPPING OR BLOCKING OF CUSTOMERS

Duan-Shin Lee and Bhaskar Sengupta

NEC USA, Inc., C&C Research Laboratories 4 Independence Way Princeton, New Jersey 08540

Abstract

Tandem queues with blocking have been used to model communication and manufacturing systems. In this paper, we consider a two station system in which the second station has a finite buffer size and the discipline is to drop or block customers when the second station is full. While the discipline of blocking is well understood, the discipline of dropping works as follows. When a customer has received service from the first station and the second station is full, this customer simply departs from the system. This discipline is likely to be encountered in Broadband ISDN systems carrying video traffic. We allow semi-Markovian arrivals as the input to the tandem queues which can be used to capture the effect of correlations in the input process. Our method of solution consists of studying the stochastic behavior of the attained waiting time of the customer in service in the first station. From this result, one can obtain quantities of interest like sojourn-time and queue-length distributions and the probability that a customer is dropped. We illustrate our method by numerical examples.

1 Introduction

A growing body of literature has appeared in the past analyzing queues in tandem with *blocking*. Two common disciplines studied are *blocking after service (BAS)* and *blocking before service (BBS)*. In this paper, we introduce a variation of this phenomenon. In this variation (which we refer to as *dropping*), when the buffer in the second queue is full, a customer who has completed service in the first queue simply leaves the system. The examination of this discipline is motivated by the need to model Broadband ISDN systems carrying video traffic. There is considerable ongoing research in the field of high speed networks of the future, which are expected to carry video traffic. This traffic is characterized by high bandwidth requirements, high correlations in interarrival times, stringent delay requirement through the network and some tolerance for loss of information. The information in the video images is coded, packetized and transported through the network. When these packets are delivered at the receiving end, the packets are used to reconstruct the video image. There are 30 still frames of video images per second transmitted from the source. For the picture reconstruction to be reasonable, the actual delay and especially the variability of the delay must be small. Since the coding scheme employed is tolerant of some loss of information, one can implement a dropping scheme of the type mentioned earlier. This dropping scheme ensures that the overall delay through the network is well behaved, since all nodes inside the network have finite buffers.

Our basic model consists of two stations. Customers (packets) arrive in station 1 and receive service in station 1 according to the first-come-first-served (FCFS) discipline. After receiving service in station 1, a customer attempts to move to station 2 which has a finite buffer size of β (including the one in service). If station 2 is full, this customer leaves the system. Otherwise, the customer receives service in station 2 according to the FCFS discipline. We assume that the service times in stations 1 and 2 are independent and exponentially distributed with parameters μ_1 and μ_2 respectively. We also extend the basic model to the case when the service times in station 1 have the phase-type distribution and the sequence of service times in station 1 could be dependent. To model the correlated interarrival times, we assume that arrivals are according to a Markov renewal process $\{A_k, M_k; k = 0, 1, \ldots\}$. Here, A_k denotes the interarrival time between the kth and $(k+1)$st arrivals and M_k is the state of the Markov chain associated with the kth customer. Let

$$H_{ij}(x) = P(A_{k+1} \leq x, M_{k+1} = j | M_k = i) \tag{1}$$

for $i, j = 1, \ldots, n$ and let $H(x)$ denote an $n \times n$ matrix whose (i,j)th element is $H_{ij}(x)$. We refer to $H(x)$ as the semi-Markov kernel for the arrival process.

On account of the semi-Markovian arrivals and dependencies in service times, the problem stated above is more general than others studied in the literature. Further, the method of analysis used here is quite different from that of others. We first characterize the distribution of *attained waiting time* at station 1 where attained waiting time is defined to be the time spent in the system by the customer in service. From this characterization, we obtain quantities of interest like queue-length and sojourn-time distributions and the probability that a customer is dropped. In addition to the generality of the model, an interesting aspect of our solution is that both queue-length and sojourn-time distributions can be characterized in a very natural manner. In contrast, most other methods used earlier give either the queue-length or the sojourn-time distributions (but not both) and usually for less general models. We use a method due to Sengupta [12] for our analysis. We note that some extensions of the basic model are dependent phase-type service instead of independent exponentially distributed service and the incorporation of BAS or BBS instead of dropping of customers. Our method of analysis can be used to solve these problems also. We sketch some of the changes needed for these extensions. Some of the earlier research in this area was done with the assumption that the server is subject to failure. Note that the assumption of phase-type service allows us to incorporate the feature of failures in a fairly simple manner.

One of the earliest papers on tandem queues with blocking was by Hunt [5]. Some papers on exact solutions of this class of problems are by Neuts [9], Latouche and Neuts [7], Gün and Makowski [3], Bocharov [1]. We state that Gün and Makowski [4] have also proposed an efficient approximation method for more complicated systems. For the special problem in which there is no intermediate waiting room, see Kelly [6], Massey [8]. The recent book by Perros and Altiok [11] contains a number of interesting papers on related problems.

This paper has three more sections. In Section 2, we show how to characterize the distribution of attained waiting time. In Section 3, we obtain other performance measures which are of practical interest. In section 4, we outline the changes needed for some extensions to our basic model. In Section 5, we obtain some numerical solutions.

2 The Distribution of Attained Waiting Time

Let X_t denote the time spent in the system by the customer in service in station 1 at time t if station 1 is non-empty. In the analysis to follow, we characterize the distribution of X_t

by first conditioning on the event that station 1 is non-empty. This is done by deleting from consideration all intervals of time during which station 1 is empty. If station 1 is serving the kth customer at time t, let $K_t = M_k$ $(1 \leq K_t \leq n)$. Further, let L_t denote the number of customers present in station 2 $(0 \leq L_t \leq \beta)$.

A sample path of X_t is shown in Figure 1. Normally, X_t increases at a rate of 1 until service completion. When a new customer enters service, X_t takes a downward jump. There are two types of downward jumps. A type 1 downward jump occurs when one service completion is followed immediately by a service commencement. If this happens, the amount of the downward jump exactly equals the inter-arrival time. In the example of Figure 1, this happens when the 1st, 2nd, 4th and 5th customers depart. A type 2 downward jump occurs if a service completion ends a busy period for station 1 and X_t jumps down to 0. This marks the beginning of a new busy period. In the example of Figure 1, this happens when the 3rd and 6th customers depart from station 1. Note that we have deleted the idle periods of station 1, so it is assumed that the next busy period begins immediately for station 1.

We first remark that $\{X_t, L_t, K_t; t \geq 0\}$ is a Markov process. At any time t (if $L_t > 0$), a service completion can occur at rate μ_2 from station 2 and the value of L_t drops by 1. This transition is not accompanied by any change in X_t. Further, at any time t, the rate of a type 1 downward jump in X_t is μ_1. This downward jump is accompanied by an increase in L_t by 1, if $L_t < \beta$. If $L_t = \beta$, L_t remains unchanged after the downward jump and a customer is dropped. For a type 1 downward jump, the distribution of the amount of downward jump is governed by the value of K_t and the semi-Markovian kernel $H(x)$. Finally, if a busy period in station 1 ends, X_t jumps to zero and an idle period of station 1 begins, during which a pure death process at a rate of the μ_2 takes place in station 2 until the instant of the next arrival. As mentioned before, this idle period of station 1 is not explicitly modeled in the process X_t.

If we map $\{L_t, K_t\}$ into a single variable N_t, then $\{X_t, N_t; t \geq 0\}$ behaves like a Markov process studied by Sengupta [12]. Clearly, there is not a unique way of mapping the two variables L_t and K_t into a single variable N_t. However, we adopt the convention that $N_t = nL_t + K_t$. Thus N_t takes integer values between 1 and $n(\beta + 1)$ and we denote the quantity $n(\beta + 1)$ by m. Also, for any two integers η and ξ we use $\eta \circ \xi$ to denote the integer $n\eta + \xi$. Finally, let $\gamma = \beta + 1$.

Using the notation of Sengupta [12] let $A_{ij}(u)$ denote the rate at which the process $\{X_t, N_t\}$ jumps from the state $(x + u, i)$ to somewhere between (x, j) and $(x + u, j)$. Here $x > 0$, $u < x$ and $i, j = 1, \ldots, m$. Let $A(x)$ denote the $m \times m$ matrix whose (i, j)th element is $A_{ij}(x)$. Further, let $A(0-) = 0$. Then from the foregoing,

$$A_{(\eta\circ\xi)((\eta-1)\circ\xi)}(0) = \mu_2 \tag{2}$$

for $\eta = 1, \ldots, \beta$ and $\xi = 1, \ldots, n$. This transition corresponds to a departure from station 2. For all other values of the subscripts, $A_{ij}(0) = 0$. Further, for $u > 0$,

$$A_{(\eta\circ\xi)((\eta+1)\circ\zeta)}(u) = \mu_1 H_{\xi\zeta}(u) \tag{3}$$

if $\eta = 0, \ldots, \beta - 1$ and $\xi, \zeta = 1, \ldots, n$. Also, for $u > 0$,

$$A_{(\beta\circ\xi)(\beta\circ\zeta)}(u) = \mu_1 H_{\xi\zeta}(u) \tag{4}$$

$\xi, \zeta = 1, \ldots, n$. For all other values of the subscripts, $A_{ij}(u) = A_{ij}(0)$ for $u > 0$. Note that (3) corresponds to a service completion (at rate μ_1) and a downward jump of less than or equal to u. This happens when a departure from station 1 occurs, there is room in station 2 for this departure and a service commencement takes place in station 1 immediately after this departure. On the other hand, (4) corresponds to the same transition as (3) except that the departure from station 1 is dropped.

Let $g_{ij}(x)$ denote the probability that in a pure death process at rate μ_2, the population goes from i to j in x time units. Then, for $i > 0$ and $x > 0$,

$$g_{ij}(x) = \begin{cases} \dfrac{e^{-\mu_2 x}\,(\mu_2 x)^{i-j}}{(i-j)!} & \text{if } 0 < j \le i \\ \displaystyle\sum_{k=i}^{\infty} \frac{e^{-\mu_2 x}\,(\mu_2 x)^k}{k!} & \text{if } j = 0\,. \end{cases} \tag{5}$$

Let $B_{ij}(x)$ denote the rate at which $\{X_t, N_t\}$ jumps from (x, i) to $(0, j)$ for $x > 0$ and $i, j = 1, \ldots, m$. Let $B(x)$ denote the $m \times m$ matrix whose (i,j)th element is $B_{ij}(x)$. Then, for $0 \le \nu \le \eta + 1 \le \beta$, $x > 0$ and $\xi, \zeta = 1, \ldots, n$, we have

$$B_{(\eta \circ \xi)(\nu \circ \zeta)}(x) = \mu_1 \int_x^{\infty} d\,H_{\xi\zeta}(u)\, g_{\eta+1,\nu}(u - x) \tag{6}$$

and for $\nu \le \beta$, $x > 0$ and $\xi, \zeta = 1, \ldots, n$, we have

$$B_{(\beta \circ \xi)(\nu \circ \zeta)}(x) = \mu_1 \int_x^{\infty} d\,H_{\xi\zeta}(u)\, g_{\beta,\nu}(u - x). \tag{7}$$

For all other values of the subscripts, $B_{ij}(x) = 0$ for $x > 0$. Note that (6) corresponds to the case in which a departure from station 1 ends the busy period in station 1 and this departure has room available in station 2. The busy period ends when the customer in service has an attained waiting time of x, which means that his sojourn-time must also equal x. The interarrival time between this customer and the next customer is u, which implies that the idle period in station 1 is of duration $u - x$. During this idle period, the population of station 2 drops from $\eta + 1$ to ν. The integration of u is over the interval (x, ∞). This is sufficient to guarantee that the busy period in station 1 ends. On the other hand, (7) corresponds to the same transition except that the departing customer from station 1 is dropped.

Before continuing further, we now show that the family of matrices $A(u)$ and $B(x)$ can be written in a Kronecker product form (denoted by the symbol $\otimes$). This notation enables us to carry out the algebra quite easily (see Neuts [10]). Let us define a $\gamma \times \gamma$ matrix Δ^0 whose (i,j)th element Δ^0_{ij} is given by

$$\Delta^0_{ij} = \begin{cases} 1 & \text{if } 1 \le j = i - 1 \le \beta \\ 0 & \text{otherwise} \end{cases} \tag{8}$$

For any integer ν, let I_ν denote the identity matrix of size ν. Then, from (2),

$$A(0) = \Delta^0 \otimes \mu_2 I_n\,. \tag{9}$$

Let Δ^1 denote a $\gamma \times \gamma$ matrix whose (i,j)th element is given by

$$\Delta^1_{ij} = \begin{cases} 1 & \text{if } 2 \le j = i + 1 \le \beta \\ 1 & \text{if } i = j = \beta \\ 0 & \text{otherwise}\,. \end{cases} \tag{10}$$

Then from (3) and (4), for $u > 0$,

$$dA(u) \;=\; \mu_1 \, \Delta^1 \otimes dH(u) \tag{11}$$

and from (9) and (11)

$$A(u) \;=\; \Delta^0 \otimes \mu_2 I_n \;+\; \mu_1 \, \Delta^1 \otimes H(u)\,. \tag{12}$$

Let $\Delta^2(x)$ denote a family of $\gamma \times \gamma$ matrices whose (i,j)th element is given by

$$\Delta^2_{i,j+1}(x) \;=\; \begin{cases} g_{i,j}(x) & \text{if } 1 \leq i \leq \beta \quad \text{and } j = 0,\ldots,i, \\ g_{\beta,j}(x) & \text{if } i = \beta+1 \quad \text{and } j = 0,\ldots,\beta \text{ and} \\ 0 & \text{otherwise}\,. \end{cases} \tag{13}$$

Then from (6) and (7), we have

$$B(x) = \mu_1 \int_x^{\infty} \Delta^2(u-x) \otimes dH(u) \tag{14}$$

Let e_ν denote a column vector of ones of dimension ν. As a check, we will now verify that $A(x)e_m + B(x)e_m$ is independent of x. This condition is necessary for the theory in the paper by Sengupta [12] to be applicable. From (12), we have

$$A(x)e_m = \mu_2 \, \Delta^0 e_\gamma \otimes e_n + \mu_1 \, e_\gamma \otimes H(x)e_n \tag{15}$$

and since $\Delta^2(x)e_\gamma \;=\; e_\gamma$,

$$B(x)e_m = \mu_1 e_\gamma \otimes \int_x^{\infty} dH(u)e_n\,. \tag{16}$$

From (15) and (16), we have

$$A(x)e_m + B(x)e_m \;=\; -De_m \tag{17}$$

where D is a diagonal matrix given by

$$D = \Delta^3 \otimes I_n \tag{18}$$

and Δ^3 is a $\gamma \times \gamma$ diagonal matrix with

$$\Delta^3_{ij} \;=\; \begin{cases} -\mu_1 & \text{if } i = j = 1 \\ -(\mu_1+\mu_2) & \text{if } 2 \leq i = j \leq \gamma \\ 0 & \text{otherwise}\,. \end{cases} \tag{19}$$

The stability condition for the Markov process $\{X_t, N_t; t \geq 20\}$ is given by

$$\alpha \int_0^{\infty} u dA(u)e_m > 1 \tag{20}$$

where α is the $1 \times m$ invariant probability vector of the generator $A(\infty)+D$, i.e., $\alpha(A(\infty)+D) = 0$ and $\alpha e_m \;=\; 1$. We note that $H(\infty)$ is the transition probability matrix of the Markov chain $\{M_k; k = 0,1,\ldots\}$. Let the $1 \times n$ vector α_2 denote its steady state distribution, i.e., $\alpha_2 H(\infty) = \alpha_2$ and $\alpha_2 \, e_n \;=\; 1$. Further, let λ^{-1} denote the stationary mean interarrival time, i.e.,

$$\lambda^{-1} = \alpha_2 \int_0^\infty u dH(u) e_n \,. \tag{21}$$

Theorem 1 *The Markov process $\{X_t, N_t; t \geq 0\}$ is positive recurrent if and only if $\lambda\mu_1^{-1} < 1$.*

Proof: Consider the $\gamma\times\gamma$ matrix $\mu_2\Delta^0 + \mu_1\Delta^1 + \Delta^3$. Clearly, from the definition of Δ^0, Δ^1 and Δ^3, the diagonal entries are less than 0 and the off-diagonal entries are greater than or equal to zero. Further,

$$(\mu_2\Delta^0 + \mu_1\Delta^1 + \Delta^3)e_\gamma = \begin{bmatrix} 0 \\ \mu_2 \\ \vdots \\ \mu_2 \end{bmatrix} + \begin{bmatrix} \mu_1 \\ \mu_1 \\ \vdots \\ \mu_1 \end{bmatrix} + \begin{bmatrix} -\mu_1 \\ -\mu_1 - \mu_2 \\ \vdots \\ -\mu_1 - \mu_2 \end{bmatrix} = 0 \,.$$

So, $\mu_2\Delta^0 + \mu_1\Delta^1 + \Delta^3$ is a generator of a Markov process and let the $1\times\gamma$ vector α_1 denote its invariant distribution, i.e., $\alpha_1(\mu_2\Delta^0 + \mu_1\Delta^1 + \Delta^3) = 0$ and $\alpha_1 e_\gamma = 1$. We now verify that α in (20) equals $\alpha_1 \otimes \alpha_2$. From the foregoing,

$$\begin{aligned} 0 &= \alpha_1(\mu_2\Delta^0 + \mu_1\Delta^1 + \Delta^3)\otimes\alpha_2 \\ &= \alpha_1\mu_2\Delta^0 \otimes \alpha_2 I_n + \alpha_1\mu_1\Delta^1 \otimes \alpha_2 H(\infty) + \alpha_1\Delta^3 \otimes \alpha_2 I_n \\ &= (\alpha_1 \otimes \alpha_2)(\mu_2\Delta^0 \otimes I_n + \mu_1\Delta^1 \otimes H(\infty) + \Delta^3 \otimes I_n) \\ &= (\alpha_1 \otimes \alpha_2)(A(\infty) + D) \text{ (using (12) and (18))} \end{aligned}$$

from which it follows that $\alpha = \alpha_1 \otimes \alpha_2$ since there is a unique invariant probability vector for the generator $A(\infty) + D$. Now using this in (20), we have

$$\begin{aligned} \alpha \textstyle\int_0^\infty u dA(u) e_m &= \mu_1 \alpha \textstyle\int_0^\infty u \Delta^1 e_\gamma \otimes dH(u) e_n && \text{(from (11))} \\ &= \mu_1(\alpha_1 \otimes \alpha_2)(e_\gamma \otimes \textstyle\int_0^\infty u dH(u) e_n) && \text{(from } \Delta^1 e_\gamma = e_\gamma) \\ &= \mu_1 \lambda^{-1} && \text{(from (21))} \end{aligned}$$

from which the result follows. □

Let us denote the steady state density of $\{X_t, N_t; t \geq 0\}$ by $\pi_i(x)$, i.e.,

$$\pi_i(x) = \lim_{t\to\infty} \frac{1}{t} \int_0^t P(u < X_u \leq u + du, N_u = i) du \,.$$

Let $\pi(x)$ denote a $1\times m$ vector whose ith element is $\pi_i(x)$.

Theorem 2 *If $\{X_t, N_t; t \geq 0\}$ is positive recurrent, then $\pi(x) = \pi(0) \exp(Tx)$ for $x > 0$, where the $m\times m$ matrix T is the minimal solution of the non-linear matrix equation*

$$T = D + \int_{0-}^\infty \exp(Tx) dA(x) \triangleq \mathcal{F}(T)$$

and $\pi(0)$ satisfies $\pi(0) = \pi(0) \int_0^\infty \exp(Tx) B(x) dx$ and $\pi(0) T^{-1} e_m = -1$.

Proof: Follows from the description of $A(x), B(x)$ and the results in Section 2 of Sengupta [12]. □

Remark Note that the method of solving for T is by iteration, i.e., let $T_0 = D$ and $T_{n+1} = \mathcal{F}(T_n)$. Then, $T_n \to T$ as $n \to \infty$ and the convergence is monotone in the sense that $T_n \leq T_{n+1}$ for all n.

3 Other Performance Measures

In this section, we find the distributions of quantities which are likely to be of interest to practitioners. In particular, we find some sojourn-time distributions, queue-length distributions and the probability that a customer is dropped.

Theorem 3 *Let $W^1(x)$ denote the distribution of sojourn-time at station 1. Then,*

$W^1(x) = 1 + \pi(0)\, T^{-1}\, \exp(Tx)e_m$.

Proof: We first note that station 1 behaves like a G/GI/1/∞ queue with exponentially distributed service times. The residual service time is exponentially distributed. So from Corollary 4.5.4 of Franken et al [2], the conditional distribution of the virtual waiting time given that it is positive is the same as that of the sojourn-time. From Sengupta [13], the virtual waiting time and the attained waiting time have the same distribution. It now follows that the density of the sojourn-time is given by $\pi(x)e_m$ and the result follows from Theorem 2. □

Theorem 4 *Let $\tilde{W}(s)$ denote the Laplace-Stieltjes Transform (LST) of the distribution of sojourn time in the two stations. Then, $\tilde{W}(s) = \pi(0)(sI_m - T)^{-1}(u(s) \otimes e_n)$ where $u(s)$ is a $\gamma \times 1$ vector whose ith element $u_i(s)$ is given by $[\mu_2/(\mu_2+s)]^i$ if $i < \gamma$ and 1 if $i = \gamma$.*

Proof: From Figure 1, we note that the downward jumps in X_t are governed by a Poisson process at rate μ_1. Therefore, by the PASTA theorem (Wolff [15]), the departures from station 1 see time averages of $\{X_t, N_t; t \geq 0\}$ just prior to their departure. This, incidentally, is a different proof of Theorem 3. Given that there are $i-1$ customers in station 2 when the departure occurs from station 1, the LST of sojourn-time distribution in station 2 is given by $u_i(s)$ for $i = 1, \ldots, \beta$. Note that for $i = \gamma$, the customer departs immediately and so, $u_\gamma(s) = 1$. Now the result of the theorem follows from the observation that the time spent in station 1 is conditionally independent of the time spent in station 2 given that we know the number in station 2 at the departure epoch from station 1. □

Theorem 5 *The probability that a randomly chosen customer is dropped (denoted by P_D) is given by the γth element of the $1\times\gamma$ vector $-\pi(0)T^{-1}(I_\gamma \otimes e_n)$.*

Proof: Note that a customer is dropped when a downward jump occurs and station 2 has β customers in it already. From arguments used in the previous theorem, the downward jumps occur according to a Poisson process and therefore, the probability that a downward jump "sees" β customers in station 2 is given by

$$P_D = \sum_{\xi=1}^{n} \int_0^\infty \pi_{\beta o \xi}(x)dx \, .$$

Clearly P_D is γth element of the $1\times\gamma$ vector $\int_0^\infty \pi(x)dx(I_\gamma \otimes e_n) = -\pi(0)T^{-1}(I_\gamma \otimes e_n)$. □

Let q_j denote a $1\times\gamma$ vector whose ith element q_{ji} is the joint probability that at a departure epoch of station 1, there are j customers in station 1 and i customers in station 2.

Theorem 6 *The joint queue-length distribution at departure epochs at station 1 is given by $q_j = -\pi(0)T^{-1}R^j(I_m - R)(I_\gamma \otimes e_n)$ where the R is the mxm matrix given by $R = \int_0^\infty \exp(Tx)(I_\gamma \otimes dH(x))$.*

Proof: Let $H^{(j)}(x)$ denote the j-fold convolution of $H(x)$ with itself. From Theorem 6 of Sengupta [14],

$$R^j = \int_0^\infty \exp(Tx)(I_\gamma \otimes dH^{(j)}(x)) \,. \tag{22}$$

Let the sojourn-time for a customer in station 1 be x. Then, the queue-length in station 1 at the departure epoch of this customer from station 1 exactly equals the number of arrivals in $(0, x)$. Further, the probability that there are j arrivals in $(0, x)$ given that the state of the Markov-renewal process for this departure is i is given by the ith element of $(H^{(j)}(x) - H^{(j+1)}(x))e_n$. These together with the fact that departures "see" time averages of the $\{X_t, N_t\}$ process, we have

$$q_j = \int_0^\infty \pi(x)\{I_\gamma \otimes (H^{(j)}(x) - H^{(j+1)}(x))e_n\}dx \,.$$

Integrating this by parts and using (22), gives the desired result. □

Corollary 1 *At station 1, the mean sojourn-time* (σ_1) *and mean number of customers* (r_1) *are given by* $\sigma_1 = \pi(0)T^{-2}e_m$ *and* $r_1 = \lambda\sigma_1$ *. At station 2, for customers who are not dropped, the mean sojourn-time* (σ_2) *and the mean number of customers* (r_2) *are given by* $\sigma_2 = \pi(0)T^{-1}(u'(0) \otimes e_n)/(1 - P_D)$ *and* $r_2 = (1 - P_D)\lambda\sigma_2$ *.*

Proof: The result for station 1 follows from Theorem 3 and Little's law. The result for σ_2 follows by first taking a derivative of $\tilde{W}(s)$ (Theorem 4) with respect to s, and letting $s \to 0+$. Then we subtract σ_1 from this and note the result is the sojourn-time in station 2 including those that were dropped. So, dividing this by $(1 - P_D)$ gives the mean sojourn-time for in station 2 for customers who were not dropped. The mean number of customers in station 2 is obtained by Little's law and the observation that $(1 - P_D)\lambda$ is the arrival rate of customers in station 2. □

4 Phase-Type Distribution and Blocking Problems

In this section, we sketch the changes that would be needed for three problems. The first is a problem in which the sequence of service times in station 1 are dependent and have phase-type distributions. The second is blocking before service and the third is blocking after service. For each of these problems, we first show how to obtain the family of matrices $A(x)$ and $B(x)$. At the end of this section, we show a general method of obtaining the sojourn-time distribution in station 1 for all three problems. This result parallels Theorem 3 in Section 3 for the problems considered. Other results presented in Theorems 4 through 6 can be obtained with a little extra effort.

Let us first generalize the problem considered to include phase-type service in station 1. Let (P, S) be a representation of the dependent phase-type distributions for the service times. The $c \times c$ matrix S denotes the sub-generator of the Markov process that governs the service times. Let S^* denote a $c \times 1$ vector satisfying $Se = -S^*$. If we let s_i^* denote the ith element of S^*, then s_i^* is the rate at which a service completion takes place from the state i. P is a $c \times c$ transition probability matrix, where P_{ij} represents the probability that the service of the next customer begins the state j given that the service of the previous customer ended in state i. In this mechanism, the starting state of service for one customer depends on the ending state of service of the previous customer through the matrix P. This introduces dependencies between successive service times. For this problem, the Markov process of interest is $\{X_t, L_t, K_t, J_t; t \geq 0\}$ where J_t is the state of the service process ($1 \leq J_t \leq c$). We let $m = (\beta + 1)nc$ and we map $\{L_t, K_t, J_t\}$ into a single variable $N_t = c(nL_t + K_t - 1) + J_t$. Finally, for three integers η, ξ and

χ we denote by $\eta \circ \xi \circ \chi$ the number $c(n\eta + \xi - 1) + \chi$. With this transformation, a knowledge of N_t is equivalent to that of L_t, K_t and J_t and *vice versa*. It is now possible to show that for $\eta = 1, \ldots, \beta$, $\xi = 1, \ldots, n$ and $\chi = 1, \ldots, c$, $A_{(\eta\circ\xi\circ\chi)((\eta-1)\circ\xi\circ\chi)}(0) = \mu_2$ and for $\tau \neq \chi$, $A_{(\eta\circ\xi\circ\chi)(\eta\circ\xi\circ\tau)}(0) = s_{\chi\tau}$. For all other values of the subscripts $A_{ij}(0) = 0$. Note that the first transition above occurs when a service completes in station 2 and the second when there is state change in the service at station 1 but no service completion actually takes place. Now, for $u > 0$, $\eta = 0, \ldots, \beta - 1$, $1 \leq \xi, \zeta \leq n$ and $1 \leq \chi, \tau \leq c$, we have $A_{(\eta\circ\xi\circ\chi)((\eta+1)\circ\zeta\circ\tau)}(u) = s_\chi^* P_{\chi\tau} H_{\xi\zeta}(u)$. Further, $A_{(\beta\circ\xi\circ\chi)(\beta\circ\zeta\circ\tau)}(u) = s_\chi^* P_{\chi\tau} H_{\xi\zeta}(u)$. For all other values of the subscripts, $A_{ij}(u) = A_{ij}(0)$ for $u > 0$. Note that the first transition above corresponds to a service completion followed by an immediate service commencement and no dropping. The second transition above corresponds to the case of a dropped customer. Now for the $B(x)$ matrix, for $x > 0$, $0 \leq \nu \leq \eta + 1 \leq \beta$, $1 \leq \xi, \zeta \leq n$ and $1 \leq \chi, \tau \leq c$, we have $B_{(\eta\circ\xi\circ\chi)(\nu\circ\zeta\circ\tau)}(x) = s_\chi^* P_{\chi\tau} \int_x^\infty dH_{\xi\zeta}(u) g_{\eta+1,\nu}(u - x)$. Finally, for $\nu \leq \beta$, $x > 0$, $1 \leq \xi, \zeta \leq n$ and $1 \leq \chi, \tau \leq c$, we have $B_{(\beta\circ\xi\circ\chi)(\nu\circ\zeta\circ\tau)}(x) = s_\chi^* P_{\chi\tau} \int_x^\infty dH_{\xi\zeta}(u) g_{\beta,\nu}(u - x)$. For all other values of the subscripts, $B_{ij}(x) = 0$ for $x > 0$. The first transition above corresponds to an end of a busy period in station 1 with no dropping of customers while the second transition corresponds to the same case but with a dropped customer. This completes our discussion of the dependent phase-type distribution in station 1.

The second problem we consider is that of blocking before service. In this discipline, the server in station 1 blocks a customer in station 1 if station 2 is full. Therefore, service commences in station 1 only if an empty position is available in station 2. We assume that the service time in station 1 is exponentially distributed with rate μ_1. For $\eta = 1, \ldots, \beta$ and $\xi = 1, \ldots, n$, we have $A_{(\eta\circ\xi)((\eta-1)\circ\xi)}(0) = \mu_2$. This transition corresponds to a departure from station 2. For all other subscripts, $A_{ij}(0) = 0$. Next, for $u > 0$, $\eta = 1, \ldots, \beta - 1$, $1 \leq \xi, \zeta \leq n$, we have $A_{(\eta\circ\xi)((\eta+1)\circ\zeta)}(u) = \mu_1 H_{\xi\zeta}(u)$. For all other values of the subscripts, $A_{ij}(u) = A_{ij}(0)$ for $u > 0$. This transition corresponds to a service completion from station 1, followed immediately by a service commencement. Finally, when a busy period ends, we have for $x > 0$, $0 \leq \nu \leq \eta + 1 \leq \beta$ and $1 \leq \xi, \zeta \leq n$, $B_{(\eta\circ\xi)(\nu\circ\zeta)}(x) = \mu_1 \int_x^\infty dH_{\xi\zeta}(u) g_{\eta+1,\nu}(u - x)$. For all other values of the subscripts, $B_{ij}(x) = 0$ for $x > 0$. This ends our description of blocking before service.

Finally, we consider blocking after service. In this discipline, a customer on reaching the head of the line in station 1, commences service. However, on service completion, the server in station 1 is blocked if there is no waiting room in station 2. To solve this problem, we need to augment the state space of L_t by the element $\beta + 1$. If $L_t = \beta + 1$, it implies that there are β customers in station 2 and the server in station 1 is blocked. For all other values of L_t between 0 and β, the server in station 1 is free to serve a customer. Note that this definition of the state space gives $m = n(\beta + 2)$. We now have for $\eta = 1, \ldots, \beta$ and $\xi = 1, \ldots, n$, $A_{(\eta\circ\xi)((\eta-1)\circ\xi)}(0) = \mu_2$ and for $\xi = 1, \ldots, n$, $A_{(\beta\circ\xi)((\beta+1)\circ\xi)}(0) = \mu_1$. The first transition above corresponds to a departure from station 2 whereas the second one corresponds to a service completion in station 1 which results in blocking. For all other values of the subscripts, $A_{ij}(0) = 0$. Now, for $u > 0$, $\eta = 1, \ldots, \beta - 1$ and $1 \leq \xi, \zeta \leq n$, we have $A_{(\eta\circ\xi)((\eta+1)\circ\zeta)}(u) = \mu_1 H_{\xi\zeta}(u)$. This corresponds to a service completion from station 1 immediately followed by a service commencement. Further, for $1 \leq \xi, \zeta \leq n$ and $u > 0$, $A_{((\beta+1)\circ\xi)(\beta\circ\zeta)}(u) = \mu_2 H_{\xi\zeta}(u)$ which corresponds to unblocking server 1 as a result of a departure from station 2. For all other values of the subscripts, $A_{ij}(u) = A_{ij}(0)$ for $u > 0$. Finally, we distinguish between two cases when the busy period ends in station 1. We first consider the case when the busy period ends due to a service completion in station 1. For $x > 0$, $0 \leq \nu \leq \eta + 1 \leq \beta$ and $1 \leq \xi, \zeta \leq n$, we have $B_{(\eta\circ\xi)(\nu\circ\zeta)}(x) = \mu_1 \int_x^\infty dH_{\xi\zeta}(u) g_{\eta+1,\nu}(u - x)$. Next we consider the case when the busy period ends because a service completion occurs in station 2 and as a result, server 1 becomes unblocked. For $x > 0$, $\nu = 0, 1, \ldots, \beta$ and $1 \leq \xi, \zeta \leq n$, we have

$B_{((\beta+1)\circ\xi)(\nu\circ\zeta)}(x) = \mu_2 \int_x^\infty dH_{\xi\zeta}(u) g_{\beta\nu}(u-x)$. For all other values of the subscripts, $B_{ij}(x) = 0$ for $x > 0$. This ends our discussion of blocking after service.

We end this section with a result that is applicable to all the problems considered so far. We find the sojourn-time distribution in station 1. The common thread in all the problems considered so far is that the ith element of the $m \times 1$ vector $(A(x) + B(x))e_m$ for any $x > 0$ represents the total rate at which state changes can take place when $N_t = i$. It is readily verified that this is independent of x in all problems considered. Let us denote this $m\times 1$ vector by $-De_m$. Further, the ith element of $A(0)e_m$ represents the rate at which state changes can take place when $N_t = i$ which also have the property that X_t does not have any accompanying change. Thus, $[-D - A(0)]e_m$ is an $m \times 1$ vector whose ith element represents the rate of departures from station 1 if $N_t = i$. Let G be an $m \times m$ diagonal matrix whose ith diagonal element g_i is the ith element of the vector $-[D + A(0)]e_m$. Let $W(x)$ represent a $1 \times m$ vector whose jth element is $W_i(x)$. Here $W_i(x)$ is the probability that the sojourn-time is less than or equal to x and the value of N_t immediately before a downward jump is i. The following theorem is now a generalization of Theorem 3.

Theorem 7 *If* $\{X_t, N_t; t \geq 0\}$ *is positive recurrent,* $W(x) = c\pi(0)T^{-1}(\exp(Tx) - I)G$ *where* $c = -[\pi(0)T^{-1}Ge_m]^{-1}$.

Proof: Immediately prior to a downward jump, X_t and N_t represent the sojourn-time of the customer who is about to depart from station 1 and the phase of the system respectively. So $W(x)$ is the distribution of X_t and N_t immediately before a downward jump. Clearly, the probability that the downward jump occurs from state $N_t = i$ is proportional to g_i. Thus, $W(x) = c\int_0^x \pi(u)\,du G$. The constant c can be evaluated from the relationship $W(x)e_m \to 1$ as $x \to \infty$. □

5 Numerical Example

In this section, we report a numerical example. The problem considered is the one solved in Section 3 and both stations have exponential service times with $\mu_1 = \mu_2$, $\beta = 5$ and $n = 2$. The elements of the semi-Markovian kernel $H(x)$ are given by:

$$
\begin{aligned}
H_{11}(x) &= p(1-\exp(-0.2x)) \\
H_{12}(x) &= \begin{cases} 0 & x < 1 \\ (1-p)(x-0.5) & 1 \leq x < 8 \\ 1-p & x \geq 8 \end{cases} \\
H_{21}(x) &= \begin{cases} 0 & x < 0.4 \\ 0.3(1-p) & 0.4 \leq x < 0.8 \\ 0.9(1-p) & 0.8 \leq x < 1.2 \\ 1-p & x \geq 1.2 \end{cases} \\
H_{22}(x) &= p(1-\exp(-x)),
\end{aligned}
$$

where p is a parameter.

Note that the distribution of interarrival times can be exponential, uniform or general discrete depending on the state of the Markov-renewal process. We first compute T through the iteration procedure outlined in the remark at the end of Section 2 (see [14] for additional details). Then we compute $\pi(0)$ by Theorem 2. The distribution and the mean of the sojourn-time, the dropping probability and the queue-length distribution can all be easily found using the results of Section 3.

First we set $p = 0.8$. In Figure 2, we show the sojourn-time distribution at station 1 for $\rho = 60\%$, 80% and 90%, where $\rho = \lambda/\mu_1$ is the utilization of station 1. Figure 3 shows three mean sojourn-times; for the system and for the two stations separately. The mean sojourn-time of the system is defined as $\sigma = \sigma_1 + (1 - P_D)\sigma_2$ which is the sojourn-time of a randomly chosen customer (whether dropped or not). So σ is the mean of the distribution whose LST is $\tilde{W}(s)$ (see Theorem 4). With this definition, the sojourn-times of the two individual stations do not add to the sojourn-time of the system. On Figure 4 we show the dropping probability as a function of the utilization ρ.

The conditional sojourn-time distribution at station 1 given the state of the semi-Markov process at arrival epochs can be easily derived from Theorem 7. Specifically, the conditional sojourn-time distribution at station 1 given that the customer arrives when the semi-Markov process is in state j is given by $Q_j(x)/Q_j(\infty)$, where $Q_j(x)$ is the jth element of the $1 \times n$ vector $W(x)e_\gamma \otimes I_n$. Let the mean sojourn-time conditioned on state j of the semi-Markov process be denoted by ϕ_j. Figure 5 shows the conditional sojourn-time distributions at station 1 for $\rho = 80\%$ and $p = 0.95$. We show the conditional mean sojourn times as functions of ρ for $p = 0.8$ and $p = 0.95$ on Figure 6. Notice that the average interarrival time for state 1 is about 10 times that for state 2. However, the conditional mean sojourn times for the two states are of the same order. This shows that the presence of the state with high arrival rate has a strong adverse effect on the performance of the state with low arrival rate. The ratio ϕ_2/ϕ_1 is plotted on Figure 7. When p is small, the ratio can be smaller than 1. Note that the ratio is not an increasing function of ρ. The ratio approaches 1 when ρ approaches 0 or 1. Note that this not unusual. When p is close to 0, the sojourn time is dominated by the service time (which is the same, whether the arrival occurs in state 1 or 2 of the semi-Markov process). On the other hand, when ρ is close to 1, the waiting time is so long that the state of the arrival has little impact on the sojourn time. It is only for moderate values of ρ where the performance significantly depends on the state of the arrival.

References

[1] P.P. Bocharov (1989) On the two-node queueing network with finite capacity, *Queueing networks with blocking*, 105-126, eds. H.G. Perros and T. Altiok, North-Holland, New York.

[2] P. Franken, D. König, U. Arndt and V. Schmidt (1982) *Queues and point processes*, Wiley, New York.

[3] L. Gün and A. Makowski (1989) "Matrix-geometric solution for two node tandem queueing systems with phase-type servers subject to blocking and failures," *Stochastic Models*, 5, 431-458.

[4] L. Gün and A. Makowski (1988) "An approximation method for general tandem queueing systems subject to blocking" *Proceedings of the First International Workshop on Queueing Networks with Blocking*, 152-171, NCSU, Raleigh, NC, May, 1988.

[5] G.C. Hunt (1956) Sequential arrays of waiting lines, *Operations Research*, 4, 674-683.

[6] F.P. Kelly (1982) "The throughput of a series of buffers," *Adv. App. Prob.*, 14, 633-653.

[7] G. Latouche and M.F. Neuts (1980) "Efficient algorithmic solutions to exponential tandem queues with blocking," *SIAM J. Alg. and Disc. Meth.*, 1, 93-106.

[8] W.A. Massey (1991) "Balanced queues in series with communication blocking," *to appear in Math Opn. Res.*.

[9] M.F. Neuts (1968) "Two queues in series with a finite intermediate waiting room," *J. App. Prob.*, 5, 123-142.

[10] M.F. Neuts (1981) *Matrix-geometric solutions in stochastic models*, John Hopkins, Baltimore, Maryland.

[11] H.G. Perros and T. Altiok (1989) *Queueing networks with blocking*, North-Holland, New York.

[12] B. Sengupta (1989) "Markov processes whose steady state distribution is matrix-exponential with an application to the GI/PH/1 queue," *Adv. App. Prob.*, 21, 159-180.

[13] B. Sengupta (1989) "An invariance relationship for the G/G/1 queue," *Adv. App. Prob.*, 21, 956-957.

[14] B. Sengupta (1990) "The semi-Markovian queue: Theory and Applications," *Stochastic Models*, 6, 383-413.

[15] R.W. Wolff (1982) "Poisson arrivals see time averages," *Operations Research*, 30, 223-231.

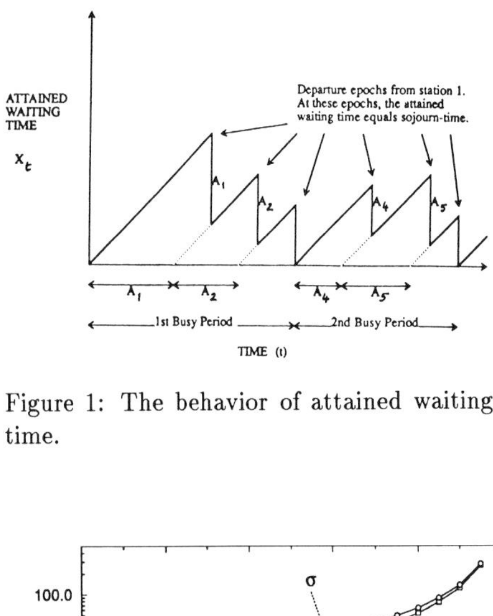

Figure 1: The behavior of attained waiting time.

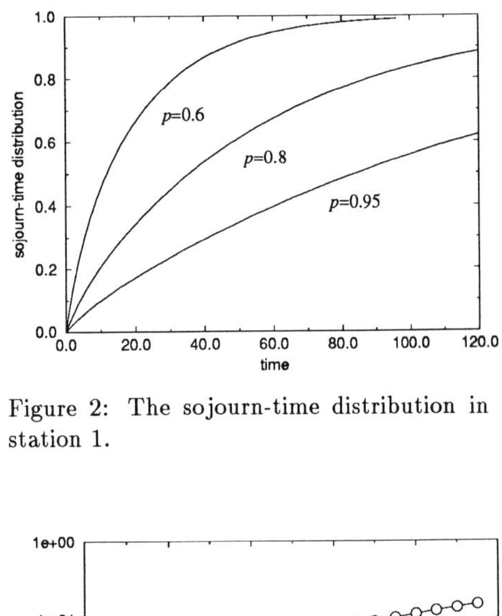

Figure 2: The sojourn-time distribution in station 1.

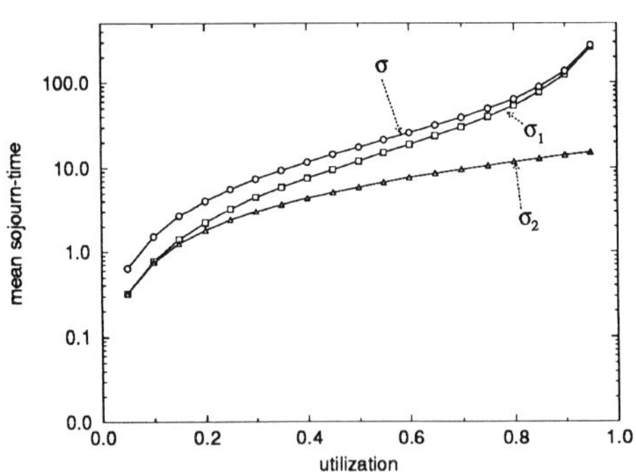

Figure 3: The mean sojourn-time in station 1 and station 2 and the mean sojourn-time in the system.

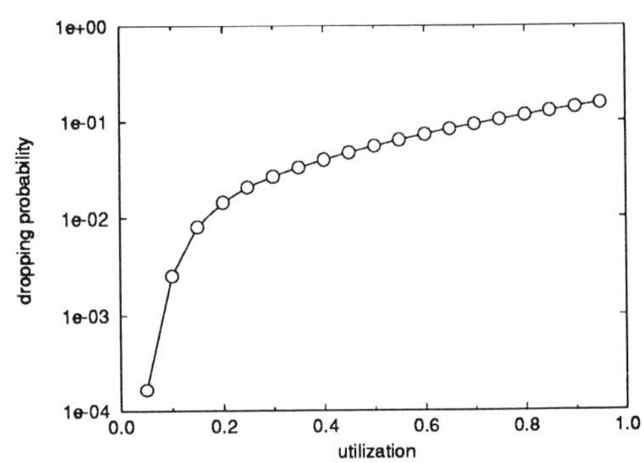

Figure 4: Dropping probability.

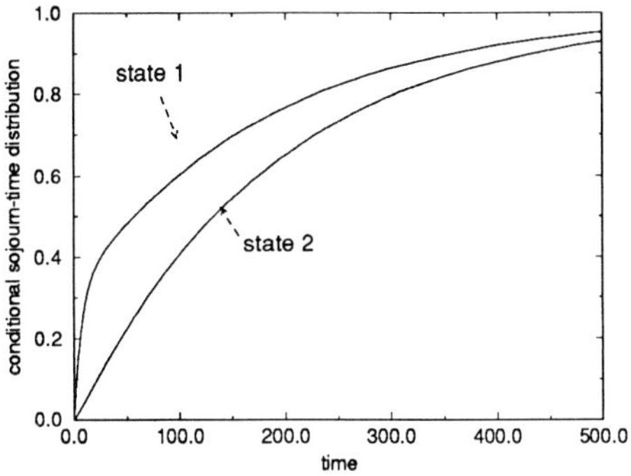

Figure 5: Conditional sojourn-time distribution.

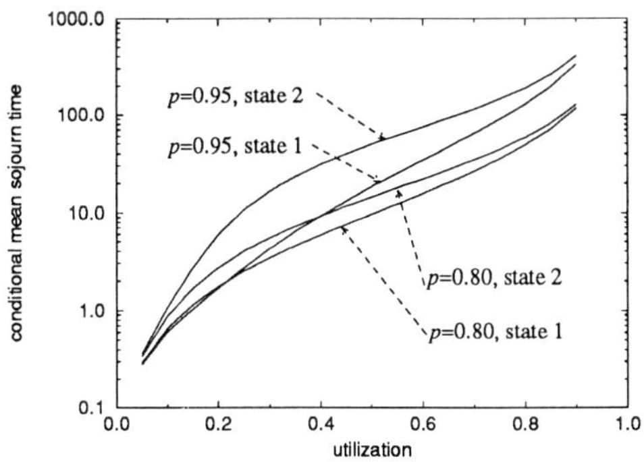

Figure 6: Conditional means sojourn time.

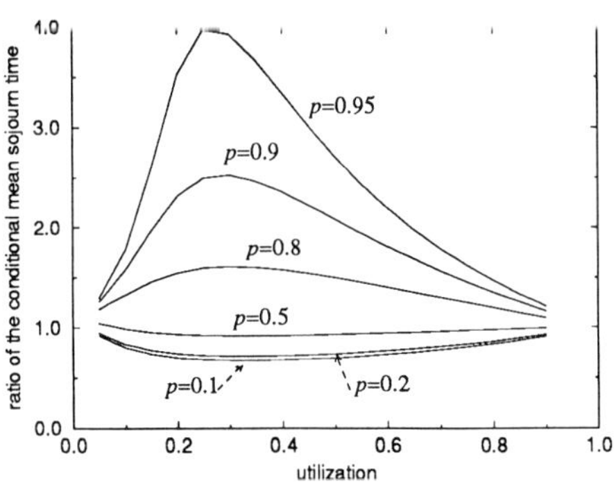

Figure 7: The ratio of the conditional mean sojourn times.

Queueing Networks with Finite Capacity
R.O. Onvural and I.F. Akyildiz (Editors)

Exact Analysis of Multiple Job Classes and Different Types of Blocking

T. Choukri

College of Computing, Georgia Institute of Technology, Atlanta, GA 30332, USA

Abstract

Queueing networks with finite buffers can be used to model communication, computer and manufacturing systems. We prove that a multi-job class open network with finite capacity queues is *pathwise-equivalent* to a closed queueing network with finite buffers. This implies particularly that the two networks have the same queue length distribution. Furthermore, the service and interarrival time distributions are assumed to be general. We prove that such networks are irreducible if the underlying networks with infinite capacities are irreducible. Finally, the steady-state distribution of a multi-job class queueing network with finite buffers is computed using the steady-state distribution of a single job class queueing network with finite buffers. Moreover, we derive the performance measures for the individual classes using only the global performance measures of the constructed queueing network. [1]

1 Introduction

Queueing networks with finite capacity queues are useful to model communication, computer and manufacturing systems [7, 8, 13]. Different types of blocking have been studied in the literature: *blocking-after-service, blocking-before-service, and repetitive-service with random or fixed destination* [2, 7, 9, 13]. Some equivalencies between queueing networks with different blocking mechanisms have been established in terms of steady-state queue length distribution [8, 9, 14]. Also some algorithms have been developed to compute performance measures in these networks [1, 2, 3, 5, 10, 11, 12]. However, the majority of the results is applicable only to single job class queueing networks with blocking. The major goal of this paper is to develop a tool to extend the known results for single job class queueing networks to multi-job class queueing networks. The steady-state distribution of a multi-job class queueing network with finite buffers is computed using the steady-state distribution of a single job class queueing network with finite buffers. Moreover, we derive the performance measures for the individual classes using only the global performance measures of the constructed single job class queueing network with finite buffers. This is an extension of the result given in [4] for BCMP networks. However the proof in [4] is based explicitly on the product-form solution for multi-job class and single job class networks. However, such proof cannot be extended to finite buffer queueing networks since a general solution does not exist currently. The constructed queueing network can easily be analyzed and performance measures for each station can be obtained by applying any single job class computational algorithm [1, 2, 3, 5, 10, 11, 12]. We prove also that a multi-job class open network with finite capacity queues is *pathwise-*

[1]This work was supported by National Science Foundation (NSF) under Grant No. CCR-90-11981.

equivalent to a closed queueing network with finite buffers. This implies particularly that the two networks have the same queue length distribution. Furthermore, the service and interarrival time distributions are assumed to be general. This extends the results given in [14] for an exponential single job class queueing network with finite capacities. This extension is based on the exploitation of the notion of residual service time distribution in spite of memoryless of exponential distributions. We prove also that such networks are irreducible if the underlying networks with infinite capacities are irreducible.

The paper is organized as follows. In section 2 we prove the equivalencies between open and closed multi-job class queueing networks with finite buffers and general service time distributions. In section 3 we extend the result about the irreducibility property in queueing networks with infinite buffers to queueing networks with finite buffers. In section 4 we construct the single job class queueing network. We show how the routing probability matrix can be determined. We state a theorem showing that both queueing networks with single job class and multiple job classes have the same queue length steady-state distribution. We show also that the steady state distribution of the multi-job class queueing network is equivalent to the queue length steady-state distribution of the single job class queueing network up to a multiplicative terms depending only on the routing probability matrix. We introduce new formulae for the computation of performance measures for individual classes in section 5. Section 6 concludes the paper.

2 Equivalencies between Open and Closed Networks with Finite Buffers

In [14], Onvural and Perros show that a single job class open queueing network with finite capacity queues is equivalent to a closed queueing network with blocking. They assume that the interarrival and service time distributions are exponential. The proof is essentially based on the memoryless property of the exponential distribution. In the following we extend this result to multi-job class queueing networks with blocking and to general interarrival and service time distributions using the notion of residual service time distribution.

We consider a multiple job class queueing network with N stations and R classes. Jobs may change class membership according to the routing probabilities $[P] = (p_{ir,js})$ where $p_{ir,js}$, for $1 \leq i,j \leq N;\ \ 1 \leq r,s \leq R$, is the probability that a job of class r completes service at station i proceeds to station j and joins class s. The probability that an external arrival first enters station j in class s is denoted by $p_{0,js}$ and the probability that a job of class r leaves the network after receiving the service at station i is denoted by $p_{ir,0}$. The transition matrix $[P]$ defines a Markovian process with states identified by (i,r). The Markov process is assumed to be decomposable into C ergodic chains, denoted by $C_1,\ldots,C_C$. We denote C_{ic} as the set of classes belonging to chain c, $(1 \leq c \leq C)$ at node i, $(1 \leq i \leq N)$. We denote also C_{open} as the set of open chains. The number of open chains will be denoted by CO. The terms $p_{0,js}$ and $p_{ir,0}$ for closed chains will be equal to zero.

We assume that for each open chain c, the jobs arrive according to a general distribution G_0^c. The service time distribution is assumed to be general and is denoted by G_i^r for class r in station i. For the buffer capacities, we consider two types of stations:

- A station i is of type CB (common buffer) if there is a common buffer with capacity B_i shared by all job classes in this station.
- A station i is of type SB if there is a specific buffer for each job class with capacities $(B_i^r)_{r \in C_i}$.

The total capacity of a station i is equal to the buffer capacities plus the number of servers denoted by m_i. The blocking mechanism for each station may be one of the following types:

- Repetitive-service-blocking with fixed destination
- Repetitive-service-blocking with random destination
- Blocking-after-service
- Blocking-before-service

In the following this network will be referred to as Γ. We assume that Γ is deadlock free [6] and all the buffer capacities $(B_i)_{i \in CB}$, $(B_i^r)_{i \in SB}$ are finite. Hereafter, we construct a closed queueing network referred to as Θ *pathwise-equivalent* to Γ: we say that two networks are *pathwise-equivalent* if there is one-to-one correspondence between their trajectories.

Theorem 2.1 *The queueing network Γ is pathwise-equivalent to a closed queueing network Θ composed of $N+CO$ stations. The Nth first stations in Θ have the same characteristics as the Nth first stations in Γ. The stations enumerated $N+1, \ldots, N+CO$ in Θ are of type CB and contain a single server each. The blocking mechanism for these stations is repetitive-service-with-random-destination, the service descipline is FCFS, and the service time distributions are $G^*_{N+c} = G^c_0$ $(1 \leq c \leq CO)$. The routing probability from a station $N+c$, $(1 \leq c \leq CO)$ is*

$$p^*_{N+c,ir} = p_{0,ir}, \quad r \in C_{ic}$$

The routing probability to a station $N+c$, $(1 \leq c \leq CO)$ is

$$p^*_{ir,N+c} = p_{ir,0}, \quad r \in C_{ic}$$

The service time distributions, the blocking mechanism and the routing probabilities for the other stations are identical in both networks. The number of jobs in Θ corresponding to an open chain c in Γ is

$$K_c = \sum_{i \in SB} \sum_{r \in C_{ic}} B_i^r + \sum_{i \in CB \& c \in C_i} B_i + \sum_{c \in C_i} m_i + 1, \quad c \in C_{open}. \tag{1}$$

Proof.
In network Γ a source for an open chain c behaves as a server with a queue containing an infinite number of jobs and evolving according to the service time distribution G^c_0. So, the server is always busy. The state of the network Γ can be represented by: for each station $(1 \leq i \leq N)$ the state representing the jobs in this station and the jobs blocked by this station, and the service residual time distributions of the jobs being served; for each source the residual service time distribution of the job being served. The constructed closed network Θ is described by the state of the stations i $(1 \leq i < N+CO)$. According to the routing probability p^* a station $N+c$ $(1 \leq c \leq CO)$, contains only jobs belonging to chain c. We identify the stations $1, \ldots, N$ in both networks and the stations $N+1, \ldots, N+CO$ in Θ with the sources in Γ. For an initial and identical

state for the two networks, let us identify each job in Γ with his corresponding job in the network Θ. For the stations 1 to N the evolution of jobs is identical for both networks since the routing probability, the service time distributions, and the blocking mechanisms are the same. There is a particular rule for the station $N+c$ in Θ and the source c in Γ. A job blocked in source c of network Γ is lost. The corresponding job in the station $N+c$ in Θ is also blocked and will receive another service according to the repetitive blocking discipline. We identify again this job with the successor of the lost job in Γ. This means that we identify the job being served in the station $N+c$ in Θ and the job in source c in Γ. This identification can be done because there is always a job in station $N+c$ due to condition (1). Moreover, each station $N+c$ $(1 \leq c \leq CO)$ in Θ can be described only by the residual service time of the job being served (as for the sources in Γ). The number of jobs in these stations can be determined by the total number of jobs in chain c and the number of jobs in chain c in stations $1 \ldots N$.

Q.E.D.

3 Irreducibility

We consider a single job class closed queueing network with N stations, K jobs and exponential service time distributions. The routing probability $P = (p_{ij})_{1 \leq i,j \leq N}$ is assumed to be irreducible. For the infinite buffer capacity case, the vector constituted by the number of jobs in each station is an irreducible Markov chain. This is due to the fact that for any stations i and j, a job can move from station i to station j without any changes in the number of jobs in the other stations. According to the irreducibility of the matrix P, there is at least a sequence $ij_1j_2 \ldots j_kj$ such that $p_{ij_1}p_{j_1j_2} \cdots p_{j_kj} > 0$. Then, a job can move with positive probability from station i to station j following the path $i, j_1, j_2, \ldots, j_k, j$. In the finite capacity network, these transitions may be not possible due to full stations. However, the move from station i to station j may be achieved with respect to the full stations.

Theorem 3.1 *We assume that the network is deadlock free and that the routing probability is irreducible. Then, the Markov process constituted by the number of jobs in each station is irreducible.*

Proof.
The proof is based on three steps. The *first step* is that from any unblocked state U_1 the network can evolve to any other unblocked state U_2. This is due to the fact that a job can be moved from a station to another without any changes in the remaining stations. Consider two stations i and j with at least one job at station i and one free buffer at station j. Since P is irreducible, there exist at least a path $i, j_1, j_2, \ldots, j_k, j$ such that $p_{ij_1}p_{j_1j_2} \cdots p_{j_kj} > 0$. A job from the last station $j^1 (i \leq j^1 \leq j_k)$ with at least one job, before station j can be moved to station j. The rule will be applied recursively to stations $j^1, j^2, \ldots$ until station $i = j^l$, where j^2 $(i \leq j^2 \leq j^1)$ is the last station with at least one job, before station j^1, and so on. At the end of these transitions, there will be one job less in station i, one job more in station j, and the same number of jobs in the intermidiate stations. The *second step* is that there is at least one unblocked state U_3 who may lead to a blocked state B_1. To illustrate this step we consider for instance, a type 1 stations and BAS blocking mechanism queueing network. Let us consider a blocked state $B_1 = (k_1, k_2, \ldots, k_{j_1-1}, (k_{j_1}, f_{j_1}), (k_{j_2}, f_{j_2}), \ldots, (k_{j_L}, f_{j_L}), \ldots, k_N)$ where k_i represents the number of jobs in station i $(1 \leq i \leq N)$, (k_{j_l}, f_{j_l}) represents the number of jobs in station j_l $(1 \leq l \leq L)$ and that this station is blocked by station f_{j_l}. This implies that $p_{j_1f_{j_1}} > 0, p_{j_2f_{j_2}} > 0, \ldots, p_{j_Lf_{j_L}} > 0$. So, the transition from

$U_3 = (k_{1,2}, \dots, k_{j_1-1}, k_{j_1}, k_{j_2}, \dots, k_N)$ (unblocked state) to B_1 is possible with positive probability. The *third step* is that any blocked state B_2 can evolve to an unblocked state U_4 due to the deadlock free assumption. Based on these three macro-steps, the network can evolve from any state to any other one.

Q.E.D.

Note that this theorem can be generalized to multiple chains as follows: if the underlying BCMP network contains R Markov chains, then the network with finite buffer capacities and deadlock free condition contains also R Markov chains.

4 Composition of Multi-Job Classes into Single Job Class in Queueing Networks with Finite Buffers

In [4] we prove that the steady-state distribution of a multi-job class BCMP network (single chain) is equivalent to a single job class BCMP network steady-state distribution. Here we extend this result to networks with finite capacity queues.

4.1 Model Description

In this section we consider a multi-job class closed queueing network with N stations and K jobs. The routing matrix, denoted by $P = (p_{ir,js})_{1 \le i,j \le N,\ 1 \le r,s \le R}$ is assumed to be irreducible. We assume also that the service does not depend on the job class. The stations have a finite buffer capacities and the blocking mechanism for each station can be either RS-FD, BAS, or BBS. The stations are assumed to be of type CB. We refer to this network as Γ. The class of jobs in station i is described by

$$\bar{K}_i = (k_{i1}, \dots, k_{ik_i})$$

where k_i is the number of jobs present at station i and $k_{ij} = 1, \dots, R$ $(1 \le j \le k_i)$ is the class index of the jth job in the order of the arrivals. The jth job blocked by station i if any, is described by $n_{ij} = (n_{ij}^1, n_{ij}^2)$ where n_{ij}^1 is the station index of the jth job blocked by station i and n_{ij}^2 is the position of that job in station n_{ij}^1. If station n_{ij}^1 is of Type 1, then $n_{ij}^2 = 1$ otherwise (i.e. n_{ij}^1 is of Type 2, 3, or 4) $1 \le n_{ij}^2 \le k_{n_{ij}^1}$. If the blocking mechanism of station n_{ij}^1 is RS-FD or BBS the destination of the n_{ij}^2 job in station n_{ij}^1 is station i. We denote by b_i the number of jobs blocked by station i, and by b_{ji} the number of jobs blocked in station j by station i $(b_i = \sum_{j=1}^{N} b_{ij})$. The blocking state of station i is described by the vector

$$\bar{N}_i = (n_{i1}, \dots, n_{ib_i})$$

The state of station i is defined by $\hat{K}_i = (\bar{K}_i, \bar{N}_i)$, and the state of the entire network is represented by the vector

$$\hat{K} = (\hat{K}_1, \dots, \hat{K}_N). \tag{2}$$

The description $\hat{K}$ is sufficiently detailed to take into account the class property. The blocked state description $\bar{N}_i$ does not give the class of the blocked jobs by station i to avoid the redundancy. The class of such job can be determined by the position in its station and the class description $\bar{K}$. As will be shown, the evolution of the number of blocked and unblocked jobs can be determined using only the state-space

$$\hat{k} = (\hat{k}_1, \ldots, \hat{k}_N) \tag{3}$$

where $\hat{k}_i = (k_i, n_{i1}^1, n_{i2}^1, \ldots, n_{ib_i}^1)$.
Hereafter, we describe a single job class closed network with the same queue length distribution as Γ. In addition, the steady-state distribution of network Γ is obtained using only the queue length distribution and an argument from combinatorics theory.

4.2 Constructing Single Job Class Queueing Network

An analysis of "to-belong-to-a-class" property is needed in order to study the queue length distribution separately of the classes. For a tagged job α in Γ, we define the Markovian chain $(Y_n, Z_n)_{n \leq 1}$ where Y_n and Z_n are the station and the class indexes of the tagged job at the nth transition. The transition matrix of the Markovian chain is also the routing probability matrix $[P]$. So, the steady-state distribution $(e_{ir})_{1 \leq i \leq N, 1 \leq r \leq R}$ of $(Y_n, Z_n)_{n \leq 1}$ is the solution of the traffic equation:

$$e_{ir} = \sum_{j=1}^{N} \sum_{s=1}^{R} e_{js} p_{js,ir} \tag{4}$$

$$1 = \sum_{i=1}^{N} \sum_{r=1}^{R} e_{ir}. \tag{5}$$

We denote by e_i and $e_{r/i}$ for the tagged job α, respectively the steady-state distribution to be at station i (just before a transition) and its steady-state distribution to be of class r given that it is at station i, i.e.,

$$e_i = P(Y_n = i) \tag{6}$$

$$e_{r/i} = P(Y_n = i, Z_n = r / Y_n = i). \tag{7}$$

Lemma 4.1 *For a tagged job α, the steady-state distribution e_i to be at station i and its steady-state distribution $e_{r/i}$ to be of class r, given it is at station i are computed from*

$$e_i = \sum_{s=1}^{R} e_{is} \tag{8}$$

$$e_{r/i} = \frac{e_{ir}}{\sum_{s=1}^{R} e_{is}}, \tag{9}$$

where (e_{is}) are computed from equations (4) and (5).

Proof.
This lemma is a direct consequence of the probability measure and the conditional probability definitions as given hereafter.

$$e_i = P(Y_n = i)$$

$$
\begin{aligned}
&= \sum_{s=1}^{R} P(Y_n = i, Z_n = s) \\
&= \sum_{s=1}^{R} e_{is}. \\
e_{r/i} &= P(Y_n = i, Z_n = r / Y_n = i) \\
&= \frac{P(Y_n = i, Z_n = r)}{P(Y_n = i,)} \\
&= \frac{e_{ir}}{\sum_{s=1}^{R} e_{is}}.
\end{aligned}
$$

Q.E.D.

Lemma 4.2 *A tagged job is routed from station i to station j, without considering its class with the following probability* $[Q] = (q_{ij})_{1 \leq i,j \leq N}$, *where*

$$
q_{ij} = \sum_{r=1}^{R} \sum_{s=1}^{R} e_{r/i} p_{ir,js}. \tag{10}
$$

Proof.
By definition, we have

$$
q_{ij} = E[Y_{n+1} = j / Y_n = i]
$$

where E denotes the expectation. According to the conditional probability definition, we have

$$
\begin{aligned}
q_{ij} &= \sum_{s=1}^{R} E[Y_{n+1} = j, Z_{n+1} = s / Y_n = i] \\
&= \sum_{r=1}^{R} \sum_{s=1}^{R} E[Y_{n+1} = j, Z_{n+1} = s / Y_n = i, Z_n = r] \frac{P(Y_n = i, Z_n = r)}{P(Y_n = i)} \\
&= \sum_{r=1}^{R} \sum_{s=1}^{R} e_{r/i} p_{ir,js}.
\end{aligned}
$$

Q.E.D.

Now we can construct the equivalent single job class queueing network which we denote by Θ. The network Θ contains N stations and K jobs and each station has the same queueing discipline (Type 1,2,3 and 4) and the same blocking mechanism (RS-FD, BBS, and BAS) as in network Γ. The routing probability matrix for network Θ is $[Q] = (q_{ij})$ given by expression (10). The service time distributions and the buffer capacities are identical for both networks.

4.3 Steady-State Distributions

In the following, the performance measures related to the network Γ (respectively network Θ) will be indexed by Γ (respectively Θ). We convene that $Y_i(l)$ is equal to i if the lth job in station i is not blocked and it is equal to the index of the destination station if the job is blocked. $Z_i(l)$ is the class of this job.

Theorem 4.1

[a] The network Γ and the network Θ have the same number of blocked and unblocked jobs steady-state distribution, i.e.

$$P^{\Gamma}(\hat{k}) = P^{\Theta}(\hat{k}). \tag{11}$$

[b] The steady-state distribution of network Γ is given by

$$P^{\Gamma}(\hat{K}) = P^{\Theta}(\hat{k}) \prod_{i=1}^{N} \prod_{l=1}^{k_i} e_{Z_i(l)/Y_i(l)} \tag{12}$$

Proof.
Firstly, the class of a tagged job α in network Γ is governed by the same rules as a single job in the network since the routing probability does not depend on the blocking scheme neither on the number of jobs. For this reason, we excluded RS-RD blocking mechanism from this model. Secondly, the class of each job evolves independently from the others. According to definitions (2) and (3) of the states $\hat{K}$ and $\hat{k}$, we have

$$P^{\Gamma}(\hat{K}) = P^{\Gamma}(\hat{k}_i, Y_i(1), Z_i(1), \ldots, Y_i(k_i), Z_i(k_i), 1 \le i \le N)$$

The conditional probability definition gives

$$P^{\Gamma}(\hat{K}) = P^{\Gamma}(\hat{k}) P^{\Gamma}(Y_i(1), Z_i(1), \ldots, Y_i(k_i), Z_i(k_i), 1 \le i \le N | \hat{k}, Y_i(1), \ldots, Y_i(k_i), 1 \le i \le N)$$

Using the independence property explained in the beginning of the proof, we obtain

$$P^{\Gamma}(\hat{K}) = P^{\Gamma}(\hat{k}) \prod_{i=1}^{N} \prod_{l=1}^{k_i} P^{\Gamma}(Y_i(l), Z_i(l) | Y_i(l))$$

The steady-state distribution of a tagged job to be of certain class given it is in a certain station (see equation (7)) implies:

$$P^{\Gamma}(Y_i(l), Z_i(l) | Y_i(l)) = e_{Z_i(l)/Y_i(l)}$$

Henceforth

$$P^{\Gamma}(\hat{K}) = P^{\Gamma}(\hat{k}) \prod_{i=1}^{N} \prod_{j=1}^{k_i} e_{Z_i(j)/Y_i(j)}$$

All we need now is to prove that $P^{\Gamma}(\hat{k}) = P^{\Theta}(\hat{k})$, which will be done using the global balance equation for both networks. Remark that $(\hat{k})$ is a Markov process in network Θ but not in Γ. For the sake of simplicity, we assume that all stations are of type 1. Three different types of transitions may lead to the state $\hat{K}$ (in network Γ):

[a] Transition from a station to another one: A job moves from station i to station j. We assume in this case that station i is neither blocked nor blocking another station ($b_i = 0$). The set of blocked stations will be denoted by Δ. The initial state has to be $T^{ir,j}(\hat{K})$ where $T^{ir,j}(\hat{K})$ is the same state as $\hat{K}$ but with one more job of class r at the the first position in station i and one less job in station j in the last position. If we denote s the class of the job in the last position in station j for $\hat{K}$ i.e. $s = k_{jk_j}$, This transition occurs with rate $\mu_i(k_i + 1)p_{ir,js}$.

[b] Deblocking transitions: A job of class r in the first postion at station $i_0 = i$ moves in class s to station j. We assume that station i_0 is blocking at least one station. We denote the consecutive blocked stations by $i_1 \ldots, i_l$, i.e. i_1 is the first station blocked by station i_0, i_2 is the first station blocked by station i_1, and so on ($l \geq 1$). The initial state $D^{i_0,i_1 \ldots,i_l r,j}(\hat{K})$ that may lead to state $\hat{K}$ is as follows: the state of station j is the same as $\hat{K}_j$ but minus the last job which is in class $s = k_{jk_j}$. The state of station $i_0 = i$ is the same as $\hat{K}_i$ but minus the last job and plus a job in class r in the first position. There is also one more station i_1 blocked by station i in the first position. The state of station $i_g (1 \leq g \leq l-1)$ is the same as $\hat{K}_g$ but minus the last job and plus a job in class $k_{i_g 0} = k_{i_{g-1}, k_{i_g}}$ in the first position. For $g \leq l-1$ there is also one more station i_{g+1} blocked by station i_g in the first position. This transition occurs with rate $\mu_i(k_i + 1)p_{ir,js}$.

[c] Blocking transitions: A job of class r in station i would move in class s to a full station j i.e. $k_j = B_j$. So, the job is blocked. In fact, station i is the last blocked station for the state $\hat{K}$ i.e. $i = n^1_{jb_j}$, and s is the class of the first job in station i i.e. $s = k_{i1}$. The initial state has to be $B^{ir,j}(\hat{K})$ where $B^{ir,j}(\hat{K})$ is the same state as $\hat{K}$ but with the first job in station i in class r in spite of class s, and minus the last blocked job in station j. This transition occurs with rate $\mu_i(k_i)p_{ir,js}$.

The global balance equation, according to the transitions described above is given by

$$\begin{aligned} P^{\Gamma}(\hat{K}) \sum_{i \notin \Delta} \mu_i(k_i) &= \sum_{i \notin \Delta, b_i = 0} \; \sum_{j s.t. k_j < B_j} \; \sum_{r=1}^{R} P^{\Gamma}(T^{ir,j}(\hat{K}))\mu_i(k_i + 1)p_{ir,js} \\ &+ \sum_{i \notin \Delta, b_i > 0} \; \sum_{j s.t. k_j < B_j} \; \sum_{r=1}^{R} \sum_{i_1 \ldots i_l} P^{\Gamma}(D^{i_0,i_1 \ldots,i_l r,j}(\hat{K}))\mu_i(k_i + 1)p_{ir,js} \\ &+ \sum_{b_j > 0, i = n^1_{jb_j}} \; \sum_{r=1}^{R} P^{\Gamma}(B^{ir,j}(\hat{K}))\mu_i(k_i)p_{ir,js} \end{aligned} \tag{13}$$

We derive from the definitions of $\hat{K}$ and $\hat{k}$ and by summing over all the classes that

$$\sum_{classes} P^{\Gamma}(\hat{K}) = P^{\Gamma}(\hat{k}). \tag{14}$$

In a similar way, we obtain by summing over all the classes but class r that

$$\sum_{classes} P^{\Gamma}(T^{ir,j}(\hat{K})) = P^{\Gamma}(T^{i,j}(\hat{k}))e_{r/i} \tag{15}$$

$$\sum_{classes} P^{\Gamma}(B^{ir,j}(\hat{K})) = P^{\Gamma}(B^{i,j}(\hat{k}))e_{r/i} \tag{16}$$

$$\sum_{classes} P^{\Gamma}(D^{ir,j}(\hat{K})) = P^{\Gamma}(D^{i,j}(\hat{k}))e_{r/i} \tag{17}$$

The notations $T^{i,j}(\hat{k})$, $B^{i,j}(\hat{k}))$ *and* $D^{i,j}(\hat{k}))$ are self-explanatory and define the states that may lead to state $\hat{k}$ in network Θ. We deduce from equations (14)-(17), and from definition (10) of q_{ij} that

$$\begin{aligned} P^{\Gamma}(\hat{k})\sum_{i\notin\Delta}\mu_i(k_i) &= \sum_{i\notin\Delta,b_i=0}\ \sum_{j s.t. k_j<B_j} P^{\Gamma}(T^{i,j}(\hat{k}))\mu_i(k_i+1)q_{ij} \\ &+ \sum_{i\notin\Delta,b_i>0}\ \sum_{j s.t. k_j<B_j}\ \sum_{i_1\ldots i_l} P^{\Gamma}(D^{i_0,i_1\ldots,i_l,j}(\hat{k}))\mu_i(k_i)q_{ij} \\ &++ \sum_{b_j>0,i=n^1_{jb_j}} P^{\Gamma}(B^{i,j}(\hat{k}))\mu_i(k_i)q_{ij} \end{aligned} \tag{18}$$

Equation (18) is in fact the global balance equation of network Θ. Thus by unicity of the solution of such equation, we deduce that

$$P^{\Gamma}(\hat{k}) = P^{\Theta}(\hat{k}).$$

Q.E.D.

In the following we assume that the performance measures such as mean number of unblocked jobs $\bar{h}_i$ and the mean number of jobs $\bar{b}_{ij}$ blocked in station i by station j for network Θ are completely determined. This step may be done using an analytical solution as in [1, 2, 3, 5, 10, 11, 12].

5 Performance Measures for Multi-Job Class Queueing Network

We denote by h_{ir} the number of unblocked jobs of class r in station i and by b_{ijr} the number of jobs of class r blocked in station i by station j. We define also the following rates for the ith station $(1 \le i \le N)$

$$\alpha_{irs} = \frac{e_{ir}}{e_{is}} \; for \; 1 \le r, s \le R. \tag{19}$$

Theorem 5.1 *The mean number of unblocked jobs $\bar{h}_{ir}$ of class r in station i and the mean number of blocked jobs $\bar{b}_{ijr}$ of class r in station i by station j are computed by*

$$\bar{h}_{i1} = \frac{\bar{h}_i}{\sum_{r=1}^{R}\alpha_{ir1}} \tag{20}$$

$$\bar{h}_{ir} = \alpha_{ir1}\bar{h}_{i1}, \; for \; 1 \le i \le N, 1 \le r \le R, \tag{21}$$

$$\bar{b}_{ij1} = \frac{\bar{b}_{ij}}{\sum_{r=1}^{R}\alpha_{ir1}} \tag{22}$$

$$\bar{b}_{ijr} = \alpha_{ir1}\bar{b}_{ij1}, \; for \; 1 \le i \le N, 1 \le r \le R. \tag{23}$$

where $\bar{h}_i$ and $\bar{b}_{ij}$ $(1 \le i, j \le N)$ are computed from the network Θ.

Proof.
According to Theorem 4.1 and to combinatorics theory, the steady-state distribution that there are h_{ir} and b_{ijr} jobs regardless of their positions in the queue is a multinomial distribution i.e.,

$$\begin{aligned} P^{\Gamma}(h_{ir}, b_{ijr} \;\; 1 \le i,j \le N \;\; 1 \le r \le R) &= P^{\Theta}(h_i, b_{ij} \;\; 1 \le i,j \le N) \\ &\times \prod_{i=1}^{N} \binom{h_{ir} \;\; 1 \le r \le R}{h_i} \\ &\times \prod_{i=1}^{N} \binom{b_{ijr} \;\; 1 \le j \le N \;\; 1 \le r \le R}{b_{ij}} \\ &\times \prod_{i=1}^{N} \prod_{r=1}^{R} (e_{r/i})^{h_{ir}} \prod_{i=1}^{N} \prod_{j=1}^{N} \prod_{r=1}^{R} (e_{r/j})^{b_{jir}} \end{aligned}$$

So the conditional expectation of h_{ir} and b_{ijr} knowing h_l and b_{lm} $1 \le l,m \le N$, are obtained by

$$E[h_{ir}/h_l, b_{lm}, 1 \le l,m \le N] = P^{\Theta}(h_l, b_{lm} \;\; 1 \le l,m \le N) h_i e_{r/i} \tag{24}$$

$$E[b_{ijr}/h_l, b_{lm}, 1 \le l,m \le N] = P^{\Theta}(h_l, b_{lm} \;\; 1 \le l,m \le N) b_{ij} e_r/j \tag{25}$$

By computing the expectation over h_i and b_{ij} in equations (24) and (25), we obtain the mean number of blocked and unblocked jobs of class r in station i

$$\begin{aligned} \bar{h}_{ir} &= S\bar{h}_i e_{r/i} \\ \bar{b}_{ijr} &= T\bar{b}_{ij} e_{r/j} \end{aligned}$$

where S and T are two constants independent of the classes. Thus, we deduce that

$$\frac{\bar{h}_{ir}}{\bar{h}_{i1}} = \frac{e_{r/i}}{e_{1/i}} = \alpha_{ir1}, \tag{26}$$

$$\frac{\bar{b}_{ijr}}{\bar{b}_{ij1}} = \frac{e_{r/j}}{e_{1/j}} = \alpha_{jr1} \tag{27}$$

These equations combined to the fact that $\bar{h}_i = \sum_{r=1}^{R} \bar{h}_{ir}$ and $\bar{b}_{ij} = \sum_{r=1}^{R} \bar{b}_{ijr}$ imply equations (20)-(23).

Q.E.D.

6 Conclusions

We introduced a new concept which allows to construct a queueing network with finite capacity queues containing a single composite job class. This constructed queueing network model has the same number of blocked and unblocked jobs distribution as the originally given multi-job class queueing network with finite capacity queues. In addition, their steady-state distributions are equal up to a multiplicative term depending only on the routing probability matrix. The performance measures of the multi-job class queueing network with finite buffers are deduced from the performance measures of the single job class queueing networks.

Acknowledgment
I am grateful to Prof. I.F. Akyildiz at Georgia Tech, USA for his suggestions and constructive comments.

References

[1] I. F. Akyildiz. Exact analysis of queueing networks with rejection blocking. *Proc. of the Workshop on Queueing Networks with Blocking, North-Holland*, pages 19–29, 1989.

[2] I. F. Akyildiz. Product form approximations for queueing networks with multiple servers and blocking. *IEEE Transactions on Computers*, 38(1):99–114, Jan. 1989.

[3] I. F. Akyildiz and H. Von Brand. Exact solutions for networks of queues with blocking-after-service. *Technical Report, Georgia Tech, College of Computing*, (GIT-CoC-91-26), April 1991.

[4] I. F. Akyildiz and T. Choukri. Composition of multi-job classes into single job class in multi-chain networks of queues. *Technical Report, Georgia Tech, College of Computing*, Nov. 1991.

[5] I. F. Akyildiz and T. Choukri. Exact product form solution for single class queueing networks with finite capacity and different types of stations. *Technical Report, Georgia Tech, College of Computing*, Nov. 1991.

[6] I. F. Akyildiz and J. Liebeherr. Optimal deadlock free buffer allocation in multiple chain blocking networks of queues. *Proc. of the Int. Conference on the Performance of Distributed Systems and Integrated Communication Networks*, Sep. 10-12 1991.

[7] I. F. Akyildiz and H. G. Perros. Special issue on queueing networks with finite capacity queues: Introduction. *Performance Evaluation*, 10(3), Dec. 1989.

[8] S. Balsamo and G. Iazeolla. Some equivalence properties for queueing networks with and without blocking. *Performance' 83, North-Holland, Amsterdam*, pages 351–360, 1983.

[9] S. Balsamo and V. De Nitto-Parsone. Closed queueing networks with finite capacities: Blocking types, product-form solution and performance indices. *Performance Evaluation*, 12(2):85–102, April 1991.

[10] Y. Dallery and Y. Frein. A decomposition method for the approximate analysis of closed queueing networks with blocking. *Proc. First International Workshop on Queueing with Blocking, North-Holland, Amsterdam*, pages 193–215, 1989.

[11] F.P. Kelly. Reversibility and stochastic networks. *Wiley, Chichester and New York*, 1979.

[12] G. Latouche and M. F. Neuts. Efficient algoritmic solutions to exponential tandem queues with blocking. *SIAM J. ALG. DISC. MATH.*, 1(1):93–106, 1980.

[13] R. O. Onvural. A survey of closed queueing networks with finite buffers. *ACM Computing Surveys*, 22(2):83–121, June 1990.

[14] R. O. Onvural and H. G. Perros. Equivalencies between Closed Queueing Networks with Finite Buffers. *Performance Evaluation*, 9:263–269, 1988/89.

Queueing Networks with Finite Capacity
R.O. Onvural and I.F. Akyildiz (Editors)

Cycle time distribution of cyclic networks with blocking

S.Balsamo[a], M. C. Clò[b] and L. Donatiello[c]

[a]Department of Mathematics, University of Modena, Modena, Italy

[b]Center for the International Development of Informatics in Academic world, Bologna, Italy

[c]Department of Mathematics, University of Bologna, Bologna, Italy

Abstract

An algorithm is provided to compute the cycle time distribution for cyclic closed exponential queueing networks with N finite capacity nodes and blocking. The blocking phenomenon is due to the nodes with finite capacities which can be used to represent systems with limited resources. Moreover a recursive algorithm is provided to evaluate the cycle time moments, and a closed form solution, both in terms of distribution and moments of the cycle time, is provided for the model under special constraints.

1. INTRODUCTION

Many real systems are characterized by finite capacity of critical resources which lead to the blocking phenomenon. The behaviour of such systems can be represented by queueing network models with finite capacity queues. When a job attempts to enter a full finite capacity queue it is rejected and the blocking phenomenon arises.

Performance evaluation of systems with finite capacity resources can be carried out by solving the queueing network model with finite capacity queues. Such models are in general difficult to examine analytically. Moreover, both exact and approximate results have been obtained in terms of average performance indexes and queue length distribution [12], while few results have been obtained in terms of the cycle time distribution [2].

In the literature, several blocking types have been considered, corresponding to different application fields. A detailed description and comparison of different blocking types have been presented in [1,12]. In this paper, we consider the blocking mechanism known as *Blocking Before Service* which is defined as follows [12]: when a node j becomes full, it blocks service in node i, provided node i is serving a job whose destination is j. Service will be resumed as soon as a departure occurs from node j. In particular, we assume that the server can be used to hold the job, that is, the blocking type

This work has been partially supported by CNR Project Research Funds

known as *BBS-SO*. This blocking mechanism has been used to model production and telecommunication systems.

The *cycle time* is defined as the time passing between two consecutive departures of a particular job (*tagged job*) from a specific node of the network. The Laplace-Stieltjes transform (*LST*) of the cycle time distribution in cyclic closed networks with infinite capacity queues has been derived by several authors both for $N \geq 2$ exponential nodes [4,6,13] and for $N = 2$ where one or both nodes can be non-exponential [3,5,8]. To the best of the authors' knowledge the only result in terms of cycle time distribution of cyclic exponential network with two finite capacity has been obtained in [2].

In this paper, we provide an algorithm to evaluate the cycle time distribution for cyclic closed exponential networks formed by $N \geq 2$ nodes, where each node can have finite capacity queue and blocking type BBS-SO. Moreover, a recursive algorithm is provided to evaluate all the moments of the cycle time. The proposed algorithms demonstrates a low order polynomial time computational complexity in the state space of the Markov process associated with the queueing network.

The algorithm is based on an equivalence relation between the steady-state queue length probability distribution at random time and at input time. This result is an extension to finite capacity networks of the arrival theorem which has been proved for queueing networks with infinite capacity in [11,14].

Note that the cyclic closed exponential network has been proved to be equivalent, in terms of queue length distribution, to the so-called dual network in [9]. However such equivalence between networks does not hold in terms of cycle time distribution and therefore it is useless for the evaluation of this performance index.

The paper is organized as follows. In Section 2 the queueing network model and the Markov process describing the tagged job evolution are defined. In Section 3 we provide recursive equations to evaluate the LST of the cycle time distribution. In Section 4 recursive equations for the cycle time moments are derived. Finally, in Section 5, we provide a closed form solution for the cycle time distribution of the network under special constraints. Conclusions and future research are presented in Section 6.

2. THE MODEL

Consider a cyclic closed queueing network with N ($N \geq 2$) nodes, in which K jobs circulate, as shown in Fig. 1. Each node i ($1 \leq i \leq N$) is a single server service center with FCFS service discipline and exponential service time distribution with mean $1/\mu_i$. Let B_i denote node i capacity. When $B_i < K$ node i has a finite capacity queue and blocking arises.

As said before, we consider the BBS-SO blocking assumption. For the cyclic topology, when a job completes service at node i, its destination is node $i \texttt{ mod } N + 1$, $1 \leq i \leq N$. Therefore, when queue i reaches its capacity B_i ($1 \leq i \leq N$), service from the sending node is blocked. This service will be resumed as soon as a job leaves node i and a space in its queue becomes available.

Cycle time is defined as the time which passes between two consecutive departures of a particular job from node N of the network. In order to evaluate the cycle time distribution, we follow an approach similar to the one used in [2]. We tag a job (*tagged job* or *job* $\mathcal{C}$) and define a transient Markov process A which describes the evolution of the tagged job in a complete walk through network nodes.

Let p denote the position of the tagged job during the cycle in the network which is defined as the number of jobs which the N-th node will serve before the cycle end. By cycle definition, just after its first departure from node N, the tagged job is in position $p = K$, i.e., K jobs will be served by node N before the next departure of job $\mathcal{C}$ from node N. As jobs receive service and leave node N, position p decreases to value zero, which corresponds to the cycle end.
The state of Markov process A contains, in addition to the tagged job position, information about the allocation of the K jobs in N network nodes. Hence the process state is defined as the $(N+1)$-vector $(n_1, \ldots, n_i, \ldots, n_N, p)$, where n_i is the number of jobs at node i $(1 \le i \le N)$ and p is the position of job $\mathcal{C}$.
State space E of process A is defined as follows:

$$E = \Big\{(n_1, \ldots, n_i, \ldots, n_N, p) \,|\, \max(0, K - \sum_{\substack{j=1 \\ j \neq i}}^{N} B_j) \le n_i \le B_i \,,\, 1 \le i \le N \,;$$

$$\sum_{i=1}^{N} n_i = K \,;\, 0 \le p \le K \Big\}$$

The topology of cyclic network allows only certain job transitions between the nodes. When a job leaves node i, it joins node $i \ \mathtt{mod}\ N + 1$ and no other transition is allowed, $1 \le i \le N$.
The transition rate matrix R of process A is thus:

$$R = \| \, R[(n_1, \ldots, n_i, \ldots, n_N, p), (n'_1, \ldots, n'_i, \ldots, n'_N, p')] \, \|$$

where $(n_1, \ldots, n_i, \ldots, n_N, p), (n'_1, \ldots, n'_i, \ldots, n'_N, p') \in E$ and

1) $R[(n_1, \ldots, n_i, \ldots, n_N, p), (n'_1, \ldots, n'_i, \ldots, n'_N, p')] = \mu_i$
if $n'_i = n_i - 1 \,,\, n'_{i+1} = n_{i+1} + 1 \,,\, p' = p \,,\, n'_j = n_j$ for $j = 1, \ldots, N \,,\, j \neq i, i+1$,
if $n_i \ge 1$ and $n_{i+1} < B_{i+1}$, for $1 \le i \le N-1$.

2) $R[(n_1, \ldots, n_i, \ldots, n_N, p), (n'_1, \ldots, n'_i, \ldots, n'_N, p')] = \mu_N$
if $n'_N = n_N - 1 \,,\, n'_1 = n_1 + 1 \,,\, p' = p - 1 \,,\, n'_j = n_j$ for $2 \le j \le N-1$,
if $n_N \ge 1$ and $n_1 < B_1$.

3) $R[(n_1, \ldots, n_i, \ldots, n_N, p), (n'_1, \ldots, n'_i, \ldots, n'_N, p')] = 0$
otherwise.

Note that only type 2) transitions lead to a decrease of tagged job position $(p' = p - 1)$ because they correspond to a completion of a job service from node N. On the contrary, type 1) transitions correspond to a completion of a job service from nodes which precede node N.

In order to compute the cycle time distribution of the tagged job we define set E_0 of initial states and set F of final states. Initial states arise at the departure time of the tagged job from node N. E_0 is a subset of space E and it is defined as follows:

$$E_0 = \Big\{ (n_1, \dots, n_i, \dots, n_N, p = K) \,|\, \max(1, K - \sum_{j=2}^{N} B_j + 1) \le n_1 \le B_1 \,; \\ \max(0, K - \sum_{\substack{j=1 \\ j \ne i}}^{N} B_j) \le n_i \le B_i \,;\, 2 \le i \le N-1 \,; \\ \max(0, K - \sum_{j=1}^{N-1} B_j) \le n_N \le B_N - 1 \,;\, \sum_{i=1}^{N} n_i = K \Big\}$$

Final states arise immediately after the successive departure of the tagged job from node N. Set F is defined as follows:

$$F = \Big\{ (n_1, \dots, n_i, \dots, n_N, p = 0) \,|\, \max(1, K - \sum_{j=2}^{N} B_j + 1) \le n_1 \le B_1 \,; \\ \max(0, K - \sum_{\substack{j=1 \\ j \ne i}}^{N} B_j) \le n_i \le B_i \,;\, 2 \le i \le N-1 \,; \\ \max(0, K - \sum_{j=1}^{N-1} B_j) \le n_N \le B_N - 1 \,;\, \sum_{i=1}^{N} n_i = K \Big\}$$

In order to reduce the computational complexity of the evaluation of the cycle time distribution, we reduce the process state space E by modifying the transition rate matrix R. Such modification is performed by ignoring those transitions between states which do not influence the cycle time completion. Let us define those transitions which can be overlooked. Let $S = (n_1, \dots, n_i, \dots, n_N, p)$ be the process state and let job $\mathcal{C}$ be in node i, as shown in Fig. 2. Then the following condition holds:

$$\sum_{j=i+1}^{N} n_j < p \le \sum_{j=i}^{N} n_j \tag{2.1}$$

where, if $i = N$, first summation is null.
If, for state S, we can determine a node with index h $(1 \le h \le i-1)$ for which the following condition holds:

$$p + \sum_{j=1}^{h} n_j \le \sum_{j=1}^{h} B_j \tag{2.2}$$

then nodes 1, 2,..., h capacitics are sufficient to contain both the jobs already in their queues, and the p jobs which will be served by node N before the end of the cycle. Therefore transitions out of nodes h, $h+1$, ..., $i-1$ do not influence completion of

cycle time. In our study, for state S, we overlook state transitions which correspond to jobs service completions from node j to node $j+1$, with $j = h, \ldots, i-1$.
As the system evolves in time, the job $\mathcal{C}$ position decreases and consequentely so does the value of index h. Hence as p decreases the number of ignored states increases. Consequently, set F of final states reduces to $F = \{S_f\}$ where

$$S_f = \left(B_1, \ldots, B_{\bar{n}}, K - \sum_{j=1}^{\bar{n}} B_j, 0, \ldots, 0, p = 0\right)$$

where $\bar{n}$ $(1 \le \bar{n} \le N-1)$ is the index which verifies:

$$\sum_{j=1}^{\bar{n}} B_j \le K < \sum_{j=1}^{\bar{n}+1} B_j.$$

3. CYCLE TIME DISTRIBUTION

In the preceding Section we have defined the Markov process A describing the evolution of the tagged job, set E_0 of initial states and the final absorbing state of cycle S_f. Moreover we have provided the condition under which some states can be ignored, as function of job position. We shall now evaluate the *Laplace-Stieltjes transform* (LST) of the cycle time distribution.
We follow an approach similar to that used in the case of networks with two nodes [2].

The cycle time distribution can be computed by evaluating the first hitting time probability distribution of Markov process A to absorbing state S_f, starting from any state in E_0.
Let $T(n_1, \ldots, n_i, \ldots, n_N, p, s)$ denote the LST of the passage time distribution from state $(n_1, \ldots, n_i, \ldots, n_N, p)$ to final state S_f. By using the matrix R definition we can write the following recursive relations for $p > 0$:

$$\begin{aligned} T(n_1, n_2, \ldots, n_N, p, s) = \quad & \quad (3.1) \\ = \sum_{j=1}^{h-1} \frac{\alpha_j(n_j, n_{j+1})}{M_{h-1} + L_i + s} & T(n_1, \ldots, n_j - 1, n_{j+1} + 1, \ldots, n_N, p, s) + \\ + \sum_{j=i}^{N-1} \frac{\alpha_j(n_j, n_{j+1})}{M_{h-1} + L_i + s} & T(n_1, \ldots, n_j - 1, n_{j+1} + 1, \ldots, n_N, p, s) + \\ + \frac{\alpha_N(n_N, n_1)}{M_{h-1} + L_i + s} & T(n_1 + 1, n_2, \ldots, n_N - 1, p - 1, s) \end{aligned}$$

when equations (2.1) and (2.2) hold, where functions M_{h-1} and L_i depend on state $(n_1,$

$n_2, \ldots, n_N, p)$ and are defined as follows:

$$M_{h-1} = \sum_{j=1}^{h-1} \alpha_j(n_j, n_{j+1}) \tag{3.2}$$

$$L_i = \sum_{j=i}^{N} \alpha_j(n_j, n_{j \bmod N+1}) \tag{3.3}$$

$$\begin{aligned} T(n_1,n_2,\ldots,n_N,p,s) &= \\ &= \sum_{j=1}^{N-1} \frac{\alpha_j(n_j,n_{j+1})}{M+s} T(n_1,\ldots,n_j-1,n_{j+1}+1,\ldots,n_N,p,s)+ \\ &+ \frac{\alpha_N(n_N,n_1)}{M+s} T(n_1+1,n_2,\ldots,n_N-1,p-1,s) \end{aligned} \tag{3.4}$$

otherwise, where:

$$M = \sum_{j=1}^{N} \alpha_j(n_j, n_{j \bmod N+1}) \tag{3.5}$$

and where, for $1 \leq i \leq N$,

$$\alpha_i(n_i, n_{i \bmod N+1}) = \delta(n_i)\, \mu_i\, \beta_{i \bmod N+1}(n_{i \bmod N+1}) \tag{3.6}$$

and

$$\beta_i(n) = \begin{cases} 0 & \text{if } n = B_i \\ 1 & \text{if } n < B_i \end{cases} \qquad \delta(n) = \begin{cases} 0 & \text{if } n = 0 \\ 1 & \text{if } n > 0 \end{cases}$$

and with

$$T(B_1,\ldots,B_{\bar{n}},K-\sum_{j=1}^{\bar{n}} B_j,0,\ldots,0,p=0,s) = 1. \tag{3.7}$$

The cycle time distribution can be evaluated as the sum of passage time distributions from any initial state to the final absorbing state, weighed by the probability of the initial state.

Let $T_{K,B_1,\ldots,B_N}(s)$ denote the LST of the cycle time distribution of a closed cyclic network with N nodes, K jobs and node finite capacities B_i, $1 \leq i \leq N$. It can be computed as follows:

$$\begin{aligned} T_{K,B_1,\ldots,B_N}(s) &= \\ &= \sum_{(n_1,\ldots,n_i,\ldots,n_N,p=K)\in E_0} P(n_1,\ldots,n_i,\ldots,n_N,K)\, T(n_1,\ldots,n_i,\ldots,n_N,K,s) \end{aligned} \tag{3.8}$$

where $P(n_1,\ldots,n_i,\ldots,n_N,K)$ is the probability for the system to be in state $(n_1, \ldots, n_i, \ldots, n_N)$ at cycle starting time, i.e., after the first departure of $\mathcal{C}$ from node N.

Then $P(n_1, \ldots, n_i, \ldots, n_N, K)$ can be evaluated as the steady-state probability of the Markov process embedded at input times at node 1.

Let $\pi(n_1, \ldots, n_N)$ be the steady-state probability of network state $(n_1, \ldots, n_N)$ at random time and let $\xi_1(m_1, \ldots, m_N)$ denote steady-state probability of state $(m_1, \ldots, m_N)$ at input time at node 1, where state $(m_1, \ldots, m_N)$ does not include the arriving job.

One can prove the following theorem [7, eq. 4.8]:

Theorem

The steady-state distribution at arrival time at node 1 is proportional to the steady-state distribution at random time:

$$\xi_1(m_1, \ldots, m_j, \ldots, m_N) = \frac{\pi(n_1, \ldots, n_j, \ldots, n_N)}{\sum_{(n'_1, \ldots, n'_N) \in \sigma_N} \pi(n'_1, \ldots, n'_N)} \tag{3.9}$$

for each state $(m_1, \ldots, m_N) = (n_1, \ldots, n_{N-1}, n_N - 1)$, where $(n_1, \ldots, n_N) \in \sigma_N$ and where

$$\begin{aligned} \sigma_N = \Big\{ (n_1, \ldots, n_i, \ldots, n_N) \,|\, & \max(0, K - \sum_{j=2}^{N} B_j) \le n_1 \le B_1 - 1 ; \\ & \max(0, K - \sum_{\substack{j=1 \\ j \ne i}}^{N} B_j) \le n_i \le B_i \, , \, 2 \le i \le N - 1 ; \\ & \max(1, K - \sum_{j=1}^{N-1} B_j) \le n_N \le B_N \, ; \, \sum_{i=1}^{N} n_i = K \Big\} \end{aligned}$$

is the set of all states where a job transition from node N to node 1 can occur. □

In addition the following equality holds:

$$P(n_1, n_2, \ldots, n_N, p = K) = \xi_1(n_1 - 1, n_2, \ldots, n_N) \tag{3.10}$$

where $(n_1, n_2, \ldots, n_N, p = K) \in E_0$ is the state at arrival time at node 1. Hence, from formulas (3.9) and (3.10), one obtains:

$$\begin{aligned} P(n_1, n_2, \ldots, n_N, p = K) &= \xi_1(n_1 - 1, n_2, \ldots, n_N) = \\ &= \frac{\pi(n_1 - 1, n_2, \ldots, n_N + 1)}{\sum_{(n'_1, \ldots, n'_N) \in \sigma_N} \pi(n'_1, \ldots, n'_N)} \end{aligned} \tag{3.11}$$

thus completing the evaluation of LST of the cycle time distribution.

Note that the result stated by the theorem is an extension to finite capacity networks of the arrival theorem which has been proved for queueing networks with infinite capacity queue nodes in [11,14]. Such result holds even if the network does not show the product-form solution for steady-state queue length distribution at random time. In the special case of product-form solution, formula (3.9) can be simplified as follows:

Corollary

If

$$K > \sum_{i=1}^{N} B_i - \min_{1 \le i \le N} B_i \tag{3.12}$$

then

$$\xi_1(\bar{m}) = \pi^*(\bar{m}) \tag{3.13}$$

$\forall \bar{m}$ and where π^* denotes the steady-state joint queue length distribution at random time of the network with $K-1$ customers and node capacities B_i^*, where $B_i^* = B_i - 1$ for $i = 1, N$ and $B_i^* = B_i$ for $2 \le i \le N-1$.

Proof: Under condition (3.12), the steady-state probability π shows product-form solution, as shown in [9], and by substituting this product-form solution in formula (3.9) one immediately obtains formula (3.13). □

Remark

We have provided a recursive scheme to evaluate the LST of the cycle time distribution. However, by using similar arguments as for networks with $N = 2$ nodes with blocking [2], one can derive an expression for the cycle time distribution in the time domain.

4. CYCLE TIME MOMENTS

In the preceding Section we have provided an equation to evaluate the LST of the cycle time distribution. In many practical applications, it is sufficient to evaluate the first few moments of the cycle time. In this Section we provide a recursive scheme to evaluate all the moments of the cycle time.

Let $E\left[(n_1, n_2, \ldots, n_N)^m\right]$ and $E\left[(n_1, n_2, \ldots, n_N, p)^m\right]$ denote the order m ($m \ge 1$) moments, respectively, of sojourn time in state $(n_1, n_2, \ldots, n_N, p)$ and of passage time from that state to final state S_f.

By applying to equations (3.1) and (3.4) standard properties of LST, the following equations hold, when $p > 0$:

$$E\big[(n_1, n_2, \dots, n_N, p)^m\big] = E\big[(n_1, n_2, \dots, n_N)^m\big] + \tag{4.1}$$
$$+ \sum_{j=1}^{h-1} \frac{\alpha_j(n_j, n_{j+1})}{M_{h-1} + L_i} \, E\big[(n_1, \dots, n_j - 1, n_{j+1} + 1, \dots, n_N, p)^m\big] +$$
$$+ \sum_{j=i}^{N-1} \frac{\alpha_j(n_j, n_{j+1})}{M_{h-1} + L_i} \, E\big[(n_1, \dots, n_j - 1, n_{j+1} + 1, \dots, n_N, p)^m\big] +$$
$$+ \frac{\alpha_N(n_N, n_1)}{M_{h-1} + L_i} E\big[(n_1 + 1, n_2, \dots, n_N - 1, p - 1)^m\big]$$

when the equations (2.1) and (2.2) are valid, and where funcions M_{h-1} and L_i are defined by formulas (3.2) and (3.3), and

$$E\big[(n_1, n_2, \dots, n_N)^m\big] = m! \frac{1}{\Big(M_{h-1} + L_i\Big)^m}$$

$$E\big[(n_1, n_2, \dots, n_N, p)^m\big] = E\big[(n_1, n_2, \dots, n_N)^m\big] + \tag{4.2}$$
$$+ \sum_{j=1}^{N-1} \frac{\alpha_j(n_j, n_{j+1})}{M} \, E\big[(n_1, \dots, n_j - 1, n_{j+1} + 1, \dots, n_N, p)^m\big] +$$
$$+ \frac{\alpha_N(n_N, n_1)}{M} E\big[(n_1 + 1, n_2, \dots, n_N - 1, p - 1)^m\big]$$

otherwise, where function M is defined by equation (3.5) and

$$E\big[(n_1, n_2, \dots, n_N)^m\big] = m! \frac{1}{M^m}$$

and

$$E\big[(B_1, \dots, B_{\bar{n}}, K - \sum_{j=1}^{\bar{n}} B_j, \dots, 0, \dots, 0, p = 0)^m\big] = 0 \tag{4.3}$$

We can compute the order m moment of the cycle time as the sum of order m moments of passage time from any initial state to the final absorbing state, weighed by the probability of the initial state.

Let $E^m_{K,B_1,\dots,B_N}$ denote the order m moment of the cycle time. Then it can be computed as follows:

$$E^m_{K,B_1,\dots,B_N} = \tag{4.4}$$
$$\sum_{(n_1,\dots,n_N,p=K) \in E_0} P(n_1, n_2, \dots, n_N, K) \, E\big[(n_1, n_2, \dots, n_N, K)^m\big]$$

where $P(n_1, n_2, \ldots, n_N, K)$ has been defined in the previous Section and it can be computed by equation (3.11).

5. Special Case

In this Section we provide a closed form solution of the LST of the cycle time distribution for a special case.

Consider a special exponential cyclic closed queueing network with blocking with N nodes and K jobs, where K is defined as follows:

$$K = \left(\sum_{i=1}^{N} B_i \right) - 1$$

For each network state, there are exactly $N-1$ nodes which are full and block the preceding $N-1$ nodes. Hence, only one node at a time, say node i, is not full and does not block the preceding node $i-1$. Therefore, there is a unique non zero transition from the state corresponding to a job movement from node $i-1$ to node i; the transition rate is $\delta(n_{i-1})\mu_{i-1}\beta_i(n_i) = \mu_{i-1}$.

The state space of the transient Markov process A can be rewritten as follows:

$$E = \left\{ (B_1, \ldots, B_i - 1, \ldots, B_N, p) \,|\, 1 \le i \le N \,;\, 0 \le p \le K \,;\, K = \sum_{i=1}^{N} B_i - 1 \right\}$$

There exists only one initial state of cycle:

$$E_0 = \left\{ (B_1, \ldots, B_i, \ldots, B_N - 1, p = K) \right\}$$

and the final state is:

$$S_f = (B_1, \ldots, B_i, \ldots, B_N - 1, p = 0)$$

First we compute the LST of passage time distribution from initial to final state. By using the recursive equations (3.1), (3.4) and (3.7), one can derive

$$T(B_1, B_2, \ldots, B_{N-1}, B_N - 1, p = K, s) = \prod_{j=1}^{N} \frac{\mu_j^K}{(\mu_j + s)^K}$$

Since the set of initial states is formed by a unique state (B_1, B_2, $\ldots$, B_{N-1}, $B_N - 1$, $p = K$), one immediately obtains

$$P(B_1, B_2, \ldots, B_{N-1}, B_N - 1, p = K) = 1.$$

Hence the LST of cycle time distribution is:

$$T_{K,B_1,\dots,B_N}(s) = \prod_{j=1}^{N} \frac{\mu_j^K}{(\mu_j + s)^K}$$

Now we compute the order m moment of the passage time distribution of the network by applying equations (4.1), (4.2) and (4.3), which lead to

$$E\big[(B_1, B_2, \dots, B_{N-1}, B_N - 1, p = K)^m\big] = m!\,K \sum_{i=1}^{N} \frac{1}{\mu_i^m}$$

Hence the order m ($m \geq 1$) moment of the cycle time distribution can be computed as follows:

$$E^m_{K,B_1,\dots,B_N} = m!\,K \sum_{i=1}^{N} \frac{1}{\mu_i^m}$$

Note that in this special case, condition (3.12) in the corollary holds and the network shows a product-form solution for the steady-state queue length distribution, and it is equivalent to a dual network defined in [9]. However, one can easily verify that the two networks are not equivalent in terms of cycle time distribution.

6. CONCLUSIONS AND OPEN PROBLEMS

We have provided a recursive scheme to evaluate the LST of the cycle time distribution for cyclic exponential closed networks with finite capacity queues and blocking. Moreover, a recursive scheme to calculate all the moments of the cycle time has been presented. Possible extension of such results concerns the identification of equivalence properties between classes of network models with different parameter values, either in terms of cycle time moments or also in terms of distribution. Another extension concerns the analysis, in terms of cycle time distribution, of other classes of networks having a different type of blocking, different topologies such as central server or star topology, and non-exponential service time distributions.

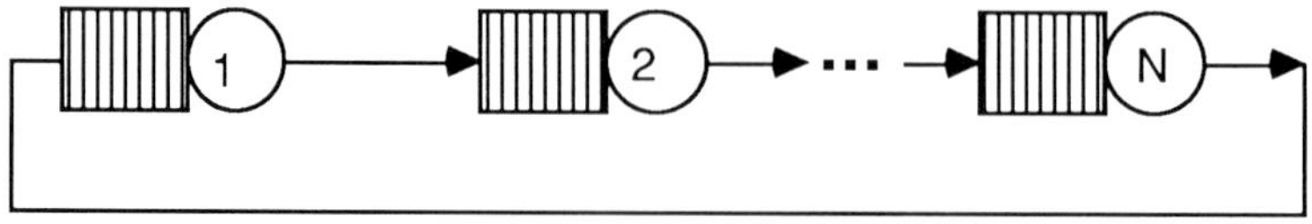

- Fig. 1 - Closed cyclic N-node network -

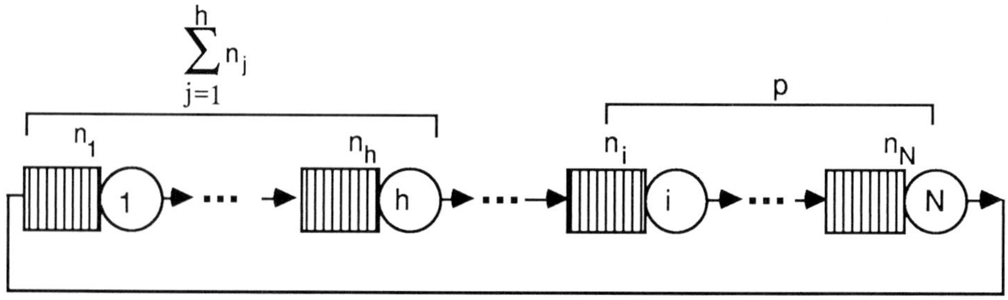

- Fig. 2 - Network state with the tagged job in node i -

7. REFERENCES

[1] S. Balsamo, V. De Nitto Personé, "Closed queueing networks with finite capacities: blocking types, product-form solution and performance indices", Performance Evaluation, vol. 10, n° 3, April 1991.

[2] S. Balsamo, L. Donatiello, "On the cycle time distribution in a two-stage cyclic network with blocking", IEEE Transactions on Software Engineering, vol. 15, n° 10, October 1989, pp. 1205-1216.

[3] O. J. Boxma, "The cyclic queue with one general and one exponential server", Adv. Appl. Prob., vol. 15, 1983, pp. 857-873.

[4] O. J. Boxma, F. P. Kelly, A. G. Konheim, "The product form for sojourn time distributions in cyclic exponential queues", J. ACM, vol. 31, n° 1, January 1981, pp. 128-133.

[5] S. Carbini, L. Donatiello, G. Iazeolla, "An efficient algorithm for the cycle time distribution in two-stage cyclic network with a non-exponential server", Teletraffic Analysis and Computer Performance Evaluation, O. J. Boxma, L. Cohen, H. C. Tijms, Eds. North-Holland, 1986, pp. 99-115.

[6] W. M. Chow, " The cycle time distribution of exponential cyclic queues", J. ACM, vol. 27, n° 2, April 1980, pp. 281-286.

[7] M. C. Clò, "Modelli stocastici per sistemi a capacità finita", Doctoral Thesis, Math. Dept., University of Modena, Italy, 1991.

[8] H. Daduna, "Two-stage queues with non-exponential servers: Steady-state and cycle time", Oper. Res., vol. 34, n° 3, 1986, pp. 445-459.

[9] W. J. Gordon, G. F. Newell, "Cyclic queueing systems with restricted length queues", Oper. Res., vol. 15, 1967, pp. 266-278.

[10] A. Hordijk, N. Van Dijk, "Adjoint processes, job local balance and insensitivity of stochastic networks." Bull: 44 session, Int. Stat. Inst. 50, 1982, pp. 776-788.

[11] S. S. Lavenberg, M. Reiser, "Stationary state probabilities at arrival instants for closed queueing networks with multiple types of customers", J. Appl. Prob., 17, 1980, pp. 1048-1061.

[12] R. O. Onvural, "Survey of closed queueing networks with blocking", ACM Computing Surveys, vol. 22, n° 2, June 1990, pp. 83-121.

[13] R. Schassberger, H. Daduna, "The time for a round trip in a cycle of exponential queues", J. ACM, vol. 30, n° 1, January 1983, pp. 146-150.

[14] K. C. Sevcik, I. Mitrani, "The distribution of queueing network states at input and output instants", J. ACM, vol. 28, n° 2, April 1981, pp. 358-371.

Queueing Networks with Finite Capacity
R.O. Onvural and I.F. Akyildiz (Editors)

Deadlock-Free Buffer Allocations in Multiple Chain Blocking Networks with Tandem Sequences *

Jörg Liebeherr
Computer Science Division,
University of California - Berkeley,
Berkeley, CA 94720, U. S. A.

Abstract

In order to verify that a blocking network with multiple routing chains is deadlock-free, all cycles in the network need to be tested whether they satisfy the deadlock freedom conditions. Networks with so-called tandem sequences are characterized by a high number of cycles. Finding all cycles in a blocking network with tandem sequences is computational impractical if the size of the network is large. It is shown that the effort of finding all cycles can be reduced by substituting each tandem sequence in a blocking network with a single station. If the substitute network is deadlock-free then the original network is deadlock-free. A method is devised that provides a deadlock-free buffer allocation with a minimum total number of buffers for blocking networks with tandem sequences. It is shown that a deadlock-free and minimal buffer allocation for networks consisting of only one tandem sequence can be directly obtained.

1 Introduction

In this work we consider the so-called blocking-after-service (BAS), also referred as type 1 or manufacturing blocking mechanism [1]. A job upon service completion at a station i attempts to enter the destination station j. If station j is full at that moment, the job is forced to wait in station i's server until it can enter destination station j. The server remains blocked for this period of time. It cannot serve other jobs waiting in the queue.

Finite station capacities and blocking can introduce a deadlock situation. In a simple example, a deadlock may occur if a job which has finished its service at station i's server wants to join station j. If station i is full the job is blocked in station i. Another job which has finished service at station j now wants to proceed to station i which is also full. It blocks the jth station. Both jobs are waiting for each other. As a result a deadlock situation arises. There are two possible solutions for the deadlock problem: Either include a strategy to handle deadlocks in the model or simply restrict oneself to cases where deadlock is impossible. We select the second approach and find conditions under which a multiple chain blocking network is deadlock-free. Kundu and Akyildiz [3] showed that queueing networks with a single routing chain are deadlock-free if the number of jobs in the network is less than the capacity of the directed cycle with minimal capacity.

* This work was supported by the National Science Foundation under Grant No. CCR-90-11981

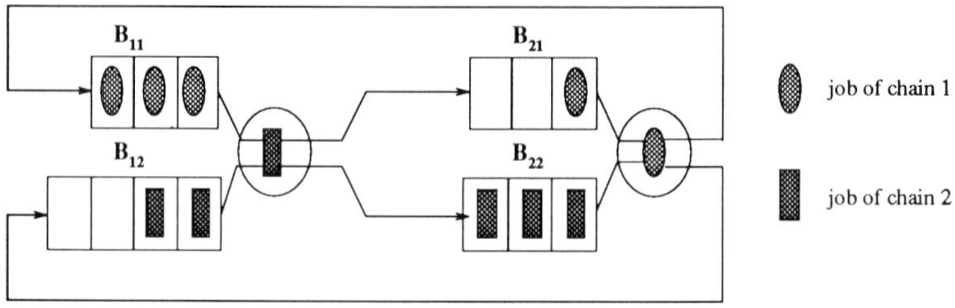

Figure 1: Blocking Network with 2 chains.

However, the conditions for deadlock freedom with multiple routing chains become very complex due to interactions between routing chains.

We demonstrate these interactions in the network model given in Figure 1. There are 2 stations and 2 routing chains in the model. Let B_{ir} denote the buffer of chain r at station i (for $i, r = 1,2$) and let K_r denote the total number of jobs in chain r ($r = 1,2$). Let the parameters be given by $K_1 = 5$, $K_2 = 6$, $B_{11} = 3$, $B_{21} = 3$, $B_{12} = 4$ and $B_{22} = 3$. Here a deadlock situation occurs, even though the condition for deadlock freedom given in [3] are satisfied for each routing chain in isolation.

In addition to determining the conditions for deadlock freedom another important issue is the allocation of station capacities in a queueing network such that deadlocks cannot occur. Akyildiz and Liebeherr [2] proved the necessary and sufficient conditions for deadlock freedom in multiple chain queueing networks with blocking. They presented an algorithm that computes a deadlock-free buffer allocation with the least number of total buffers. The algorithm generates the set of so-called buffer cycles. Finally, the deadlock-free allocation with the minimum total number of buffers is obtained by applying linear programming techniques.

However, for certain network topologies it is computationally impractical to compute all buffer cycles. In this study we show for the class of blocking networks with so-called *tandem sequences* that the effort to find a deadlock-free allocation with minimum number of buffers can be reduced.

The remaining sections of this paper are organized as follows. In section 2 we specify the class of queueing networks which is considered in this study. We review the deadlock freedom conditions and the buffer allocation algorithm presented in [2]. In section 3 we are concerned with finding a minimal deadlock-free buffer allocation for blocking networks with tandem sequences. If the entire network constitutes a single tandem sequence we prove that a minimal deadlock-free buffer allocation can be directly obtained. We show that for blocking networks with tandem sequences a minimal deadlock-free buffer allocation can be obtained from a substitute network with a simpler structure. In section 4, we present an example that applies our results. We summarize our results in section 5.

2 Deadlock Freedom in Multiple Chain Blocking Networks

2.1 Model Description

We consider a closed queueing network Γ with the following properties:

a) The system consists of N stations and R disjoint routing chains. Stations and routing chains are referred to by their indexes. Let $\mathcal{N}$ and $\mathcal{R}$ denote the set of stations and routing chains, respectively. Let $\mathcal{R}_i \subseteq \mathcal{R}$ denote the set of routing chains visiting station i. Let $\mathcal{N}_r \subseteq \mathcal{N}$ denote the set of stations visited by routing chain r.

b) Each station i has a single server. The service time distribution and the scheduling discipline of a station is arbitrary, but non-preemptive.

c) Each station keeps separate buffers for jobs from different routing chains. B_{ir} denotes the buffer at station i (excluding the server) for jobs from routing chain r $(i \in \mathcal{N},\ r \in \mathcal{R}_i)$. Each buffer may accommodate only a finite number of jobs. Let Φ be an assignment of capacities to the buffers of Γ, i.e.,

$$\Phi : \{B_{ir} | i \in \mathcal{N}, \quad r \in \mathcal{R}_i\} \rightarrow \{0, 1, 2, \ldots\ldots\} \tag{1}$$

Note that a buffer can have infinite capacity ($\Phi(B_{ir}) = \infty$) or no capacity at all ($\Phi(B_{ir}) = 0$).

d) The number of jobs in the system is fixed at

$$\underline{K} = (K_1, K_2, \ldots, K_R) \tag{2}$$

where

$$K_r = \sum_{i \in \mathcal{N}_r} k_{ir}, \qquad \text{for } r = 1, \ldots, R \tag{3}$$

represents the total number of jobs in routing chain r, and k_{ir} is the number of jobs of chain r in the i-th station (in buffer and server). The state of the network is represented by a vector $\underline{k} = (\underline{k}_1, \underline{k}_2, \ldots, \underline{k}_N)$ where $\underline{k}_i$ is a vector $(k_{ir_1}, k_{ir_2}, \ldots, k_{ir_{|\mathcal{R}_i|}})$ with $r_u \in \mathcal{R}_i$ $(u = 1, 2, \ldots, |\mathcal{R}_i|)$. The total capacity of station i is computed by $\sum_{r \in \mathcal{R}_i} \Phi(B_{ir}) + 1$.

e) A job of routing chain r which has received service in station i proceeds to the station j with probability $p_{ij,r}$ (for $i, j \in \mathcal{N}$, $r \in \mathcal{R}_i$ and $r \in \mathcal{R}_j$). The number of jobs in the buffer of station j for jobs of chain r cannot exceed its capacity $\Phi(B_{jr})$. If B_{jr} is *saturated*, i.e., $k_{jr} = \Phi(B_{jr})$, the job of chain r is blocked at buffer B_{jr} and has to remain in the server of station i until a place in B_{jr} becomes available. This is the BAS (blocking-after-service) blocking mechanism.

2.2 Deadlock Freedom Conditions

The following definitions are needed to state the conditions for deadlock freedom of a network Γ.

Definition 1 *A buffer cycle (of length $t-1$ $(1 \leq t-1 \leq |\mathcal{N}|)$) is a sequence of buffers $C = (B_{\sigma_1\rho_1}, B_{\sigma_2\rho_1}, \ldots, B_{\sigma_{t-1}\rho_{t-1}}, B_{\sigma_t\rho_t})$ such that*

$$\begin{aligned} &\sigma_u \in \mathcal{N} && (1 \leq u \leq t) \\ &\sigma_t = \sigma_1 && \\ &\rho_u \in R_{\sigma_u}, \quad \rho_{u+1} \in R_{\sigma_u}, \quad p_{\sigma_u\sigma_{u+1},\rho_{u+1}} > 0 && (1 \leq u < t). \end{aligned} \tag{4}$$

Let $\mathcal{C}$ denote the set of all buffer cycles in Γ.

Definition 2 *The set of stations with buffers from routing chain r in buffer cycle C is defined by*

$$S_r^{(C)} = \{i \mid B_{ir} \in C, i \in \mathcal{N}, r \in \mathcal{R}_i\} \tag{5}$$

The necessary and sufficient deadlock freedom conditions for multiple chain blocking networks are given by the following theorem [2].

Theorem 1 (DLF condition) *A multiple chain queueing network* Γ *is* deadlock-free *(DLF) if and only if for all $C \in \mathcal{C}$ with $C = (B_{\sigma_1\rho_1}, B_{\sigma_2\rho_1}, \ldots, B_{\sigma_{t-1}\rho_{t-1}}, B_{\sigma_t\rho_t})$ there exists a routing chain $r \in \mathcal{R}$ such that:*

$$\sum_{1<u\leq t \wedge \sigma_u \in S_r^{(C)}} \Phi(B_{\sigma_u r}) + |S_r^{(C)}| > K_r \tag{6}$$

In other words, each buffer cycle must contain at least one routing chain which cannot both saturate the buffers of its chain in the cycle and have blocked jobs waiting at the saturated buffers. If the DLF condition (equation (6)) is satisfied for a buffer cycle C they are clearly satisfied for all cycles which include the buffers of cycle C as a subset. Therefore, it is not required to test the DLF condition for all buffer cycles to guarantee deadlock freedom of the network. This is formalized in the following Lemma [2].

Lemma 1 *Given a multiple chain queueing network Γ and the set of all buffer cycles $\mathcal{C}$ of Γ. For any $C', C'' \in \mathcal{C}$ let $C' \subseteq C''$, if the set of buffers in C' is a subset of the set of buffers in C''. Let $\tilde{\mathcal{C}} \subseteq \mathcal{C}$ be such that*

$$(\forall C', C'' \in \tilde{\mathcal{C}})(C' \not\subseteq C''). \tag{7}$$

Then, Γ is deadlock-free if the DLF condition (equation (6)) is satisfied for all $C \in \tilde{\mathcal{C}}$.

2.3 Deadlock-Free Buffer Allocation Algorithm

Once the conditions for a deadlock-free network are known, the problem of finding a deadlock-free allocation with the least number of buffers can be investigated. We define *optimal buffer allocations* to be deadlock-free buffer allocations with the least total number of buffers.

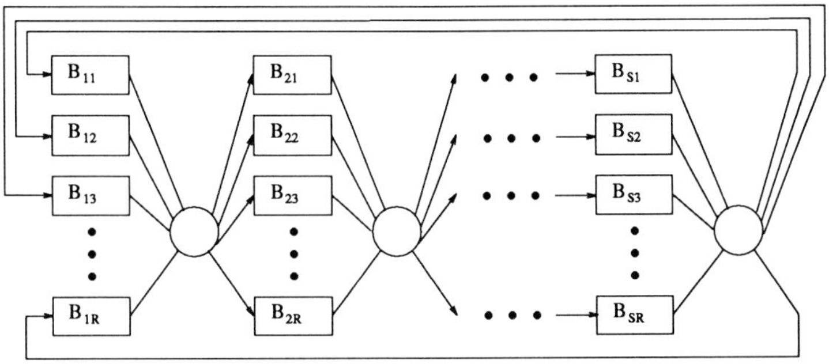

Figure 2: Tandem Network.

Definition 3 *A buffer allocation Φ^* for a queueing network Γ is optimal if:*

(i) Φ^* satisfies the DLF condition (equation (6)) for all $C \in \mathcal{C}$,

(ii) Φ^* is minimal, i.e., no deadlock-free allocation Φ for Γ allocates less capacities to the buffers of Γ than Φ^*:

$$\sum_{i \in \mathcal{N}} \sum_{r \in \mathcal{R}_i} \Phi^*(B_{ir}) \leq \sum_{i \in \mathcal{N}} \sum_{r \in \mathcal{R}_i} \Phi(B_{ir}) \tag{8}$$

In [2], an algorithm was presented that gives an optimal buffer allocation for a blocking network Γ with K_r jobs in each chain r. The computational effort to obtain an optimal buffer allocation is dependent on the number of buffer cycles in the network. In the following section we show that the optimization procedure can be made computationally efficient even if the number of cycles in the network grows large.

3 Special Network Topologies

Even though the optimal buffer allocation algorithm is applicable to all network topologies, the computational complexity of the algorithm increases with the size of $\tilde{\mathcal{C}}$, the set of cycles where no cycle is a subset of another. For some networks, the size of $\tilde{\mathcal{C}}$ can make the optimization algorithm impractical. As a worst case consider the network in Figure 2. Here, $\tilde{\mathcal{C}}$ is identical with the set of all buffer cycles in the network. Note that in the network in Figure 2 the set $\tilde{\mathcal{C}}$ contains R^S cycles. In the following we show that obtaining an optimal buffer allocation can be simplified, if we take advantage of structural properties of the network. Here we consider two types of networks, so-called *tandem networks* which have a topology as shown in Figure 2, and networks with so-called *tandem sequences*, i.e., networks that contain tandem networks as subnetworks.

Definition 4 *Given a multiple chain queueing network* Γ.

1. *A* tandem sequence *is a sequence of stations* $(\kappa_1, \kappa_2, \ldots, \kappa_{s-1}, \kappa_s)$ *such that*

$$\begin{array}{ll} \mathcal{R}_{\kappa_v} = \mathcal{R}_{\kappa_w} & (1 \leq v, w \leq s) \\ (\forall r \in \mathcal{R}_{\kappa_v})(p_{\kappa_v \kappa_{v+1}, r} = 1) & (1 \leq v < s) \\ (\forall r \in \mathcal{R}_{\kappa_v})(p_{i \kappa_{v+1}, r} = 0) & (1 \leq v < s, i \neq \kappa_v). \end{array} \tag{9}$$

2. *A* tandem network Γ *with* S *stations is a tandem sequence* $(\kappa_1, \kappa_2, \ldots, \kappa_S, \kappa_{S+1})$ *where* $\kappa_1 = \kappa_{S+1}$.

Next we show that finding an optimal buffer allocation can be simplified if the network is a tandem network or contains tandem sequences. In the next subsection we prove that optimal buffer allocations for tandem networks with blocking can be obtained without running any optimization procedure. Then we prove for blocking networks with tandem sequences that the efforts to find an optimal buffer allocation can be greatly reduced.

3.1 Optimal Buffer Allocation for Closed Tandem Networks

In the next Lemma we show that an optimal buffer allocation for a tandem network assigns nonzero buffer spaces to the buffers of only one station.

Lemma 2 *For a tandem network* Γ *with* S *stations and* R *routing chains, the following buffer allocations* $\hat{\Phi}_i$ $(i \in \mathcal{N})$ *are optimal:*

$$\hat{\Phi}_i(B_{jr}) = \begin{cases} K_r & \text{if } j = i \text{ and } \sum_{\rho \in \mathcal{R} \setminus \{r\}} K_\rho \geq S \text{ and } \sum_{r=1}^{R} K_r \geq S \\ K_r - S + \sum_{\rho \in \mathcal{R} \setminus \{r\}} K_\rho & \text{if } j = i \text{ and } \sum_{\rho \in \mathcal{R} \setminus \{r\}} K_\rho < S \text{ and } \sum_{r=1}^{R} K_r \geq S \\ 0 & \text{if } j = i \text{ and } \sum_{r=1}^{R} K_r < S \\ 0 & \text{if } j \neq i \end{cases} \quad \text{for } j \in \mathcal{N} \text{ and } r \in \mathcal{R}. \tag{10}$$

Proof:
(1) $\hat{\Phi}_i$ *is a deadlock-free allocation*
Let us first consider the case: $\sum_{r=1}^{R} K_r < S$. In this case, no state exists where all stations of Γ are occupied. Hence, for each cycle C there exists a chain $r' \in \mathcal{R}$ with:

$$|S_{r'}^{(C)}| > K_{r'}, \tag{11}$$

which satisfies equation (6). For the other cases, consider $i = \kappa$ $(1 \leq \kappa \leq S)$. In a tandem network, each buffer cycle C has to contain a buffer from station κ, say $B_{\kappa\tau}$ $(\tau \in \mathcal{R})$. Since:

$$\begin{array}{lll} \hat{\Phi}_\kappa(B_{\kappa\tau}) = K_\tau & \wedge\ |S_\tau^{(C)}| > 0, & \text{if } \sum_{\rho \in \mathcal{R} \setminus \{\tau\}} K_\rho \geq S, \\ \hat{\Phi}_\kappa(B_{\kappa\tau}) = K_\tau - \sum_{\rho \in \mathcal{R} \setminus \{\tau\}} K_\rho & \wedge\ |S_\tau^{(C)}| > S - \sum_{\rho \in \mathcal{R} \setminus \{\tau\}} K_\rho, & \text{if } \sum_{\rho \in \mathcal{R} \setminus \{\tau\}} K_\rho < S. \end{array} \tag{12}$$

We obtain for either case:

$$\hat{\Phi}_\kappa(B_{\kappa\tau}) + |S_\tau^{(C)}| > K_\tau. \tag{13}$$

This satisfies the DLF condition for chain τ in Cycle C.

(2) $\hat{\Phi}_i$ is minimal
Since buffer allocation $\hat{\Phi}_i$ is a deadlock-free allocation we can use it as an upper bound for any minimal allocation. To prove the minimality of allocations $\hat{\Phi}_i$ for chain r ($r \in \mathcal{R}$) we will show that any other allocation which allocates the same total capacities to the buffers as $\hat{\Phi}_i$ will contain a deadlock. We have to distinguish four cases:

(a) $\sum_{r=1}^{R} K_r < S$
In this case, no capacity is allocated to any buffer, i.e., $\hat{\Phi}_i$ is minimal.

(b) $K_r \geq S$ and $\sum_{\rho \in \mathcal{R}\setminus\{r\}} K_\rho \geq S$
Assume there exists an allocation Φ^* different from any $\hat{\Phi}_i$ with $\sum_{i \in \mathcal{N}_r} \Phi^*(B_{ir}) = K_r$. Without loss of generality we may assume that $R = 2$ with two chains r and r'. Let m ($m < S$) be such that m buffers in chain r ($B_{\kappa_1 r}, B_{\kappa_2 r}, \ldots, B_{\kappa_m r}$) satisfy:

$$\sum_{s=1}^{m} \Phi^*(B_{\kappa_s r}) + m > K_r. \tag{14}$$

Since $K_r > S$ we have:

$$\sum_{s=m+1}^{S} \Phi^*(B_{\kappa_s r}) + (S - m) < K_r, \tag{15}$$

and $\sum_{s=1}^{S} \Phi^*(B_{\kappa_s r}) \leq K_r$ implies:

$$\sum_{s=m+1}^{S} \Phi^*(B_{\kappa_s r}) < m. \tag{16}$$

Because of (15) chain r' must satisfy:

$$\sum_{s=1}^{m} \Phi^*(B_{\kappa_s r'}) + m > K_{r'}. \tag{17}$$

This allows to perform an induction over m.

Induction start: $m = 1$.
Then Φ^* is equal to allocation $\hat{\Phi}_{\kappa_1}$. But this violates the assumption that Φ^* is different from any allocation $\hat{\Phi}_i$.

Induction assumption:
Assume that for $m = l - 1$ no deadlock-free allocation Φ^* exists with $\sum_{i \in \mathcal{N}_r} \Phi^*(B_{ir}) = K_r$ which satisfies (14).

Induction step: $m = l$
Consider the set of all cycles which contain the buffer sequence ($B_{\kappa_1 r}, B_{\kappa_2 r}, \ldots, B_{\kappa_{l-1} r}, B_{\kappa_l r'}$):

$$\mathcal{C}' = \{(B_{\kappa_1 r}, B_{\kappa_2 r}, \ldots, B_{\kappa_{l-1} r}, B_{\kappa_l r'}, B_{\kappa_{l+1} \rho_{l+1}}, B_{\kappa_{l+2} \rho_{l+2}}, \ldots, B_{\kappa_S \rho_S}) \mid \rho_s \in \{r, r'\}\} \tag{18}$$

Assume a state where the jobs of chain r occupy all buffer space in $(B_{\kappa_1 r}, B_{\kappa_2 r}, \ldots, B_{\kappa_{l-1} r})$ and occupy servers $(\kappa_S, \kappa_2, \ldots, \kappa_{l-2})$. Assume further that jobs from chain r' occupy $B_{\kappa_l r'}$ and the server of station κ_{l-1}. Note that due to the induction assumption:

$$\sum_{s=1}^{l-1} \Phi^*(B_{\kappa_s r}) + (l-1) \leq K_r \tag{19}$$

and

$$\Phi^*(B_{lr'}) + 1 \leq K_{r'} \tag{20}$$

If the remaining jobs from class r and r' can be distributed over the remaining $S - l$ stations such that all these stations are blocked, then a deadlock occurs. Since any cycle $C \in \mathcal{C}'$ must be deadlock-free, the following condition must hold:

$$S - l > (K_r - \sum_{s=1}^{l-1} \Phi^*(B_{\kappa_s r}) - (l-1)) + (K_{r'} - B_{lr'} - 1) - z. \tag{21}$$

with $z = \sum_{s=l+1}^{S} \Phi^*(B_{\kappa_s \rho_s})$ $(\rho_s \in \{r, r'\})$. Let z_r $(z_{r'})$ be the sum of buffer capacities of chain r (r') which are allocated to buffers $B_{\kappa_{l+1}\rho}, B_{\kappa_{l+2}\rho}, \ldots, B_{\kappa_S \rho}$ of C, i.e.,

$$z_r = \sum_{s=l+1}^{S} (\Phi^*(B_{\kappa_s r}) \cdot \eta(C, \kappa_s, r)) \qquad z_{r'} = \sum_{s=l+1}^{S} (\Phi^*(B_{\kappa_s r'}) \cdot \eta(C, \kappa_s, r')) \tag{22}$$

with:

$$\eta(C, \kappa_s, r') = \begin{cases} 1 & \text{if } B_{\kappa_s r} \in C \\ 0 & \text{otherwise} \end{cases} \tag{23}$$

Note that $z_r, z_{r'} < l$ (equation (16)). Then for chain r:

$$\sum_{s=1}^{l} \Phi^*(B_{\kappa_s r}) + l - z_r = K_r. \tag{24}$$

We obtain for equation (21):

$$S - l > K_r - (K_r - l + z_r - B_{lr}) - (l-1) + K_{r'} - B_{lr'} - 1 - z, \tag{25}$$

We can choose a cycle C from $\mathcal{C}'$ such that $z_r + z \leq l$. Then :

$$S > K_{r'} + B_{lr'} - B_{lr}. \tag{26}$$

Now, consider the set of cycles

$$\mathcal{C}'' = \{(B_{\kappa_1 r'}, B_{\kappa_2 r'}, \ldots, B_{\kappa_{l-1} r'}, B_{\kappa_l r}, B_{\kappa_{l+1}\rho_{l+1}}, B_{\kappa_{l+2}\rho_{l+2}}, \ldots, B_{\kappa_S \rho_S}) \mid \rho_s \in \{r, r'\}\} \tag{27}$$

As before, we can construct cycles which have to satisfy the condition:

$$S > K_r + B_{lr} - B_{lr'}. \tag{28}$$

Adding up (26) and (28), we obtain:

$$2S > K_r + K_{r'} \tag{29}$$

But this is a contradiction, since $K_r \geq S$ and $K_{r'} \geq S$. Therefore, for $m = l$, Φ^* is not deadlock-free.

(c) $K_r \geq S$ *and* $\sum_{\rho \in \mathcal{R}\setminus\{r\}} K_\rho < S$
Since in this case

$$|S_r^{(C)}| > S - \sum_{\rho \in \mathcal{R}\setminus\{r\}} K_\rho \tag{30}$$

we can construct a tandem network $\overline{\Gamma}$ from Γ with $\overline{S}$ stations and $\overline{K}_\rho$ jobs in chain ρ with:

$$\begin{aligned} \overline{S} &= \sum_{\rho \in \mathcal{R}\setminus\{r\}} K_\rho \\ \overline{\mathcal{R}} &= \mathcal{R} \\ \overline{K}_r &= K_r - (S - \sum_{\rho \in \mathcal{R}\setminus\{r\}} K_\rho) \\ \overline{K}_\rho &= K_\rho \qquad\qquad (\rho \in \mathcal{R} \setminus \{r\}) \end{aligned} \tag{31}$$

Per construction, Γ contains a deadlock if $\overline{\Gamma}$ contains a deadlock. Since $\sum_{\rho \in \mathcal{R}\setminus\{r\}} K_\rho \geq \overline{S}$, case (b) applies.

(d) $K_r < S$ *and* $\sum_{\rho \in \mathcal{R}\setminus\{r\}} K_\rho \geq S$
Since in this case

$$\sum_{\rho \in \mathcal{R}\setminus\{r\}} |S_\rho^{(C)}| > S - K_r \tag{32}$$

we can construct a tandem network $\overline{\Gamma}$ from Γ with $\overline{S}$ stations and $\overline{K}_\rho$ jobs in chain ρ with:

$$\begin{aligned} \overline{S} &= K_r \\ \overline{\mathcal{R}} &= \mathcal{R} \\ \overline{K}_r &= K_r \\ \overline{K}_\rho &= K_\rho - k_\rho \qquad (\rho \in \mathcal{R} \setminus \{r\}) \end{aligned} \tag{33}$$

with:

$$\sum_{\rho \in \mathcal{R}\setminus\{r\}} k_\rho = S - K_r \tag{34}$$

Per construction, Γ contains a deadlock if $\overline{\Gamma}$ contains a deadlock. Since $K_r \geq \overline{S}$, case (b) applies. □

If $R = 1$, then $\sum_{\rho \in \mathcal{R}\setminus\{r\}} K_\rho = 0$. Therefore, an optimal buffer allocation for a single chain network can be obtained from Lemma 2 by:

$$\hat{\Phi}_i(B_j) = \begin{cases} K - S & \text{if } j = i \\ 0 & \text{if } j \neq i \end{cases} \tag{35}$$

From the proof of Lemma 2, we can follow immediately:

Corollary 1 *A buffer allocation Φ^* for a tandem network with S stations and R routing chains with $\sum_{\rho\in\mathcal{R}\setminus\{r\}} K_\rho \geq S$ (for $1 \leq r \leq R$) is optimal, if and only if there exists an $i \in \mathcal{N}$ such that for all $j \in \mathcal{N}$ and $r \in \mathcal{R}_j$:*

$$\Phi^*(B_{jr}) = \hat{\Phi}_i(B_{jr}) \tag{36}$$

3.2 Optimal Buffer Allocation for Networks with Tandem Sequences

With Lemma 2 we are able to directly give an optimal buffer allocation without running any optimization. However, the Lemma is only applicable for the special class of tandem networks. Next we show how to take advantage of the regularity of networks which contain a tandem sequence, i.e, contain a network like the network in Figure 2 as a subnetwork. The following Lemma prepares the so-called *Reduction Lemma* which will be presented at the end of this section. The *Reduction Lemma* allows to perform the allocation algorithm in [2] on a network where each tandem sequence is substituted by a single station. The optimal buffer allocation from the reduced network can be used to obtain an optimal allocation for the original network.

Lemma 3 *Given an optimal buffer allocation Φ^* for a network Γ. If Γ contains a tandem sequence of S stations $(\kappa_1, \kappa_2, \ldots, \kappa_{S-1}, \kappa_S)$, the following buffer allocation $\hat{\Phi}$ is optimal:*

$$\hat{\Phi}(B_{ir}) = \begin{cases} \sum_{v=1}^{S} \Phi^*(B_{\kappa_v r}) & \text{if } i = \kappa_1 \\ 0 & \text{if } i = \kappa_v \text{ and } 1 < v \leq S \\ \Phi^*(B_{ir}) & \text{otherwise} \end{cases} \quad \text{for } i \in \mathcal{N}, r \in \mathcal{R}_i. \tag{37}$$

Proof: $\hat{\Phi}$ is per definition minimal. It remains to be shown that $\hat{\Phi}$ yields a deadlock-free network. Assume the network with allocation $\hat{\Phi}$ is in a deadlock. The deadlock has to be along a cycle which includes one buffer from each of stations $(\kappa_1, \kappa_2, \ldots, \kappa_{S-1}, \kappa_S)$. Otherwise, Γ has a deadlock under Φ^*. Note that Φ^* is equal to $\hat{\Phi}$ for buffers of stations not in the tandem sequence. Let the buffer cycle which contains a possible deadlock be given by:

$$\hat{C} = (B_{\sigma_1\rho_1}, B_{\sigma_2\rho_2}, \ldots, B_{\kappa_1\tau_1}, B_{\kappa_2\tau_2}, \ldots, B_{\kappa_{S-1}\tau_{S-1}}, B_{\kappa_S\tau_S}, \ldots, B_{\sigma_{t-1}\rho_{t-1}}, B_{\sigma_t\rho_t}) \tag{38}$$

Consider the tandem network Ψ obtained from Γ by eliminating the stations not belonging to the tandem sequence. Let $\mathcal{N}_\Psi$, $\mathcal{R}_\Psi$, $K_{\Psi,r}$ denote the set of stations, the set of classes and the number of jobs, respectively, in Ψ, with:

$$\begin{aligned} \mathcal{N}_\Psi &= \{B_{\kappa_1\tau_1}, B_{\kappa_2\tau_2}, \ldots, B_{\kappa_{S-1}\tau_{S-1}}, \ldots, B_{\sigma_S\rho_S}\}, \\ \mathcal{R}_\Psi &= \mathcal{R}_{\kappa_1}, \\ K_{\Psi,r} &= K_r - (\sum_{1<u\leq t \wedge \sigma_u \in \overline{S}_r^{(\hat{C})}} \Phi^*(B_{\sigma_u r}) + |\overline{S}_r^{(\hat{C})}|), \end{aligned} \tag{39}$$

where:

$$\overline{S}_r^{(C)} = S_r^{(C)} \setminus \{\kappa_1, \kappa_2, \ldots, \kappa_{S-1}, \kappa_S\}. \tag{40}$$

Any allocation Φ for network Γ defines an allocation for Ψ. Per definition of Ψ, allocation $\hat{\Phi}$ for Ψ violates the DLF condition for cycle $\tilde{C}$ with:

$$\tilde{C} = (B_{\kappa_1\tau_1}, B_{\kappa_2\tau_2}, \ldots, B_{\kappa_{S-1}\tau_{S-1}}, B_{\sigma_S\rho_S}), \tag{41}$$

if $\widehat{C}$ in Γ under $\hat{\Phi}$ violates the DLF condition. According to Lemma 2, we can find a minimal and deadlock-free buffer allocation for Ψ with:

$$\hat{\Phi}(B_{ir}) = 0, \qquad \text{if } i = \kappa_v \text{ and } 1 < v \leq S. \tag{42}$$

Therefore, if $\hat{\Phi}$ contains a deadlock, then the assumption that Φ^* is optimal for Γ does not hold.
□

Lemma 4 (Reduction Lemma) *Given a network Γ with a tandem sequence $(\kappa_1, \kappa_2, \ldots, \kappa_{S-1}, \kappa_S)$ with $r \in \mathcal{R}_{\kappa_1}$. Obtain a network Γ' by substituting the tandem sequence in Γ by a single station $\tilde{\kappa}$. Let $\tilde{\Phi}$ be a minimal allocation for Γ'. Define for all $r \in \mathcal{R}_{\tilde{\kappa}}$*

$$M_{\tilde{\kappa}r} = \max\{ \sum_{\rho \in \mathcal{R}\setminus\{r\}} (K_\rho - \sum_{1<u\leq t \wedge \sigma_u \in S_r^{(C)}} \tilde{\Phi}(B_{\sigma_u\rho}) - |S_\rho^{(C)}|) \mid C \text{ Cycle in } \Gamma',\ B_{\tilde{\kappa}r} \in C\}. \tag{43}$$

If the condition holds that:

$$M_{\tilde{\kappa}r} + \tilde{\Phi}(B_{\tilde{\kappa}r}) \geq S \qquad \text{for all } r \in \mathcal{R}_{\tilde{\kappa}}, \tag{44}$$

then the following buffer allocation Φ^* is optimal for Γ:

$$\Phi^*(B_{ir}) = \begin{cases} \tilde{\Phi}(B_{\tilde{\kappa}r}) & \text{if } i = \kappa_1 \text{ and } M_{\tilde{\kappa}r} \geq S-1 \\ \tilde{\Phi}(B_{\tilde{\kappa}r}) - (S-1) + M_{\tilde{\kappa}r} & \text{if } i = \kappa_1 \text{ and } M_{\tilde{\kappa}r} < S-1 \\ 0 & \text{if } i = \kappa_v \text{ and } 1 < v \leq S \\ \tilde{\Phi}(B_{ir}) & \text{otherwise} \end{cases} \tag{45}$$
$$\text{for } i \in \mathcal{N}, r \in \mathcal{R}_i.$$

Proof:
(1) Φ^* is a deadlock-free allocation
We only have to consider cycles which include buffers from stations of the tandem sequence. Assume a cycle C' in Γ' which includes buffer $B_{\tilde{\kappa}\tau_1}$ ($\tau_1 \in \mathcal{R}_{\tilde{\kappa}}$):

$$C' = (B_{\sigma_1\rho_1}, B_{\sigma_2\rho_2}, \ldots, B_{\tilde{\kappa}\tau_1}, \ldots, B_{\sigma_{t-1}\rho_{t-1}}, B_{\sigma_t\rho_t}) \tag{46}$$

C' corresponds to a set of cycles in Γ:

$$\mathcal{C}'' = \{(B_{\sigma_1\rho_1}, B_{\sigma_2\rho_2}, \ldots, B_{\kappa_1\tau_1}, B_{\kappa_2\tau_2}, \ldots, B_{\kappa_S\tau_S}, \ldots, B_{\sigma_{t-1}\rho_{t-1}}, B_{\sigma_t\rho_t}) \mid \tau_v \in \mathcal{R}_{\tau_1}, 1 \leq v \leq S\} \tag{47}$$

(a) $M_{\tilde{\kappa}r} \geq S-1$
Then for all $C'' \in \mathcal{C}''$ and for all $r \in \mathcal{R}$ the following holds:

$$\sum_{1<u\leq t \wedge \sigma_u \in S_r^{(C')}} \Phi^*(B_{\sigma_u r}) = \sum_{1<u\leq t \wedge \sigma_u \in S_r^{(C'')}} \tilde{\Phi}(B_{\sigma_u r}), \tag{48}$$

and:

$$|S_r^{(C')}| \leq |S_r^{(C'')}|, \tag{49}$$

Since C' is deadlock-free in Γ' with $\tilde{\Phi}$, it follows from (48) and (49) that all $C'' \in \mathcal{C}''$ are deadlock-free in Γ with Φ^*.

(b) $M_{\tilde{\kappa}r} < S - 1$
Recall from (45) that in this case $\Phi^*(B_{\tilde{\kappa}r}) < \tilde{\Phi}(B_{\tilde{\kappa}r})$. If the DLF condition for cycle C' in Γ' is satisfied by chain $\rho \neq \tau_1$, then equations (48) and (49) hold for chain ρ. But then, the DLF condition for $C'' \in \mathcal{C}''$ in Γ is satisfied by chain $\rho \neq \tau_1$. Now assume that chain τ_1 satisfies the DLF condition for cycle C' in Γ', i.e.,

$$\sum_{1<u\leq t \wedge \sigma_u \in S_{\tau_1}^{(C')}} \Phi^*(B_{\sigma_u \tau_1}) + |S_{\tau_1}^{(C')}| > K_{\tau_1} \tag{50}$$

Since $M_{\tilde{\kappa}r} < S - 1$, at least $((S-1) - M_{\tilde{\kappa}r})$ of the servers of stations $(\kappa_2, \kappa_3, \ldots, \kappa_S)$ have to be occupied by jobs of chain τ_1 if a deadlock persists in a cycle $C'' \in \mathcal{C}''$. But then:

$$|S_{\tau_1}^{(C'')}| \geq |S_{\tau_1}^{(C')}| + ((S-1) - M_{\tilde{\kappa}\tau_1}) \tag{51}$$

Note that from (45):

$$\sum_{1<u\leq t \wedge \sigma_u \in S_{\tau_1}^{(C')}} \Phi^*(B_{\sigma_u \tau_1}) = \sum_{1<u\leq t \wedge \sigma_u \in S_{\tau_1}^{(C'')}} \tilde{\Phi}(B_{\sigma_u \tau_1}) - ((S-1) - M_{\tilde{\kappa}\tau_1}). \tag{52}$$

From equations (51) and (52) we see immediately, that chain τ_1 satisfies the DLF condition for all $C'' \in \mathcal{C}''$ of Γ.

(2) Φ^* *is minimal*
Because of Lemma 3 we only need to consider allocations which assign capacities to station κ_1 and no capacities at all to stations κ_v $(1 < v \leq S)$ in Γ. Since $\tilde{\Phi}$ is optimal for Γ', network Γ' will have a deadlock if we assign less capacity than $\tilde{\Phi}(B_{\tilde{\kappa}\tau_1})$ to buffer $B_{\tilde{\kappa}\tau_1}$. Take a cycle $\widehat{C}$ in Γ' with, which satisfies the DLF condition for chain $\tilde{\kappa}$ if a capacity of $\tilde{\Phi}(B_{\tilde{\kappa}\tau_1})$ is assigned to $B_{\tilde{\kappa}\tau_1}$, but does not so if the capacity is $(\tilde{\Phi}(B_{\tilde{\kappa}\tau_1}) - 1)$. $\widehat{C}$ has the same structure as the cycle in (46). Let $\widehat{\widehat{\mathcal{C}}}$ be the set of cycles in Γ which correspond to $\widehat{C}$ in Γ'. $\widehat{\widehat{\mathcal{C}}}$ has the same structure as set $\mathcal{C}'$ in (47). For a cycle $\widehat{\widehat{C}} \in \widehat{\widehat{\mathcal{C}}}$ we distinguish two cases:

(a) $M_{\tilde{\kappa}r} \geq S - 1$
In this case, the servers of stations $(\kappa_2, \kappa_3, \ldots, \kappa_S)$ can be occupied by jobs of chains from $\mathcal{R} \setminus \{r\}$ such that no $\rho \in \mathcal{R} \setminus \{r\}$ satisfies the DLF condition for cycle $\widehat{\widehat{C}}$. Then:

$$\sum_{1<u\leq t \wedge \sigma_u \in S_{\tau_1}^{(\widehat{\widehat{C}})}} \Phi^*(B_{\sigma_u \tau_1}) = \sum_{1<u\leq t \wedge \sigma_u \in S_{\tau_1}^{(\widehat{C})}} \tilde{\Phi}(B_{\sigma_u \tau_1}), \tag{53}$$

and:

$$|S_{\tau_1}^{(\widehat{\widehat{C}})}| = |S_{\tau_1}^{(\widehat{C})}|, \tag{54}$$

Therefore, any allocation with:

$$\Phi^*(B_{\kappa_1 \tau_1}) < \tilde{\Phi}(B_{\tilde{\kappa}\tau_1}) \tag{55}$$

does not satisfy the DLF condition for chain τ_1, and thus, a deadlock persists for all $\hat{\hat{C}} \in \hat{\hat{\mathcal{C}}}$.

(b) $M_{\tilde{\kappa}r} < S - 1$

Here $M_{\tilde{\kappa}\tau_1}$ of the servers of stations $(\kappa_2, \kappa_3, \ldots, \kappa_S)$ can be occupied by jobs of chains from $\mathcal{R} \setminus \{r\}$ without satisfying the DLF condition for any chain $\rho \in \mathcal{R} \setminus \{r\}$. If condition (44) holds, then the remaining $((S-1) - M_{\tilde{\kappa}\tau_1})$ servers can be occupied by jobs from chain τ_1. Then:

$$\begin{aligned}
&\sum_{1<u\leq t \wedge \sigma_u \in S_{\tau_1}^{(\hat{\hat{C}})} \setminus \{\kappa_1\}} \Phi^*(B_{\sigma_u \tau_1}) + \Phi^*(B_{\kappa_1 \tau_1}) + |S_{\tau_1}^{(\hat{\hat{C}})}| \\
&= \sum_{1<u\leq t \wedge \sigma_u \in S_{\tau_1}^{(\hat{C})} \setminus \{\kappa_1\}} \Phi^*(B_{\sigma_u \tau_1}) + (\tilde{\Phi}(B_{\tilde{\kappa}\tau_1}) - (S-1) + M_{\tilde{\kappa}\tau_1}) + (|S_{\tau_1}^{(\hat{C})}| + (S-1) - M_{\tilde{\kappa}\tau_1}) \\
&= \sum_{1<u\leq t \wedge \sigma_u \in S_{\tau_1}^{(\hat{C})} \setminus \{\tilde{\kappa}\}} \tilde{\Phi}(B_{\sigma_u \tau_1}) + \tilde{\Phi}(B_{\tilde{\kappa}\tau_1}) + |S_{\tau_1}^{(\hat{C})}|
\end{aligned} \tag{56}$$

But then, any allocation with:

$$\Phi^*(B_{\kappa_1 \tau_1}) < \tilde{\Phi}(B_{\tilde{\kappa}\tau_1}) - (S-1) + M_{\tilde{\kappa}\tau_1} \tag{57}$$

does not satisfy the DLF condition for chain τ_1. We follow from the construction of $\hat{\hat{\mathcal{C}}}$ that network Γ contains a deadlock for all $\hat{\hat{C}} \in \hat{\hat{\mathcal{C}}}$. □

4 Example

In Figure 3 we show a multiple chain queueing network with a tandem sequence of length $S = 5$. Obtaining an optimal buffer allocation algorithm for networks which include a tandem sequence is a formidable task because of the high number of cycles. Even after eliminating those cycles that are a superset of other cycles according to Lemma 1, the network in Figure 3 contains 2561 cycles that have to be considered for the buffer allocation algorithm. With the *Reduction Lemma* (Lemma 4) we can replace the tandem sequence, i.e., stations $(6, 7, 8, 9, 10)$, by a single station. The reduced network contains only 11 cycles which have to be considered for the optimization. Since the *Reduction Lemma* is valid only if the condition in equation (44) is satisfied, we have to show that the reduction is valid. Note that none of the cycles $(C_1, C_2, \ldots, C_{11})$ includes buffers from all routing chains and $K_r > S$ for $r = 1, 2, 3, 4$. Therefore, condition (44) is clearly satisfied for all routing chains.

Applying the optimization algorithm in [2] to the reduced network yields an optimal buffer allocation $\tilde{\Phi}$. If we let $B_{\{6,7,8,9,10\},r}$ (for $r = 1, 2, 3, 4$) denote the chain-r-buffer of the station which replaces the tandem sequence in the reduced network we obtain an

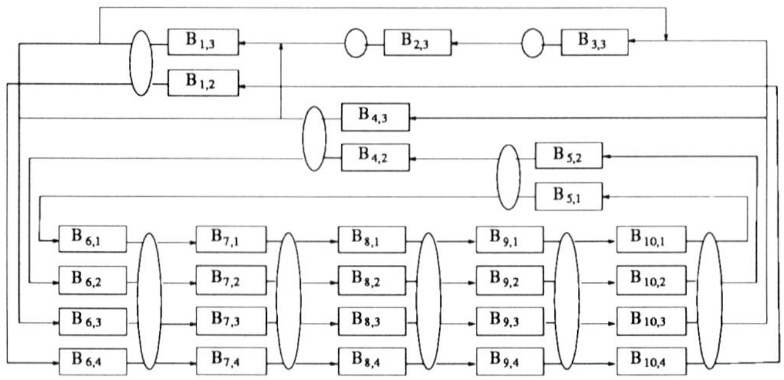

Figure 3: Multiple Chain Queueing Network with a Tandem Sequence.

optimal allocation for the network in Figure 3 by:

$$\Phi^*(B_{ir}) = \begin{cases} \tilde{\Phi}(B_{\{6,7,8,9,10\},r}) & \text{if } i = 6 \\ 0 & \text{if } i \in \{7,8,9,10\} \\ \tilde{\Phi}(B_{ir}) & \text{otherwise} \end{cases}$$

5 Conclusions

We investigated deadlock-free buffer allocations in blocking network with multiple routing chains. In order to show that a network is deadlock-free, all cycles in the network need to satisfy the deadlock freedom condition. Networks with so-called tandem sequences are characterized by a high number of cycles. Therefore, finding all cycles becomes computational impractical even for small networks. We showed that the problem of finding all cycles can be dealt with by substituting tandem sequences of a queueing network with single stations. The so-called *Reduction Lemma* guarantees that an optimal deadlock-free buffer allocation for the substitute network is an optimal deadlock-free buffer allocation for the original network. In addition, an optimal deadlock-free buffer allocation for networks consisting of only one tandem sequence can be directly obtained without applying any optimization procedure.

References

[1] I. F. Akyildiz and H. G. Perros, "Queueing Networks with Finite Capacity Queues", *Special Issue in Performance Evaluation*, Vol. 10, No. 4, December 1989.

[2] I. F. Akyildiz and J. Liebeherr, "Optimal Deadlock Free Buffer Allocation in Multiple Chain Blocking Networks of Queues", *Proc. IFIP TC7/WG 7.3 International Conference on Performance of Distributed Systems and Integrated Communication Networks*, Kyoto, Japan, September 1991, pp. 217-236.

[3] S. Kundu and I. F. Akyildiz, "Deadlock Free Buffer Allocation in Closed Queueing Networks", *Queueing Systems*, Vol. 4, No. 1, January 1989, pp. 47 - 56.

Queueing Networks with Finite Capacity
R.O. Onvural and I.F. Akyildiz (Editors)

Approximate analysis of a discrete-time tandem network of cut-through queues with blocking and bursty traffic[1]

Todd D. Morris[2] and Harry G. Perros[3]

[2]Bell Northern Research, P.O. Box 3511, Station C, Ottawa, Canada K1Y 4H7

[3]Computer Science Department, and Center for Communications and Signal Processing, North Carolina State University, Raleigh, N.C. 27695-8206

Abstract

A discrete-time tandem queueing network of cut-through queues is presented. The model allows finite capacity queues, blocking, and bursty traffic. A new bursty arrival process, IBP(k), for cut-through traffic is introduced. The tandem network is analyzed using single-node decomposition. The results obtained are approximate and validation tests have shown that the model has good accuracy.

1. INTRODUCTION

The Asynchronous Transfer Mode (ATM) is the preferred solution for high-speed networks. ATM provides the capabilities of switching bursty traffic such as voice, video and bulk files. Several ATM switch designs have been proposed so far. In general, packets are routed through an ATM switch using circuit or packet like switching techniques. An alternative approach is cut-through switching. In this case, a packet is broken into k subpackets and the subpackets proceed consecutively through the switch. That is, only a single subpacket need arrive at a stage of a switch before the subpacket is forwarded to the next stage of the switch. The operation of forwarding subpackets before the entire packet arrives is termed a *cut*. In order to avoid duplication of routing information for each subpacket, the k subpackets proceed without interruption through the switching structure. If no contention is encountered, then cut-through switching proceeds like circuit switching. Cut-through switches, like packet switches, provide synchronous switching with a back-pressure mechanism which prevents packet and subpacket loss within the switch. This back-pressure is in fact the familiar *blocking-before-service* mechanism.

[1] Supported in part by DARPA under grant no. DAEA 18-90-c-0039.

Cut-through switching was introduced and analyzed by Kermani and Kleinrock [1]. Poisson arrivals, exponential message lengths, infinite buffers, and uniform destination probabilities were assumed. A cut-through node was modelled by an M/M/m system. For simplicity, *partial cuts* were ignored. A *full cut* occurs when the appropriate output port of a switching element is free at the time a packet destined for that respective port enters the switching element. Therefore, the packet need not be buffered at all. A partial cut occurs when an incoming message is blocked for some period of time smaller than the total arrival time of the message or packet. The message should be able to make a partial cut, and thus, continue to the output port before it has arrived entirely. Kermani and Kleinrock in [2] compared cut-through switching to circuit and packet switching. It was found that cut-through switching degrades to packet switching under heavy loading (i.e. no cuts at all). With very light loading, cut-through switching performs like circuit switching since full cuts always occur. Shin and Un [3] analysed a cut-through M/D/1 queue. Full and partial cuts were allowed. Ilyas and Mouftah [4], [5] and Abo-Taleb and Mouftah [6] considered a tandem network of cut-through queues where each queue was represented by an M/G/1 queue. Abo-Taleb and Mouftah [7] extended this work to G/G/1 queues. No validation results were reported for all these models. Marsan et al [8] modelled an input buffered multi-stage interconnection network employing cut-through switching. The input port queues were modelled by an M/M/1/k queue. The network was represented by a tandem configuration of switching elements. Simulation results were compared to the model and the model was found to have a good accuracy. Exponential packet lengths, Poisson arrivals, and uniform destination distributions were assumed for both the model and simulation. In general, most cut-through switching models assume infinite queues although this assumption has yet to be justified through simulation. In one model, finite queues were assumed. However, packet lengths were assumed to be exponentially distributed, which greatly simplifies the model state description. None of the reported models incorporate bursty traffic; Poisson traffic is always assumed.

In this paper, we present a discrete-time tandem network with cut-through queues, blocking-before-service and bursty arrivals. Discrete-time queueing networks with blocking have only been considered within the context of production systems (see for instance Gershwin [9]). Recently, Morris and Perros [10] analyzed a buffered Banyan ATM switch as a network of discrete-time nodes with blocking-before-service. The model considered here differs from that reported in [10] since we assume cut-through traffic. To the best of the authors' knowledge such a model has not as yet been reported in the litterature.

The paper is organized as follows. In section 2, we describe the structure of the discrete-time tandem network under study, and we introduce a new arrival process, IBP(k), for cut-through traffic. The tandem network is analyzed using single-node decomposition. Each queue is analyzed in isolation using numerical techniques. The decomposition algorithm is presented in section 3, and its validation is discussed in section 4. Finally, the conclusions are given in section 5.

2. THE TANDEM NETWORK OF CUT-THROUGH QUEUES

Let us consider a tandem configuration of finite capacity queues. We assume that the customers in the queueing network are fixed size packets. Each packet is formed by a fixed number, k, of subpackets. All subpackets have the same size. The first subpacket contains the routing information, and in view of this, it will be referred to as the *header subpacket*. The remaining k-1 subpackets represent the rest of the packet, and they will be referred to as the *body subpackets*. We number the body subpackets from 1 to k-1.

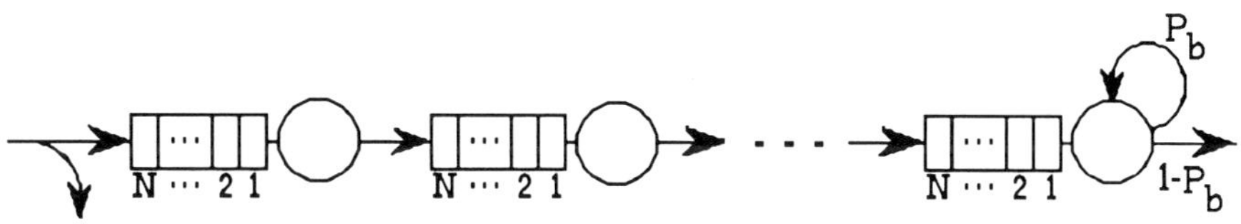

Figure 1: A Tandem Network of Cut-Through Queues

The subpackets in each queue are served individually. In particular, each server is associated with a slotted time axis. Service is constant, and it begins at the beginning of a slot and ends at the end of the same slot. These slots are referred to as *service slots*. The length of a slot is equal to the time required to send (i.e. transmit and propagate) one subpacket from one queue to another. The servers in the queueing network are not synchronized. That is, they do not start and end service at the same time. However, they all have the same service slots. The arrival of subpackets to a queue is not synchronized with the service slots of the server. A subpacket will arrive sometime during a service slot. If the server is idle, the subpacket will not be served until the beginning of the next service slot. The basic operation of a cut-through queue is as follows:

1. If a subpacket arrives to the first queue at a time when the first queue has at least one free subpacket space, the subpacket enters the queue. If the queue is full then the subpacket is lost. This loss of the subpacket causes the entire packet to get lost. That is, any subpackets already in the first queue which belong to the same packet are removed from the queue. (Note that as long as the first queue can contain at least one entire packet, i.e. N≥1, these other subpackets will always be found in the first queue.) Also, the remaining subpackets belonging to the lost packet that have not yet arrived will get lost upon arrival irrespective of whether or not there is space available in the queue.
2. Once a subpacket is buffered in the first queue, the subpacket proceeds in a FIFO manner through the queue.
3. When a subpacket reaches the head of the queue, it will be forwarded to the next queue in one service slot if the downstream queue has sufficient space for the subpacket. If there is not sufficient free space, the subpacket will wait until the next service slot. If at that time there exists a free space in the downstream queue, the subpacket will be sent out. Else, it will wait for another service slot. This will repeat until the subpacket enters the next queue. This mechanism prevents subpacket loss within the tandem network. It is identical to the blocking-before-service mechanism that has

been considered in queueing networks with blocking. We note that all queues function so that a full queue will accept another subpacket if the subpacket at the head of that queue is departing at the same time.

A tandem network of cut-through queues is shown in figure 1, where N is the capacity of a queue expressed in packets. That is, each queue can accommodate kN subpackets. We note that this tandem configuration was motivated by the buffered Banyan ATM switch, where all queues have the same capacity. The approximation algorithm presented in this paper is equally applicable to the case where each queue has a different capacity. We note that the last server of the queueing network cannot get blocked. Since all servers work at the same rate, the analysis of the queueing network becomes trivial! In view of this, we assume that the last server provides a geometrically distributed service with parameter P_b. We now proceed to describe the arrival process in detail.

2.1. The cut-through arrival process

The arrival process to the first queue of the tandem cut-through network is assumed to be bursty. Bursty traffic can be effectively modelled by an Interrupted Bernoulli Process (IBP) (see for instance Morris and Perros [10] and Nilsson, Lai, and Perros [11]). An IBP is associated with a slotted time axis, where each slot is assumed to be long enough to contain one packet. There is a geometrically distributed period during which no arrivals occur (idle state), followed by a geometrically distributed period during which arrivals occur in a Bernoulli fashion (active state). Given that the process is in the active state respectively idle state at slot t, it will remain in the same state in the next slot t+1 with probability p respectively q. During the active state, a slot contains a packet with probability α. The quantity α is also known as the peak bandwidth, i.e., the rate of arrivals during the active period. Let t_A (t_I) be the time until the next packet arrival given that the IBP process is currently in the active (idle) state. The interarrival time of packets for the IBP process can thus be represented as (also see [11]):

$$t_A = \begin{cases} 1 & ,p\alpha \\ 1+t_A & ,p(1-\alpha) \\ 1+t_I & ,1-p \end{cases} \quad \text{and} \quad t_I = \begin{cases} 1+t_I & ,q \\ 1 & ,(1-q)\alpha \\ 1+t_A & ,(1-q)(1-\alpha) \end{cases}$$

An explanation of this representation is as follows: if the arrival process is currently in the active period, A, then the time until the next arrival, t_A, will be one slot with probability $p\alpha$, $1+t_A$ slots with probability $p(1-\alpha)$, and $1+t_I$ slots with probability 1-p. Similar reasoning can be applied to the time until the next arrival given that the IBP is currently in the idle state (t_I). The z-transform of the probability distribution of the interarrival time between packets is given by

$$E\{z^{t_A}\} = \frac{z\alpha[p+z(1-p-q)]}{(1-\alpha)(p+q-1)z^2 - [q+p(1-\alpha)]z + 1} .$$

The mean interarrival time, denoted by $1/\rho$, is given by

$$\frac{1}{\rho} = \frac{2\text{-}p\text{-}q}{\alpha(1\text{-}q)} .$$

The average arrival rate, ρ, is otherwise known as the average bandwidth. The squared coefficient of variation, C^2, of the time between successive arrivals is

$$C^2 = 1+\alpha\left(\frac{(1\text{-}p)(p+q)}{(2\text{-}p\text{-}q)^2} - 1\right) .$$

For $\alpha = 1$, parameters p and q can be obtained by the following expressions:

$$p = 1 - \frac{2}{C^2(A+1)^2 + A + 1} \quad \text{and} \quad q = 1 - A(1\text{-}p) , \quad \text{where } A = \frac{\rho}{1\text{-}\rho} .$$

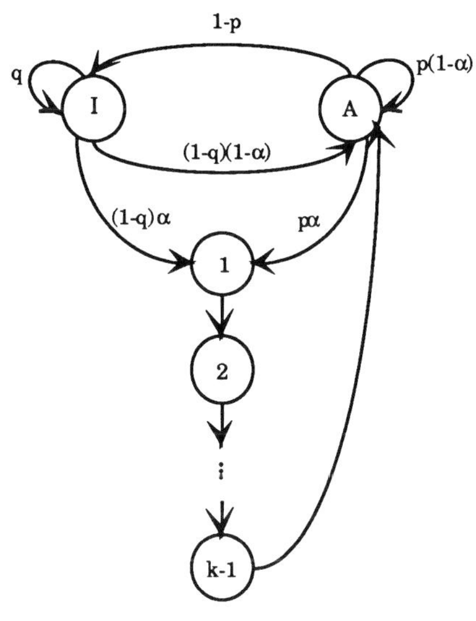

Figure 2: The Markov chain of the IBP(k)

The IBP process can be used to model bursty sources in a packet switched environment. With cut-through switching, the bursty source must not only provide a bursty interarrival time between packets, but the process must also function at the subpacket level. Thus, when the header subpacket of a new packet arrives, the body subpackets follow during the next k-1 slots. Figure 2 provides the Markov chain of this arrival process which we shall refer to as an

IBP(k). In this proces we assume that a slot is long enough to contain one subpacket. As in the IBP case, there is an idle period and an active period. The idle period is geometrically distributed with parameter q. That is, if slot t is idle, then with probability q, slot t+1 will also be idle, and with probability 1-q, slot t+1 will be active. During the active period, a slot contains a header subpacket with probability α. If a slot contains a header subpacket, then the next k-1 slots will contain body subpackets belonging to the same packet. States 1 to k-1 in figure 2 are associated with the arrival of these k-1 body subpackets. We shall refer to these states as *subpacket counting states*. If the IBP(k) is in first subpacket counting state, then a header subpacket has just arrived. If it is in the ith subpacket counting state, then i subpackets have already arrived. Therefore, the last subpacket arrives during the transition from state k-1 to state A. The time between the last body subpacket and the next header subpacket is distributed identically to the interarrival distribution of an IBP. That is, once the last body subpacket arrives, the next slot is active with probability p or idle with probability 1-p. If the next slot is active, then a new header subpacket arrives with probability α. The original IBP is an IBP(1) process under this terminology. Also, note that a slot of this arrival process is identical to a service slot.

It is of interest to determine the mean interarrival time of packets and the squared coefficient of variation of the interarrival time for this IBP(k) process. The interarrival time of packets is the same as the interarrival time of header subpackets. This is the time before the next header subpacket arrival given that the IBP(k) is in state 1. Given that a header subpacket arrives, the distribution of the interarrival time between header subpackets is the convolution of the constant delay k-1, associated with the subpacket counting states, and the distribution of the interarrival time of an IBP. Using the previous definitions for t_A and t_I, and by representing the time before the next header subpacket arrival, given that a header subpacket just arrived, by t_1 we have $t_1 = k-1+t_A$. In other words, with a probability of one, the interarrival time between header subpackets is a constant delay of k-1 added to the interarrival time of a standard IBP. The z-transform of the probability distribution of the interarrival time of packets is given by

$$E\{z^{t_1}\} = z^{k-1}\, E\{z^{t_A}\}.$$

The mean interarrival time of packets (or header subpackets) for the IBP(k) process is

$$\frac{1}{\rho} = k-1+\frac{2-p-q}{\alpha(1-q)} ,$$

and the squared coefficient of variation, C^2, of the time between successive arrivals is

$$C^2 = \frac{(p+q)^2(1-\alpha) + 4(1-\alpha-p-q) - \alpha(p-5)(p+q)}{(2-p-q-\alpha(1-k)(1-q))^2} .$$

For $\alpha = 1$, parameters p and q can be determined as follows:

$$q = \frac{\frac{C^2}{\rho^2} - (\frac{1}{\rho} - k)(1 - \frac{1}{\rho} + k)}{\frac{C^2}{\rho^2} + (\frac{1}{\rho} - k)(1 + \frac{1}{\rho} - k)} \quad \text{and} \quad p = 2 - q - (1 + \frac{1}{\rho} - k)(1 - q)\ .$$

The squared coefficient of variation, C^2, of the time between successive arrivals is used in this paper as a measure of burstiness. In view of this, it is interesting to relate the C^2 of the IBP process to that of the IBP(k) process. Every packet arrival in the IBP process takes one slot whereas a packet in the IBP(k) process requires k slots to arrive. This affects the value of p. We distinguish between the parameters p and q of an IBP and IBP(k) with the subscripts $_{IBP}$ and $_{IBP(k)}$ respectively. If each subpacket is treated as a separate entity as in the IBP process and $\alpha = 1$, we have:

$$p_{IBP} = 1 - \frac{1 - p_{IBP(k)}}{k} \quad \text{and} \quad q_{IBP} = q_{IBP(k)}.$$

By substituting p_{IBP} and q_{IBP} into the formula for the squared coefficient of variation C^2 of the time between successive arrivals for the IBP process, the relationship of C^2 for the IBP and IBP(k) processes can be obtained. We refer to the C^2 for the IBP and IBP(k) as C^2_{IBP} and $C^2_{IBP(k)}$ respectively. Also, $\rho_{IBP(k)}$ refers to the throughput of header subpackets in an IBP(k). We have:

$$C^2_{IBP} = kC^2_{IBP(k)} + (1 - 2k\rho_{IBP(k)} + k^2\rho^2_{IBP(k)})(k - 1).$$

Note that the maximum value for $\rho_{IBP(k)}$ is 1/k. Therefore, for bursty traffic where C^2 is relatively high,

$$C^2_{IBP} \sim kC^2_{IBP(k)}\ .$$

3. THE APPROXIMATION ALGORITHM

The tandem network under study was analyzed by decomposing it into individual cut-through queues. Each cut-through queue was then analyzed independently of the other queues. In order to analyze a cut-through queue in isolation, we need to know its arrival process and the blocking probability π that a subpacket at the head of a queue is blocked at the beginning of a service slot due to the downstream queue being full. The arrival process is obtained using information from the the upstream queue. As it will be seen below, it suffices to characterize the inter-departure time of header subpackets from the upstream queue. The service time of a packet is represented as follows. Since each subpacket may be blocked by the downstream queue with probability π, the service time of a subpacket is represented by a geometric distribution (see [10] and [11]). That is, at the beginning of a service slot, a subpacket is blocked with probability π or is forwarded to the next queue with probability $1-\pi$. If it is blocked, the subpacket repeats the same procedure at the beginning of the next service slot. The total service time that a packet receives, therefore, can be represented by k identical geometric distributions, as shown in figure 3. We shall refer to this distribution as Geo(k). The arrival process to an intermediate queue and the blocking probability π are calculated below in section 3.1

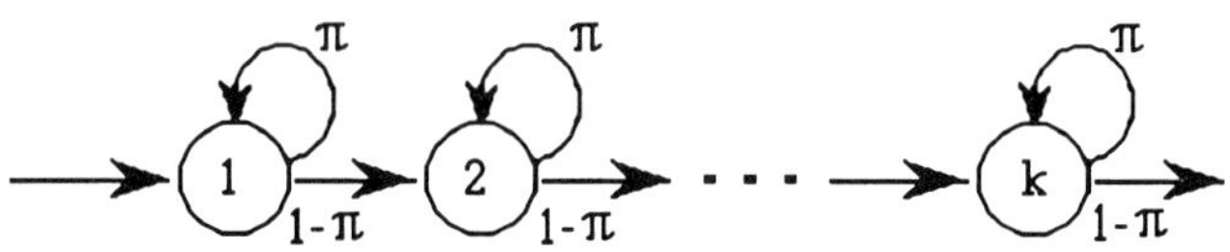

Figure 3: The Geo(k) Service Process

The state of a cut-through queue can be described by the variables $\mathbf{n}=\{n_0,n_1,n_2\}$. Variable n_0 gives the state of the arrival process and it takes the following values: 0 if the arrival process is in the idle state, 1 if the arrival process is in the active state, 2 if a header subpacket is blocked, and i+2 if i subpackets have already arrived. Variable n_1 gives the total number of whole and partial packets in the queue. At any time, there may be a packet that has partially entered the queue. The exact number of its subpackets that have entered can be determined from n_0. Also, there may be a packet that has been partially transmitted out. The exact number of its remaining subpackets can be determined from n_2. For a queue which can hold N packets, n_1 takes values from 0 to N+1. Variable n_2 gives the state of the Geo(k) server, and it takes the following values: 0 if no packet is in service (i.e., the queue is empty), or i if the packet is in the ith geometric service. If $n_2=1$, then the header subpacket is in service. A transition occurs from $n_2=1$ to $n_2=2$ when the header subpacket departs from the queue. Finally, if a packet has completed the ith service, that means that i subpackets of the packet have departed from the queue, and the remaining k-i subpackets are still in the queue.

Each cut-through queue is analyzed numerically as follows. We first generate all the states of the system, and subsequently we generate the rate matrix $\mathbf{Q}$. The stationary probability vector $\mathbf{x}$ is then obtained by solving the system of linear equations $\mathbf{xQ}=\mathbf{0}$. The above state description results in $3+k(k+1)(N+1)$ states. Furthermore, the transition probability matrix has a block tridiagonal structure in which repeating blocks are made up of $k(k+1)$ states. The system of linear equations $\mathbf{xQ}=\mathbf{0}$ was solved using Gaussian elimination. Obviously, there are other solution techniques such as matrix-geometric or block Gauss-Seidel that may be more efficient. However, the particular Gaussian elimination routine that was employed was a very fast routine optimized for a Cray Y-MP. This routine also checked for the stability and reliability of the results. For large matrices, block based numerical techniques are far more suitable, and they can also be performed on a standard workstation.

3.1. The arrival process and the blocking probability π

The arrival process to an intermediate queue is obtained by simply characterizing the time elapsing from the time the last body subpacket of a packet enters the queue to the time before the header subpacket of the next packet arrives to the queue. When the last subpacket of a packet enters the queue, the arrival process is in the active state ($n_0=1$). In the next slot the source will remain active ($n_0=1$) or become idle ($n_0=0$) with probability p and 1-p respectively. If it remains active, a header subpacket will arrive during the slot with probability α, or the slot will be empty with probability 1-α. If the source is idle, then in the next slot it will remain idle with probabilty q or it will become busy with probability 1-q. If it becomes busy, a header subpacket will arrive during the same slot with probability α. Given that a header subpacket arrives, it may enter the queue or it may get blocked ($n_0=2$) depending on the state of the queue, i.e. the values of n_1 and n_2. Subsequently, the body subpackets will enter the queue according the state of the queue. The states and transition probabilities of this arrival process are similar to those given in figure 3, except that the blocking probability is not π, but it depends on the state of the queue at the moment when a subpacket arrives.

The values for p, q, and α are obtained from the previous upstream queue as follows. Let us consider the set of all states of the queue from which it is possible to have a header subpacket departure. If the queue is in one of these states, then the departure process is assumed to be in its active state. We shall refer to these states as the *active states*, and their set will be referred to by the symbol S_1. Probability p can then be obtained as the probability that the system will be in one of the active states or in state $\{n_0,n_1>0,2\}$ in the next slot, given that it is currently in an active state. Probability q can be obtained similarly by considering the set of all states of the cut-through queue from which it is not possible to have a subpacket departure. (i.e. the queue is empty). If the queue is in one of these states, then the departure process is in its idle state. We shall refer to these states as the *idle states*, and their set will be referred to by the symbol S_2. The calculation of p and q requires the classification of the states of the queue into active and idle states. However, this is not a straight forward task. For example, if α is set so that $\alpha<1$, then certain idle states will have to be classified as active states. Below, we present two different models for characterizing p, q and α, namely model 1 and 2.

In model 1 we simplify the problem of selecting appropriate active and idle states by setting $\alpha=1$. Thus, any state of the cut-through queue from which it is possible to have a header subpacket departure is an active state, and all states in which the queue is empty are idle states. Thus, the set of active states,

S_1, consists of any state $\{n_0,n_1,n_2\}$ where $n_1>0$ and $n_2=1$. Also, define $S_1'=S_1\cup\{n_0,n_1>0,2\}$. The set of idle states, S_2, consists of any state $\{n_0,n_1,n_2\}$ where $n_1=0$ and thus $n_2=0$. (Note that the set $S_1\cup S_1'\cup S_2$ does not include all the possible states $\{n_0,n_1,n_2\}$ since certain subpacket counting states have not been included in S_1, S_1', and S_2.) Let $p(\mathbf{n})$ be the steady-state probability that the cut-through queue is in state $\mathbf{n}$, and let $t(\mathbf{n}\rightarrow\mathbf{n}')$ be the transition probability from state $\mathbf{n}$ to state $\mathbf{n}'$. Then, the probabilities p and q that define the arrival process to the next dwonstream queue can be calculated as follows:

$$p = \frac{\sum_{\mathbf{n}\in S_1} p(\mathbf{n})\left(\sum_{\mathbf{n}'\in S_1'} t(\mathbf{n}\rightarrow\mathbf{n}')\right)}{\sum_{\mathbf{n}\in S_1} p(\mathbf{n})} \quad \text{and} \quad q = \frac{\sum_{\mathbf{n}\in S_2} p(\mathbf{n})\left(\sum_{\mathbf{n}'\in S_2} t(\mathbf{n}\rightarrow\mathbf{n}')\right)}{\sum_{\mathbf{n}\in S_2} p(\mathbf{n})}.$$

Model 1 assumes that $\alpha=1$. The same model was also used in [10] for a different queueing network. It was noted in [10] that assuming $\alpha=1$ did not give good results in the case where each queue has two or more bursty arrival sources. In view of this, it was anticipated that this assumption will work well for the tandem configuration studied in this paper. In order to test this, we also developed a new model (hereafter referred to as model 2) where $\alpha<1$. In model 2, in order to achieve an $\alpha<1$, certain idle states must be reclassified as active states. In model 1, the set of idle states S_2 consists of states $\{0,0,0\}$ and $\{1,0,0\}$. Thus, in model 2, it makes sense to add the state $\{1,0,0\}$ to the active set of states S_1. The values of p and q can thus be obtained as in model 1. The value of α can be obtained by matching throughputs as in [10]. Model 2 was implemented along with model 1. Experimental evidence showed that model 2 performed almost identically to model 1. In fact, α was very close to one for model 2.

The blocking probability π can be directly obtained from the steady-state probabilities $p(\mathbf{n})$ of the next downstream queue. Let S_I be the set of states of the queue in which the source to the queue is in the idle state, S_A be the set of states of the queue in which the source to the queue is in the active state, S_B be the set of states of the queue in which the source to the queue is in the block state, and S_C be the set of states of the queue in which the source to the queue is in any of the subpacket counting states. Then, the blocking probability of the next upstream queue π' is:

$$\pi' = \frac{(1-q)\alpha\pi p(0,N,1)+p\alpha\pi p(1,N,1)+\pi p(2,N,1)+\sum_{n_0=3}^{k+1}\pi p(n_0,N+1,n_2=n_0-1)}{\sum_{\mathbf{n}\in S_C} p(\mathbf{n}) + \sum_{\mathbf{n}\in S_B} p(\mathbf{n}) + \sum_{\mathbf{n}\in S_A} \alpha p(\mathbf{n}) + \sum_{\mathbf{n}\in S_I} \alpha(1-q)p(\mathbf{n})}.$$

3.2. Summary of the approximation algorithm

The tandem network was analyzed by decomposing it into the individual cut-through queues. Each cut-through queue was then analyzed individually assuming that the arrival process and the blocking probability are known. An iteration consists of visiting each queue beginning at the first queue and continuing sequentially to the last queue. Within an iteration, each queue i was analyzed using a blocking probability calculated from queue i+1 in the previous iteration and an arrival process calculated from queue i-1 during the current iteration from the current iteration. The first and last queues were analyzed similarly with some modifications, since the arrival process to the first queue is known, and the last queue is never blocked. The main steps of the approximation algorithm are as follows:

Step 0: For each cut-through queue assume initial values for the blocking probability π. Set i to the first queue of the network (i=1).
Step 1: Analyze numerically queue i.
Step 2: For queue i, calculate the new blocking probability, and the new arrival process to its downstream queue using either model 1 or 2.
Step 3: If queue i is not the last queue of the network, then set i=i+1 and go to step 1. Else, this iteration is completed. Go to step 4.
Step 4: The convergence test compares $p(\mathbf{n})$ for each queue from one iteration to the next. The tolerance used was ten decimal place accuracy. Stop, if convergence has occurred. Else, set i=1 and go to step 1.

An alternative iterative scheme that can be employed to analyze this queueing network is the scheme that is commonly used to analyze continuous-time tandem queueing networks with blocking. This scheme consists of a forward and a backward pass. In the forward pass, the queues are analyzed from the first queue to the last queue and the departure process from each queue is updated. In the backward pass, the queues are analyzed from the last queue back to the first queue and the blocking probabilities are updated. This scheme has been found empirically to be more CPU consuming than the scheme presented above.

4. VALIDATION

The approximation model described in the previous section was implemented on a Cray Y-MP. (We note that the use of a Cray is not necessary. The approximation algorithm can efficiently run on any workstation.) The main performance measures obtained by the model were the packet loss probability that occurs at the first queue, the subpacket blocking probability between queues, the queue length distribution at each queue, and the throughput which is defined as the probability that a slot contains a header subpacket.

The approximate results were compared to simulation data. Due to lack of space, we do not give any numerical comparisons between the approximate and simulation results. (The interested reader is referred to Morris and Perros [13] and also to Morris [12], where extensive numerical comparisons are given.) Instead, we summarize below the cases where the algorithm was observed to have a good accuracy, and also the cases where the algorithm did not perform as well.

In the case where the arrival process to the first queue is Bernoulli i.e., non-bursty, the approximate results for the queue length distribution and blocking probabilities were found to have a good accuracy over the entire range of ρ, the arrival rate of header subpackets.

In the case where the queues have a small size (for instance, k=2, N=2) and the input process is bursty, the queueing network almost works in an on-off mode. (In this paper, burstiness is measured in terms of the squared coefficient of variation of the inter-arrival time of packets.) During idle periods, the network is empty, and during bursts the network quickly fills and remains full. In general, the approximation algorithm was found to have good accuracy when the queue sizes are small and the input process is bursty. When the input process is extremely bursty (i.e. the squared coefficient of variation C^2 of the inter-arrival time of header subpackets is greater than 200), the packet loss probability is over estimated and the squared coefficient of variation between successive header arrivals through the network is not as accurate.

Now let us consider the case where the queues are large (for instance, k=8 and N=4). At high load conditions, the queue length distribution is not estimated accurately. (The relative error for the mean queue length, for instance, may be more than 10%.) This can be directly attributed to the way blocking is represented in the model. With highly bursty traffic, the queueing network will empty during the idle periods, and fill during the active burst periods. At the beginning of the burst, the source will see a relatively empty system. It will take a substantive amount of traffic to fill the queues before blocking takes place. In the approximation algorithm, however, a blocking probability is applied to every subpacket regardless whether it is the beginning of a burst period or not. Thus, at the beginning of a burst period, the queues fill faster than in reality. This results in an over estimation of the Prob{Queue Full} by the model. The mean queue length is similarly affected. This phenomenon is accentuated for large queue sizes since it takes longer to fill large queues at the beginning of a burst period. This problem is alleviated in lighter traffic or when the arrival process is non-bursty.

In general, the accuracy of the header packet throughput, the packet loss, and the blocking probability at each queue are estimated quite accurately. Also, the approximation algorithm gives good results for the squared coefficient of variation of the effective interarrival time of header subpackets to each queue, except when the input process is extremely bursty.

The CPU time required for the execution of the approximation algorithm was primarily dependent on ρ, P_b, k, and N. In general, the number of iterations until convergence depended upon ρ and P_b, whereas the execution time of an iteration depended on k and N. Typically, as $k\rho+P_b$ approached 1, the number of iterations greatly increased.

5. CONCLUSIONS

In this paper, we presented an approximation algorithm for the analysis of a discrete-time tandem network of cut-through queues. Finite buffers with blocking-before-service, fixed packet sizes, and bursty traffic was assumed. A new arrival process, IBP(k), was presented and analyzed. The tandem network was analyzed by decomposing it into individual queues. Each cut-through queue was then analyzed numerically in isolation assuming that its arrival and service processes are known. Two different methods for characterizing the arrival process to a cut-through queue were presented. The parameters of the arrival and service processes of the cut-through queues were obtained using an iterative scheme. Using this algorithm, the queue length distribution, the packet loss, the blocking probability between queues, and the throughput can be obtained for different traffic parameters, subpacket-to-packet ratios, and queue sizes. Validation tests showed that the approximation algorithm has a good accuracy.

The analytic model provided good accuracy, but improvement is still possible. One possible improvement, is to use a state dependent blocking probability. If a queue contains more than one subpacket, then all downstream queues must be full. Therefore, in this case, the blocking probability of that queue should be equal to P_b, the parameter of the geometric service time of the last queue. The value of the blocking probability when the queue length is equal to one subpacket can be determined if the average blocking probability π is known, which is in fact what the current approximation algorithm calculates. This new blocking probability would resemble reality better than the blocking probability in the current algorithm.

Further generalization of the approximation algorithm for cut-through queues should allow for multiple input processes and multiple destinations. This generalization would permit the analysis of complex switching elements in a multistage buffered interconnection network.

ACKNOWLEDGEMENT

We would like to thank the North Carolina Supercomputing Center for providing us with Cray time.

REFERENCES

1 P. Kermani and L. Kleinrock, "Virtual cut-through: a new computer communication switching technique," *Computer Networks*, vol. 3, pp. 267-286, Sept. 1979.

2 P. Kermani and L. Kleinrock, "A tradeoff study of switching systems in computer communication networks," *IEEE Trans. on Computers*, vol. c-29, no. 12, pp. 1052-1059, Dec. 1980.

3 B. Shin and C. Un, "Performance analysis of a quasi-M/D/1 cut-through switching network with noisy channels," *IEEE Trans. on Communications*, vol. com-34, no. 9, pp. 882-889, Sept. 1986.

4 M. Ilyas and H.T. Mouftah, "End-to-end flow control with a new hybrid switching in computer networks," in *Proc. of GLOBCON '82*, pp. 366-370, 1982.

5 M. Ilyas and H.T. Mouftah, "Quasi cut-through: a new hybrid switching technique for computer communication networks," *IEE Proceedings*, vol. 131, pt. E, no. 1, pp. 1-9, Jan. 1984.
6 A. Abo-Taleb and H.T. Mouftah, "Delay analysis for interfered paths under general cut-through switching," *Canadian Journal of Electrical and Computer Engineering*, vol. 13, no. 2, pp. 80-84, 1988.
7 A. Abo-Taleb and H.T. Mouftah, "Delay analysis under a general cut-through switching technique in computer networks," *IEEE Trans. on Communications*, vol. com-35, no. 3, pp. 356-359, March 1987.
8 M. Marsan, G. Balbo, G. Chiola, A. Ciccardi, and G. Conte, "Estimating the average delay in a Delta interconnection network operating according to the cut-through packet switching technique," in *Symposium of Distributed and Parallel Systems*, pp. 481-500, Kyoto, Japan, 1988.
9 S.B. Gershwin, An efficient decomposition method for the approximate evaluation of tandem queues with finite storage space and blocking, *Oper. Res.* vol. 35, pp. 291-305, 1987.
10 T. Morris and H.G. Perros, "Performance modelling of a multi-buffered Banyan switch under bursty traffic," *Proc. of INFOCOM '92*, Florence, Italy, May 1992, 436-445.
11 A.A. Nilsson, F.-Y. Lai, and H.G. Perros, "An approximate analysis of a bufferless NxN synchronous Clos ATM switch," in: Cohen and Pack. eds., *Queueing, Performance, and Control in ATM (ITC-13)*, North-Holland, 1991, pp. 39-46.
12 T. Morris, "The design and analysis of buffered multistage interconnection networks," Ph.D. Thesis, Dept. of Electrical and Computer Engineering, North Carolina State University, July 1991.
13. T. Morris and H.G. Perros, "Approximate analysis of a discrete-time tandem network of cut-through queues with blocking and bursty traffic", to appear in a special issue of the *Performance Evaluation J.* comprising a selection of papers from these proceedings.

Queueing Networks with Finite Capacity
R.O. Onvural and I.F. Akyildiz (Editors)

ATM-based Broadband Switching System Network Performance under Bursty Traffic Environment

C. Chen, A. Arutaki, and G. Frank

Advanced Switching Laboratory NEC America, Inc., Irving, Texas

ABSTRACT

This paper introduces a mathematical model for estimating the performance of an ATM-based switching system network. Network planning and node capacity assignment in an ATM-based network is complicated by the bursty nature of traffic generated by its applications. Capturing burstiness with a mathematical model is the key to properly analyzing an ATM Switching System's performance. Three major sources of ATM traffic burstiness are:

1) Excessive arrival rate;
2) Sudden changes in message length;
3) Combinations of the above two.

ATM network performance modeling is further complicated by

- Wide ranges of message lengths due to wide varieties of traffic classes
- Unknown traffic service distributions due to lack of empirical data

To address the above-mentioned issues, we propose an ATM burst model based on Poisson message arrival and Gamma distribution as a model for message length. Poisson-Gamma is chosen because of the bursty nature of ATM traffic, and its capability to approximate other distributions with great accuracy. The Bayes estimation method is proposed to estimate performance parameters for its ability to incorporate engineering judgements, which is especially important due to the scarcity of empirical data on ATM traffic at the present time.

Mean cell delays at a single node, and an extension to the network level, are obtained assuming infinite buffer capacity. An algorithm for computing cell loss probability at a single node is presented. Bayes estimators for message arrival rate and mean cell length are also obtained.

1. Introduction:

An ATM switch network is designed to handle high-bandwidth traffic with diverse types of traffic patterns. A clear understanding of the traffic characteristics of applications is crucial to effectively utilize switch network capacity and guarantee high-quality services.The major difficulty in evaluating ATM switch network performance is the bursty nature of traffic sources due to

- Excessive message arrival relative to that allocated during set-up time

- Sudden change of message length
- The combination of both excessive message arrival and message length changes.

Modeling ATM network performance is further complicated by the wide range of message lengths due to numerous varieties of traffic classes, which renders the conventional Poisson-Exponential model inadequate. This paper proposes an ATM traffic model based on Poisson message arrival and Gamma distribution as a model for message length. A Poisson-Gamma model is chosen due to its extensive range of functional shapes which can describe a variety of burst patterns. In particular, the popular Exponential model is a special case of Gamma with shape parameter equal to one; furthermore, other distributions can be approximated as a finite mixture of Gamma, with an arbitrary degree of accuracy. We feel that the rich modeling power is most important for modeling the great variety of ATM traffic patterns. However, the Poisson-Gamma model does introduce technical challenges for analyzing ATM switch performance, especially at the network level, due to its violation of conditions constituting a Jackson network; hence, existing results built around the Jackson network model are not readily applicable. New methods must be developed.

The contributions of this paper are threefold:

- Mean cell delay for a single node is obtained; the result is extended to the network level. Both results are based on the assumption of infinite buffer capacity
- An algorithm for computing cell loss probability at a single node is presented
- Bayes estimators for traffic parameters such as message arrival rate and mean message length are also obtained. The Bayes method is employed due to its ability to incorporate engineering judgements and empirical data to reach accurate estimates. This technique is especially powerful when data is hard to gather, which is precisely the case in the current state of ATM technology.

2. Statement of the Problem:

Consider a network of M ATM switches $\{S_i\}$, each with N input ports and N output ports, with each output port of each S_i connected either to an input port of another S_j or a terminating point; customer-premises equipment (CPE), say. An example interconnection graph is depicted in Figure 1 below.

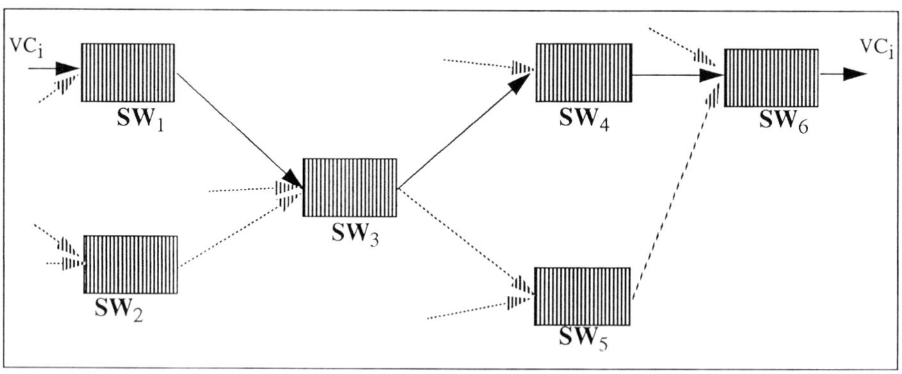

Figure 1: ATM Switch Network

Messages of a connection (called virtual circuit) reach an initial ATM node and travel through a preassigned path toward a terminal node connected to the message destination. Notice that during normal operating conditions, a message will not be routed through a given node more than once.

Figure 2 below depicts an $N \times N$ ATM switch which is present on each network node.

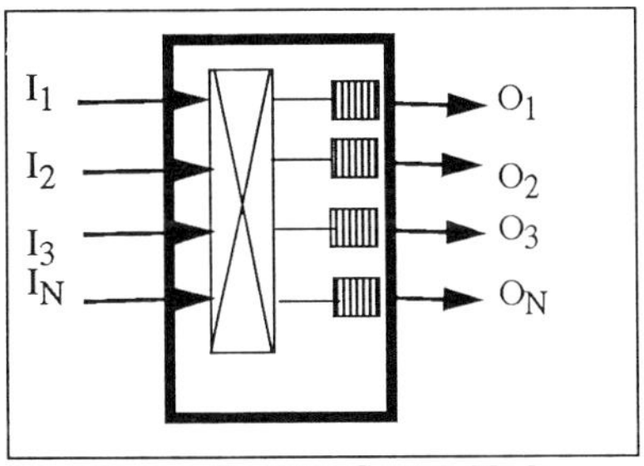

Figure 2: ATM Switch Node

3. Terminology:

C: Link capacity of an output port (measured in bit/second)

c_0: Time for processing one ATM cell $c_0 = \frac{8\left(\frac{b}{\text{byte}}\right) \times 53 \left(\frac{\text{byte}}{\text{cell}}\right)}{C}$

S_i: ATM switch node i

Q_{ij}: Unique output port queue in node i connecting to node j

L_{ij}: Link between node i and node j

$VC_i(j)$: Virtual circuit i which originates at node j

P_i: An ordered set of queues through which VC_i is routed in a predetermined fashion

$N_i(t)$: Number of messages generated by VC_i during time interval (0,t]

Y_i: Total number of ATM cells generated by VC_i during time interval (0,t]

A_{ij}: Set of virtual circuits which are routed through Q_{ij} during interval (0,t]

T_{ij}: Total traffic at Q_{ij} during time interval (0, t]

T_0: Processing time for cells belonging to the same message and ahead of arriving cells.

T_s: Residual service time: time to complete the cell currently in active service

T_b: Waiting time for the arriving cell.

4. Assumptions:

The analysis is based on the following assumptions:

1. P_i forms an acyclic graph (VC_i visits each node at most once)
2. VCs that belong to the same traffic class have the same bit-rate.
3. For each node, messages which travel toward the same adjacent node are routed through the same output port.
4. For each output port, the service discipline is First Come First Served
5. Transmission delay, propagation delay, and message-to-cell conversion time are negligible
6. No internal blocking is done within the ATM switch.
7. Message arrivals from the same or different virtual circuits are stochastically independent
8. The $E(T_{jk}) \cdot c_0 < 1$.

5. Single Node Traffic Model:

Let $VC_i(j)$ be the ith virtual circuit at node j and let $N_i(t)$ be the number of messages generated by VC_i during time interval (0,t]. Let a_i be the bit rate of VC_i with duration D_i. Assuming N_i is $\text{Poisson}(\lambda_i)$ and $D_i \sim \text{Gamma}(\alpha_i, \beta_i)$, each message of VC_i has length

$$L_i = a_i \cdot D_i \sim \text{Gamma}\left(\alpha_i, \frac{\beta_i}{\gamma_i}\right), \text{ where } \quad \gamma_i = \frac{a_i}{8 \times 53};$$

Notice that L_i is measured in number of cells.

The total number of cells generated by VC_i during (0,t] is Y_i with mean and variance as follows:

$$E(Y_i) = \frac{\lambda_i \cdot \alpha_i \cdot \gamma_i}{\beta_i}, \text{ and } \qquad \text{var}(Y_i) = \frac{\lambda_i \cdot \alpha_i \cdot \gamma_i^2}{\beta_i^2}$$

The Gamma model is chosen because the potential occurrence of wide ranges of traffic classes with different bit-rates and burst duration makes the exponential model impractical. Furthermore, the Gamma model includes the exponential model as a special case with shape parameter $\alpha = 1$, and is robust in the sense that all other distributions $F(X)$ can be approximated by Gamma distributions; namely:

$$\lim_{n \to \infty} F_n(x) = F(x)$$

$$F_n(x) = \sum_{i=1}^{n} \mu_i \cdot \text{Gamma}(\alpha_i, \beta_i); \quad \sum_{i=1}^{n} \mu_i = 1; \quad (\mu_i \geq 0)$$

6. Single Node Analysis

6.1. Queueing Delay for Single Output Port (Infinite Buffer Case)

The queueing delay for cells consists of T_s, T_b, and T_o. We analyze each quantity as follows: Let waiting time for a cell from VC_i be $W_i = T_b + T_s$; then cell delay is

$$T_i = W_i + T_o + c_0 .$$

Notice that since the service discipline is FCFS, T_b consists of service time for cells which are already in the queue (buffer) at the time the cell arrives at Q_{jk}.

The mean residual service time T_s is

$$E(T_s) = \frac{1}{2} \cdot c_0 \cdot \text{Prob(server is busy)} + 0 \cdot \text{Prob(server is idle)}$$

Thus,

$$E(T_s) = \frac{1}{2} \cdot c_0 \cdot \frac{E(T_{jk})}{\text{Link Capacity}} = \frac{1}{2} \times \sum_{i \in A_{jk}} \frac{\lambda_i \cdot \alpha_i \cdot \gamma_i}{\beta_i} \cdot c_0^2$$

Where T_{jk} is the total traffic at the queue Q_{jk}.

Let t_1 be the arrival time for the message which contains the cell under consideration and let t_2 denote the time the first cell of this message receives service. Then, T_b can be defined as $T_b = t_2 - t_1$. Little's law implies that

$$E(T_b) = \sum_{i \in A_{jk}} EQ_i \cdot c_0 = \sum_{i \in A_{jk}} \left(\frac{\lambda_i \cdot \alpha_i}{\beta_i}\right) \cdot E(W_i) \cdot c_0^2$$

Thus, we have

$$E(W_i) = \left(\sum_{i \in A_{jk}} f_i \cdot E(W_i)\right) + T_s \qquad (1)$$

The total amount of traffic at the queue Q_{ij} is

$$T_{jk} = \sum_{i \in A_{jk}} Y_i \text{; where}$$

The mean delay of cells from VC_i at Q_{jk}, denoted $E(T_i^{jk})$, is

$$E(T_i^{jk}) = E(w_i) + \frac{1}{2} \cdot \lambda_i \cdot \frac{\alpha_i \cdot \gamma_i}{\beta_i} \cdot c_0 + c_0 \text{, where}$$

$$E(w_i) = \frac{T_s}{\left(1 - \sum_{i \in A_{jk}} f_i\right)} \text{, with}$$

$$f_i = \frac{\lambda_i \cdot \alpha_i \cdot \gamma_i \cdot c_0^2}{\beta_i} \text{, and } T_s = \frac{1}{2} \cdot \sum_{i \in A_{jk}} \frac{\lambda_i \cdot \alpha_i \cdot \gamma_i}{\beta_i} \cdot c_0^2 \cdot$$

6.2. Cell Loss Probability With Given Finite Buffer Capacity B

6.3.

This section evaluates cell loss probability assuming single switch node. Namely, all cells of VC_i originate at switch node j passing through the sliding window for VC_i and arrived at output port Q_{jk}. Let T_{jk} be the total traffic during time interval I_n at Q_{jk}, and B_{jk} be the buffer size for Q_{jk}. Let E_{jk} be the number of cells discarded at Q_{jk} due to queue length exceeding buffer size. Then the actual traffic (number of cells) passing through $Q_{jk}$is $U_{jk} = T_{jk} - E_{jk}$. This section will analyze the cell loss probability for Q_{jk} with buffer capacity B_{jk} and total traffic T_{jk}.

For a given buffer capacity B_{jk}, cells will be lost if $T_{jk} = \sum_{i=1}^{N_{jk}} Y'_i \geq \frac{C}{8 \times 53} + B_{jk}$, where N_{jk} is the number of virtual circuits routed trough Q_{jk} during the interval I_n. Thus, the loss probability of a cell at Q_{jk} is

$$\mathrm{Prob}\left(\sum_{i=1}^{N_{jk}} Y'_i \geq \frac{1}{c_0} + B_{jk}\right) \qquad (2)$$

We propose using Chernoff Theorem to evaluate cell loss probability.

The moment generating function of Y'_i is $M_{Y'_i}(s) = \dfrac{e^{\lambda_i T \cdot \left(\left(\frac{\upsilon_i}{\upsilon_i - s}\right)^{\alpha_i} - 1\right)}}{p_{io}}$; where

$p_{io} = 1 - q_{io}$;, and q_{io} is the cell loss probability due to sliding window control. Thus,

$$\mu_{T_{jk}}(s) = \sum_{i=1}^{N_{jk}} \left(\lambda_i \cdot T \cdot \left(\left(\frac{\upsilon_i}{\upsilon_i - s}\right)^{\alpha_i} - 1\right)\right) - \sum_{i=1}^{N_{jk}} \log(p_{io})$$

Let $l_{jk} = \frac{1}{c_o} + B_{jk}$;then the cell loss probability at Q_{jk} is:

$$\mathrm{Prob}\,(T_{jk} \geq l_{jk}) \leq \left(e^{-\left(s^* \cdot l_{jk} - \mu_{T_{jk}}(s^*)\right)}\right) \qquad (3)$$

where s^* is the solution of the following equation:

$$l_{jk} = \frac{d}{ds}\left(\sum_{i=1}^{N_{jk}} \lambda_i \cdot T \cdot \left(\left(\frac{\upsilon_i}{\upsilon_i - s}\right)^{\alpha_i} - 1\right) - \sum_{i=1}^{N_{jk}} \log(p_{io})\right) \qquad (4)$$

Subject to the following constraints:

$0 < s < \upsilon_i$ for i = 1, 2,...., N_{jk}.

Equation (4) can be simplified into equation (5) below

$$l_{jk} = \sum_{i=1}^{N_{jk}} \lambda_i \cdot T \cdot \upsilon_i^{\alpha_i} \cdot \alpha_i \cdot (\upsilon_i - s)^{-(\alpha_i + 1)} \qquad (5)$$

Subject to the following constraints: $0 < s < \upsilon_i$ for i = 1, 2,...., N_{jk}

Notice that equation (5) is a non-linear equation in s. In general, there is no closed-form solution for s; hence, an efficient approximation algorithm is needed. We now propose such an algorithm and prove that it converges to a unique solution. First, we need the following lemmas, presented without proof.

Lemma 1: If s^* is a solution for optimization problem (5), then, s^* is also the solution for the following optimization problem (6); further, s^* is the unique solution for both problems.

Maximize:

$$(l_{jk} \cdot s) - \left(\sum_{i=1}^{N_{jk}} \lambda_i \cdot T \cdot \left(\left(\frac{\upsilon_i}{\upsilon_i - s} \right)^{\alpha_i} - 1 \right) - \sum_{i=1}^{N_{jk}} \log(p_{io}) \right) \qquad (6)$$

Subject to the following constraints:

$0 < s < \upsilon_i$ for i = 1, 2,...., N_{jk}.

Lemma 2: The objective function of (10) is a concave function in s.

As a direct consequence of Lemma 2, if s^* maximizes the optimization problem (6), s^* must be unique. However, there are still two possible locations for s^*: (1) the interior of the feasible region, or (2) on the boundary of constraints. The following lemmas ensure that (1) is the case, and propose an initial approximation for the algorithm which follows.

Lemma 3: Let s^* optimize the objective function of problem (10), s^* satisfies the following:

$0 < s^* < \min \upsilon_i$, for i = 1, 2, 3,...., N_{jk}.

and hence, s^* is the solution of the following equation (11):

$$\frac{d}{ds}\left((l_{jk} \cdot s) - \sum_{i=1}^{N_{jk}} \lambda_i \cdot T \cdot \left(\left(\frac{\upsilon_i}{\upsilon_i - s^*} \right)^{\alpha_i} - 1 \right) + \sum_{i=1}^{N_{jk}} \log(p_{io}) \right) = 0$$

Lemma 4: Let s_0 be the solution of the following optimization problem:

$$\text{Max:} \quad (l_{jk} \cdot s) - (\lambda_i \cdot T) \cdot \left(\left(\frac{\upsilon_i}{\upsilon_i - s} \right)^{\alpha_i} - 1 \right) + \log (p_{io}) \qquad (7)$$

Subject to the following constraint:

$$0 \leq s \leq \upsilon_i$$

$$\text{Then,} \quad s_0 = \upsilon_i - \left(\frac{\lambda_i \cdot T \cdot \alpha_i \cdot \upsilon_i^{\alpha_i}}{l_{jk}} \right)^{\frac{1}{(\alpha_i + 1)}} \qquad (8) \quad .$$

Then s^*, the solution of equation (5), can be obtained by the following algorithm (A), which computes a sequence of points $\{s_n\}$ which converge to s^* sufficient quickly.

Here we choose an accuracy criterion of 0.0001.

Algorithm A:

Let $io \in A_{jk}$ be the index such that $\upsilon_{io} = \min \{\upsilon_i\}$ for $i = 1, 2, \ldots, n_{jk}$.

Step 1: Let $k = 0$; $s_0 = \upsilon_0 - \left(\frac{\lambda_{i0} \cdot T \cdot \alpha_{i0} \cdot \upsilon_{i0}^{\alpha_{i0}}}{l_{jk}} \right)^{\frac{1}{(\alpha_{i0} + 1)}}$; let

$b_0 = \upsilon_o$ and $a_o = 0$

Step 2: compute the slope of the objective function in equation (6) at s_k using the following equation:

$$q(s_k) = l_{jk} - \sum_{i=1}^{N_{jk}} \lambda_i \cdot T \cdot \upsilon_i^{\alpha_i} \cdot \alpha_i \cdot (\upsilon_i - s_k)^{-(\alpha_i + 1)}$$

if $|q(s_k)| < 0.0001$ go to Step 4;

Step 3: For k = 0, If $q(s_0) \geq 0$, let $s_1 = \frac{s_0 + b_0}{2}$; and $a_1 = s_0$, $b_1 = b_0$

If $q(s_0) < 0$, let $s_1 = \frac{s_0 + a_0}{2}$, and $a_1 = a_0$, $b_1 = s_0$

For $k > 0$

Case I:

if $q(s_k) < 0$

$$s_{k+1} = \frac{s_k + a_k}{2}$$; and $a_{k+1} = a_k$, $b_{k+1} = s_k$ go to Step 2;

Case II:

If $q(s_k) \geq 0$, then, let

$$s_{k+1} = \frac{s_k + b_k}{2}$$; and $a_{k+1} = s_k$, $b_{k+1} = b_k$ go to step 2;

Step 4: Let $s^* = s_k$ and stop.

Theorem 1: The cell loss probability at Q_{jk} is q_{jk} where

$$q_{jk} \cong e^{-\left\{ s^* \cdot l_{jk} - \left\{ \sum_{i=1}^{N_{jk}} \left\{ \lambda_i \cdot T \cdot \left\{ \left(\frac{\upsilon_i}{\upsilon_i - s^*} \right)^{\alpha_i} - 1 \right\} \right\} - \sum_{i=1}^{N_{jk}} \log(p_{i0}) \right\} \right\}}$$

where s^* is obtained by Algorithm A.

For practical application, we may estimate s^*, α_i, υ_i, p_{io} by their corresponding Bayes estimators

$\hat{s}^*$, $\hat{\alpha}_i$, $\hat{\upsilon}_i$, $\hat{p}_{io}$ as obtained in section 5.1.

6.4. Application Example

In this subsection the above analysis is applied to a possible implementation: an ATM switch in a DS3 network. The question is: How many virtual circuits can be attached to a single ATM switch, and what is the corresponding cell delay? Figure 4 shows the application scenario.

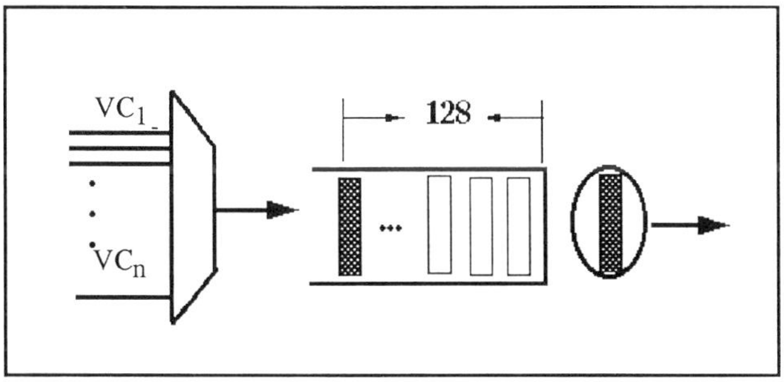

Figure 4: ATM Switch Application

The subsequent analysis is based on the following assumptions:

- Messages arrivals are statistically independent
- Buffer size: 128 cells
- Service discipline: FCFS
- All nodes connected via DS3 lines are identical
- Customers do not violate designated traffic thresholds

We make the following additional assumptions:

- average load per VC_i:

Message arrival: $N_i \sim \text{Poisson}\,(\lambda_i)$

Message Length: $L_i \sim \text{Gamma}\,(\alpha_i, \beta_i)$

Average Load/sec per VC_i: $\lambda_i \cdot \frac{\alpha_i}{\beta_i}$

For selected average message lengths $l = \frac{\alpha_i}{\beta_i}$ of 1, 10, and 20 cells and message arrival rate λ_i of 1 per second, the number of VCs multiplexed through an ATM switch output port and the corresponding cell delay are shown in Figure 5. Notice that message length affects both the delay time and the queue length (number of cells in the buffer).

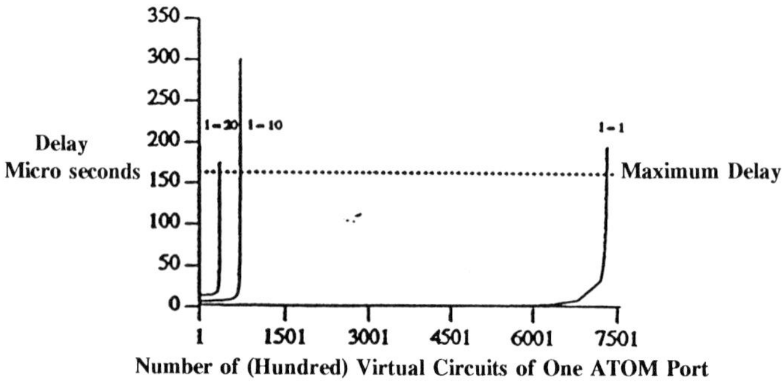

Figure 5. Average Cell Delay (Varying message lengths)

In Figure 6, we keep the message length constant (l=10 cells/message) and we vary the message arrival

rate λ_i from 1 to 4 message(s) per second.

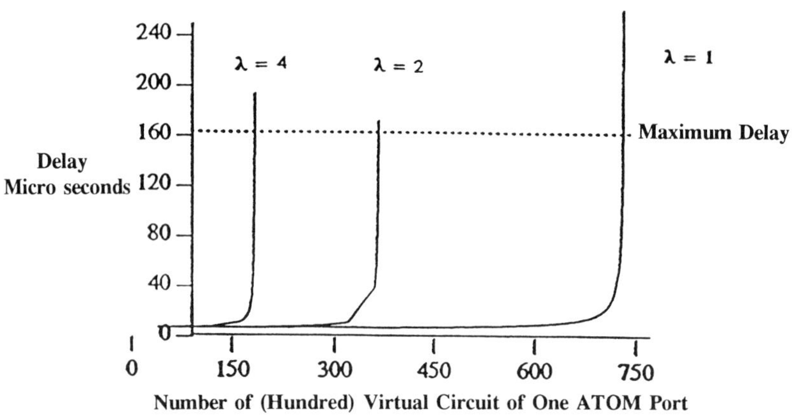

Figure 6. Average Cell Delay (Varying message arrival rates)

As the figure shows, the arrival rate will change the maximum number of virtual circuits drastically but not the minimum delay time.

Certain message arrival rates and message lengths (and combinations of both) can cause traffic bursts and result in excessive cell delay and cell loss. This implies that traffic planning analysis taking the bursty nature of traffic into consideration is indispensable. More importantly, this analysis identifies the need for ATM switching systems to provide traffic monitoring and cell admission control mechanisms during run-time to ensure that traffic bursts will be properly handled, thereby ensuring quality of service.

7. Extension to a Network of ATM Switches

In the preceding sections we described a method to calculate cell delay and cell loss probability for the Poisson - Gamma traffic model at an ATM node. It is the aim of this section to extend the Poisson-Gamma model to analyze the performance of an ATM switch network.

7.1. Equivalent Jackson Network Approximation:

Consider a virtual circuit VC_i which generates traffic according to Poisson (λ_i), with bit-rate a_i and distribution of burst duration for a message $D_i \sim \text{Gamma}(\alpha_i, \beta)$. Then the total amount of traffic (number of ATM

cells) generated by VC_i during time interval (0,t] is Y_i, where the moment generating function of Y_i is

$$M_{Y_i}(s) = e^{\lambda_i \cdot \left(\left(\frac{\theta_i}{\theta_i - s} \right)^{\alpha_i} - 1 \right)}, \text{ where } \theta_i = \frac{\beta}{a_i \cdot \frac{1}{8 \times 53}}, \text{ and}$$

$$E(Y_i) = \frac{\lambda_i \cdot \alpha_i \cdot \gamma_i}{\beta}.$$

Proposition 1: Consider a virtual circuit with message arrival process $\text{Poisson}(\rho_i)$, bit rate 1, and burst duration $\text{Exponential}(\beta)$, and let Z_i denote the number of cells generated by the virtual circuit during the time interval (0, t]. If $\rho_i = \lambda_i \cdot \alpha_i \cdot \gamma_i$, then,

$$E(Z_i) = \frac{\lambda_i \cdot \alpha_i \cdot \gamma_i}{\beta_i}$$

As a consequence of Proposition 1, we propose approximating Poisson-Gamma traffic by a Poisson-Exponential traffic pattern by properly adjusting the Poisson arrival rate according to Proposition 1. Notice that the mean of the total amount of traffic generated by the original traffic pattern has been preserved by the approximating process.

7.2. End-to-End Cell Delay of an ATM Network with Infinite Buffer Capacity

As a direct consequence of Proposition 1, an ATM network with Poisson-Gamma traffic arrival pattern can be approximated by a network with Poisson-exponential in the sense that the mean total number of cells generated within interval (0, t] is preserved. Assuming infinite buffer capacity on each queue, the approximating network is of Jackson type and the end-to-end mean cell delay is the sum of delays at the queues through which the cells are routed. To evaluate the mean cell delay at each queue on the routing path, similar arguments to those stated in section 6.1 for the single node case can readily be applied by adjusting the following parameters:

- Replace message arrival $\text{Poisson}(\lambda_i)$ with $\text{Poisson}(\rho_i)$, and
- Cell length distribution $\text{Gamma}\left(\alpha_i, \frac{\beta_i}{\gamma_i}\right)$ with $\text{Gamma}(1, \beta)$.

If the buffer capacities are finite at each queue on the ATM network, the output process will be very complex; proper methods of evaluating cell delay are under study at this time.

8. Conclusion:

This report proposes a traffic model for an acyclic network of ATM switches. Methods of computing cell delay and cell loss probability based on the Poisson-Gamma model are presented. To simplify the network performance analysis, a method of traffic approximation based on the Poisson-Exponential model is suggested, where the approximating model preserves the mean of total traffic of the original traffic model. Cell loss probability based on the approximating model is also obtained for the finite buffer case. By assuming infinite buffer capacity and using the approximating model, the network is approximated by a Jackson network, and the network performance is obtained. For the case of finite buffer capacity, both cell loss probability at each queue of the network and end-to-end cell probability for cells routed through a set nodes will be reported in the forth-coming paper.

9. Acknowledgments

The authors would like to thank the following persons for valuable advice and supports.

Mr. H. Ito (GM, ISDD, NEC Corp.), Dr. T. Yamaguchi (AGM, ISDD, NEC Corp.), Mr. K. Akiba (SM, ISDD, NEC Corp.), and Mr. H. Yabe (GM, ASL, NECAM).

10. References

1 H. Suzuki, H. Nagano, T. Suzuki, T. Takeuchi, S. Iwasaki "Output-buffer Switch Architecture for Asynchronous Transfer Mode," ICC 1989.

2 "Broadband Aspects of ISDN", T1S1.5 Baseline Document, August 1990.

3 K. Suzuki, T. Suzuki, A. Itoh, C. Itoh, "An ATM Switching System — Development and Evaluation," NEC Research and Development, 1991.

4 M. DeGroot, "Optimal Statistical Decision." McGraw-Hill, 1970.

5 Erfani, S. and Malek, M. "Dynamic Access Capacity Management In a Multiservice Packet-Mode Environment", Network Management and Control, Plenum, New York, 1990.

6 Hyman D. Tabatabai, A. and T. V. Lakshman "Statistical Analysis and Simulation Study of Video Teleconference Traffic in ATM Networks" Globecom 91.

Queueing Networks with Finite Capacity
R.O. Onvural and I.F. Akyildiz (Editors)

The upper bounds for performance measures of a finite capacity polling system under bursty arrivals

Y. Frank Jou[a], Arne A. Nilsson[a], and Fuyung Lai[b]

[a] Center for Communications and Signal Processing, Department of Electrical and Computer Engineering, North Carolina State University, Raleigh, N.C. 27695-7914

[b] IBM, V57/B660 P.O. Box 12195, Research Triangle Park, N.C. 27709

Abstract

This paper focuses on the upper bounds for both the mean delay and the loss probability that bursty arrivals incur in an ATM switching system which can be modeled as a finite capacity polling system with nonexhaustive cyclic service. The arrival process to each input port of the system is assumed to be bursty and is modeled by an Interrupted Bernoulli Process(IBP). We compute the upper bounds for this polling system by considering a cell multiplexer with the same arrival processes and equal queue capacity. Under the ATM environment, the mean delay obtained from this multiplexer cannot only serve as an upper bound but also render a fairly accurate estimation for the mean delay of the polling system. For the cell loss probability, we consider a multiple urn model with uniform occupancy distribution which will guarantee the upper bound. Also, a heuristic method is proposed to give better estimation for cases which have medium to high cell loss rate. These analytic results are compared against the simulation results of the polling system to validate this approach.

1 INTRODUCTION

Numerous high speed networking approaches have been proposed to meet the stringent requirements of the broadband integrated networks. Among these approaches, the Asynchronous Transfer Mode (ATM) technology has been selected by international standards bodies as the basis for future broadband ISDN facilities. ATM is a packet oriented transfer mode based on statistical multiplexing in which the information is transported in short, fixed length blocks referred to as cells. ATM provides the means to transporting different types of highly bursty traffic such as voice, video images and bulk files. The bandwidth flexibility, the capability to handle all services in a uniform way, and the possible use of statistical multiplexing are advantageous features of ATM.

In this paper, we study the performance of an ATM switch architecture which is shown in figure 1 [1]. This ATM switch architecture is constructed by connecting self-routing switching modules (SRMs) in a three-stage link configuration which is called multi-stage self-routing network (MSRN). Each stage of MSRN consists of eight self-routing switching

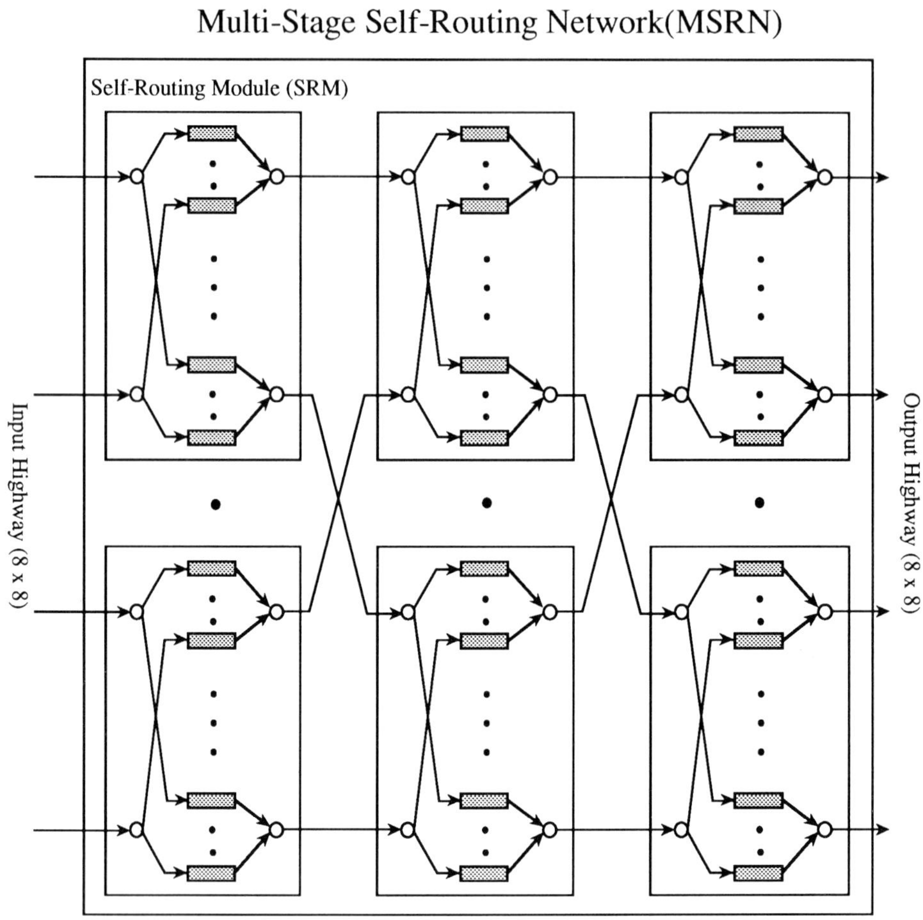

Figure 1: ATM Switching Architecture.

modules. Each module is an 8 $\times$ 8 crossbar switch which has a finite buffer associated with each crosspoint. The cells in each buffer will be transmitted in a cyclic order.

To study the performance of this SRM, we model it by a polling system with cyclic service. In the literature, multiqueue systems served by a single server have been the subject of numerous investigations (see [2], [3] and references therein). Various polling strategies like cyclic or priority service and different types of service disciplines, e.g. exhaustive, gated, or limited service have been considered. In most of these investigations, the input processes are assumed to be Poisson, and the queues of the polling system to have infinite capacity. In order to include more realistic modeling elements in the class of polling systems, we consider bursty arrival processes as inputs, and finite buffer capacity in the polling system.

In this paper, instead of solving this finite capacity polling system directly, we present a model to compute the upper bounds for mean delay and cell loss probability. This model will assume symmetric traffic load, zero switchover time, and 'limited - 1' service [4]. In Section 2 we describe in detail the models we propose. This approach will require analysis of multiplexed bursty traffic and a multiple urn model with uniform occupancy distribution which are presented in Section 3 and Section 4, respectively. In Section 5 we show some numerical results validated by computer simulations. Finally, we make some concluding remarks.

2 MODEL DESCRIPTION

In this section we describe in detail the switch architecture, the arrival process, and the queueing models which we propose.

2.1 Switch architecture

Figure 1 shows the configuration of the MSRN. The MSRN is constructed by connecting SRMs in a three-stage link configuration. In this configuration, there are multiple paths between a first-stage SRM and a third- stage SRM. This allows the traffic flow to be routed efficiently between the input highway and the output highway, and reduces the delay in the switching network. Also, this configuration is inherently reliable because a faulty second stage SRM can be bypassed. Each stage of MSRN consists of eight self-routing switching modules. Each module is an 8×8 crossbar switch which has a finite buffer associated with each crosspoint. The SRMs consist of a cell distributor at each inlet, and FIFO buffers storing cells temporarily for resolving outlet contention at each outlet. Cells are assigned to paths (links between SRMs) so that each link carries an equal amount of traffic.

2.2 Arrival process

Since most of the traffic sources that an ATM network supports are bursty, a Poisson process may no longer be suitable for describing the network traffic. For instance, interactive data and compressed video generate cells at a near-peak rate for a very short period of time. Immediately, following a near peak rate such a source may become inactive, thus generating no cells. With this scenario, the usual approximation of arrival process by a Poisson process will fail to capture the bursty nature of input traffic and may result in a quite dramatic error in the performance estimation. Kuehn[5] has shown that the system behavior is much more sensitive to arrival processes than to service process. Therefore, we propose to use the Interrupted Bernoulli Process(IBP) which is the generalization of the Poisson process in a discrete time system.

An IBP is governed by a Markov chain of two states, an active state and an idle state. The duration of stay in these two states are geometrically distributed with different parameters. Arrivals occur in a Bernoulli fashion when the process is in the active state. No arrivals occur if the process is in the idle state. Given that the process is in the active

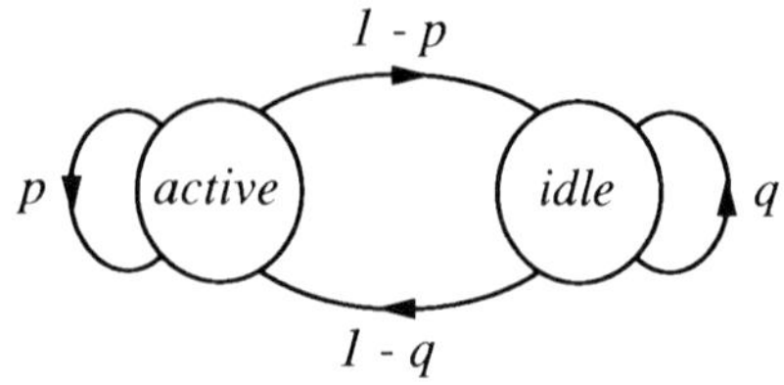

$$\pi_A = \frac{1-q}{2-p-q}$$

$$\pi_I = \frac{1-p}{2-p-q}$$

Figure 2: The Markov chain for an IBP

state (or idle state) at slot i, it will remain in the same state in the next slot $i+1$ with probability p (or q), or will change to the idle state (or active state) with probability $1-p$ (or $1-q$). The transitions between the active and idle states are shown in figure 2, where π_A and π_I are the probabilities that an arrival process is in the active and idle states, respectively. During the active state, a slot contains a cell with probability α. Here we assume α equals 1 which will generate the highest bursty effects.

Letting t be the interarrival time of a cell; it can be shown that the z-transform of the probability distribution of the interarrival time is

$$A(z) = \frac{z\alpha[p + z(1-p-q)]}{(1-\alpha)(p+q-1)z^2 - [q+p(1-\alpha)]z + 1}.$$

The mean interarrival time $E\{t\}$ and the squared coefficient of variation of the time between successive arrivals, C^2 are as follows:

$$\begin{aligned} E\{t\} &= \frac{2-p-q}{\alpha(1-q)} \\ C^2 &= \frac{Var(t)}{E\{t\}^2} \\ &= 1 + \alpha\left[\frac{(1-p)(p+q)}{(2-p-q)^2} - 1\right]. \end{aligned}$$

The average arrival rate, i.e. the probability that any slot contains a cell, λ is

$$\lambda = \frac{\alpha(1-q)}{2-p-q}.$$

By varying p and q, we can have different traffic loads and at the same time change the burstiness of the arrival process.

2.3 Queueing models

Based on the structure of SRM, we evaluate its performance by using a multiqueue system as shown in figure 3. Under the assumption of symmetric traffic, the arrival processes to the multiqueue system will be characterized also as IBPs with the same

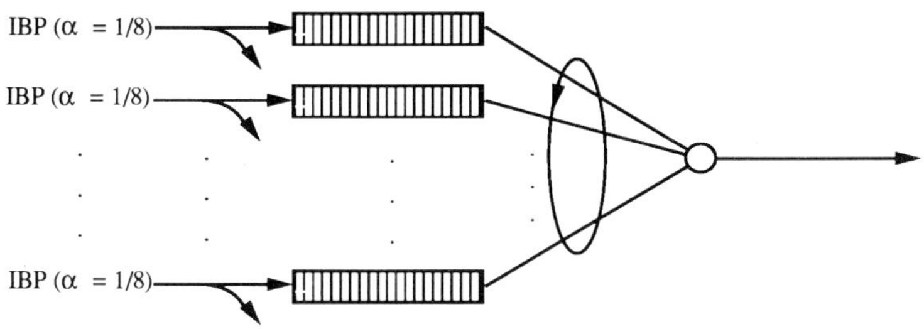

Figure 3: Multiqueue System Served by a Single Server.

parameters of p and q as of the original arrival processes. However, given in the active state, the probability of having a cell arriving in the polling system will be only one eighth.

It is recognized that the analysis of a finite capacity polling system with bursty arrivals will involve a great deal of mathematical complexity. In this paper, instead of investigating the polling system itself, we propose a method to find the upper bounds for mean delay time and cell loss probability that cells incur in this polling system.

Based on the requirement of the cell loss probability being less than 10^{-9} in the ATM environment, it is suitable to find the mean delay time which cells incur in a SRM by studying a multiplexer which has the same queue capacity as the original multiqueue system. For the extreme case of an infinite buffer where cell loss does not occur, the polling system will be equivalent to the multiplexer as far as the mean delay is concerned. Even though the service disciplines are different for these two systems, it is known that the mean delay is independent of service scheduling as long as the order of service is not chosen according to the service time required for different customers. Since we have deterministic service time in our model, the mean delay will be the same for both systems if the buffer sizes are infinite. Therefore, when the queue capacities are finite, the discrepancy of mean delays between the polling system and the cell multiplexer is only due to the different cell loss rates that cells incur in these two systems. It is obvious that the cell loss probability in the multiplexer is much smaller than in the polling system. In fact, the cell loss rate obtained from the multiplexer is a lower bound to that obtained from the polling system. Therefore, the mean delay time that cells incur in the multiplexer will be longer than in the polling system and can serve as an upper bound. It will be shown in Section 5 that if the cell loss probability is smaller than 10^{-5}, the discrepancy will be less than one percent. Based on the requirement of the cell loss rate being much smaller than 10^{-5}, this approach will not only provide the conservative bound but also render a very accurate estimation. Hence, this approach is suitable in the practical application.

Consider again the case with an infinite buffer. We call the distribution of the total number of cells in the multiqueue system as the aggregate queue length distribution. Based upon the conservation law and zero switchover time assumption, we know that the aggregate queue length distribution of the polling system will be the same as the queue length distribution in the cell multiplexer. From this observation we approximate the aggregate queue length distribution of the polling system by computing the queue length distribution of the cell multiplexer when the buffer sizes are finite. The queues in the polling system can be further described by a multiple finite capacity urn model. Given the number of cells waiting in the system, it is assumed that the occupancy of each queue is independent from each other and the cells are uniformly distributed in any queue, i.e., each position in a queue is equally likely to be occupied. From this model we can compute the weighting of the cell occupancy configuration which could cause cell loss and the cell loss probability will follow. Notice that given N cells in the system, the occupancy of these cells in reality will more likely be evenly distributed among these queues. This is because that in general the server will visit the queues with longer queue sizes more frequently than the queues with shorter queue sizes. Therefore, given the number of cells in the polling system exceeding a single queue capacity, the occupancy configuration which has at least one full queue is less likely to occur. Hence, the assumption of uniform occupancy will give us a conservative estimate which could serve as an upper bound for cell loss probability of the polling system.

3 ANALYSIS OF A CELL MULTIPLEXER

The characteristics of a simple multiplexing model are analyzed to investigate cell loss and delay at a finite buffer when bursty traffic is loaded in the buffer. This multiplexer model is shown in figure 4-a. It is assumed that both input and output links have the same speed and operate synchronously. Cells arriving from N input lines are stored in a FIFO queue with capacity Q_m. A cell is considered as an arrival only after the last bit of this cell has been received. In our model, we assume that arrivals can only occur at the beginning of each slot and departures leave the system at the end of each slot. This arrangement, as illustrated in figure 4-b, is called an early arrival system according to Hunter [6]. During a slot period, one cell may arrive on each input link and one cell may be transmitted given that the system is not empty.

Next, the cell arrival process is analyzed for a cell arriving from all inputs in a slot time. Both numbers of active input links k and arrival cells m in a unit time could vary from zero to N. The probability that the number of active input links is k in a unit time is

$$P_K(k) = \binom{N}{k} \pi_A^k \pi_I^{N-k},$$

where π_A and π_I denote the probability that an input link is in an active or idle state.

If the number of active input links is k, then m cells will arrive in a unit time with the conditional probability of

$$P_{M|K}(m|k) = \binom{k}{m} \alpha^m (1-\alpha)^{k-m},$$

and the probability distribution of the number of arriving cells is given by

$$P_M(m) = \sum_{k=0}^{N} P_K(k) P_{M|K}(m|k)$$

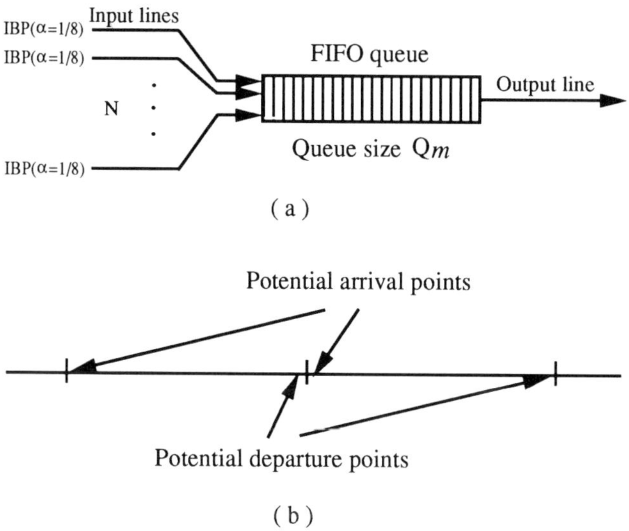

Figure 4: (a) Cell multiplexer, (b) Potential arrival and departure points.

$$= \binom{N}{m} (\alpha\pi_A)^m (1-\alpha\pi_A)^{N-m}.$$

The state transition probability of having k active lines in a slot and k' active lines in the next slot is given by

$$P_{K'|K}(k'|k) = \sum_{j=0}^{k} \binom{k}{j} p^j (1-p)^{k-j} \binom{N-k}{k'-j} q^{N-k-k'+j} (1-q)^{k'-j}.$$

Now, we define a two dimensional state variable (K,Q) such that the queue length becomes Q as the result of having M cells arrive in a slot given that K input lines are active. The state probability $P_{K,Q}(k,q)$ can be obtained by a numerical solution of the following steady state equations:

$$P_{K,Q}(k',q') = \sum_{k=0}^{N} \sum_{m'=0}^{k'} \sum_{q=0}^{Q_m} P_{K,Q}(k,q) P_{K'|K}(k'|k) P_{M|K}(m'|k'),$$

$$\sum_{k=0}^{N} \sum_{q=0}^{Q_m} P_{K,Q}(k,q) = 1.$$

From $P_{K,Q}(k,q)$, we can sum over K and find the queue length distribution $P_Q(q)$. The probability distribution of system delay will be found by using Little's result. The cell loss probability is obtained through the summation of the probabilities of the queue length overflowing the maximum queue size Q_m as follows:

$$Ploss = \frac{\sum_{k=0}^{N}\sum_{k'=0}^{N}\sum_{q=0}^{Q_m} P_{K,Q}(k,q)P_{K'|K}(k'|k)P_{M'|K'}(m'|k')(q-1+m'-Q_m)}{\lambda},$$

where $q-1+m' > Q_m$ and λ is the total arrival rate to the multiplexer.

4 MULTIPLE URN MODEL

4.1 Model description

We approximate the aggregate queue length distribution of the polling system by the queue length distribution of the cell multiplexer obtained from the last section. In order to compute the upper bound for cell loss probability of the polling system, it remains to explore the total number of ways to place r indistinguishable balls into n distinguishable boxes given that the capacity of each box is m. This is equivalent to finding the total number of solutions which satisfy the following equation:

$$r_1 + r_2 + \ldots + r_k + \ldots + r_n = r, \qquad 0 \le r_k \le m. \tag{1}$$

Every n-tuple of integers satisfying this equation describes a possible configuration of occupancy numbers. With indistinguishable balls two distributions are distinguishable only if the corresponding n-tuples $(r_1, r_2, \ldots, r_n)$ are not identical. It is shown [7] that given r $\le$ m (i.e. no capacity limit), the number of distinguishable distributions is

$$A_{n,r} = \binom{n+r-1}{r} = \binom{n+r-1}{n-1}.$$

Define $B_{n,r,k}$ as the number of ways of having at least k_i balls in i^{th} box (no limit). Then $B_{n,r,k}$ can be easily found as the following

$$B_{n,r,k} = \binom{n+r-k-1}{r-k} = \binom{n+r-k-1}{n-1}, \qquad k = \sum_{i=1}^{n} k_i.$$

From the principle of inclusion and exclusion we can obtain the total number of different solutions of equation (1) as

$$\begin{aligned} T_{n,r,m} &= \sum_{i=0}^{n}(-1)^i \binom{n}{i}\binom{n+r-i(m+1)-1}{n-1} \\ &= \sum_{i=0}^{n}(-1)^i \binom{n}{i} B_{n,r-i(m+1),0}, \end{aligned}$$

where $B_{n,r,0}$ is equal to $A_{n,r}$ which denotes the total number of solutions given no limit imposed on the capacity of boxes. $B_{n,r-(m+1),0}$ represents the number of configurations in which at least one of the n boxes contains no less than $m+1$ balls. This condition violates the limit and should be subtracted from $B_{n,r,0}$. The rest of the terms is just to compensate the over subtraction that $B_{n,r-(m+1),0}$ introduces.

We now proceed to find $C_{n,r,m}$, the number of ways of having at least one full box for this multiple urn model. If the total number of balls r is less than m, $C_{n,r,m}$ equals zero. For $m \leq r < 2m$, it is easy to verify that

$$C_{n,r,m} = nB_{n-1,r-m,0}.$$

Likewise, when $2m \leq r < 3m$

$$\begin{aligned} C_{n,r,m} &= \binom{n}{2} B_{n-2,r-2m,0} + \binom{n}{1}\left[T_{n-1,r-m,m} - \binom{n-1}{1} B_{n-2,r-2m,0}\right] \qquad (2) \\ &= \binom{n}{2} T_{n-2,r-2m,m} + \binom{n}{1}\left(T_{n-1,r-m,m} - C_{n-1,r-m,m}\right), \end{aligned}$$

where the first term on the right hand side of equation (2) denotes the number of ways to have two full boxes, and the second term gives the number of ways of having one full box. Proceeding in the same manner, we conclude that for $km \leq r < (k+1)m$

$$C_{n,r,m} = \sum_{i=1}^{k} \binom{n}{i} \left(T_{n-i,r-i*m,m} - C_{n-i,r-i*m,m}\right),$$

where $C_{n,r,m}$ is obtained recursively.

Consider again the cell loss probability of the polling system. Given r cells in the system, the (conditional) probability, $P_{loss}(r)$, that a typical arriving cell sees a full queue is

$$\begin{aligned} P_{loss}(r) &= P(\textit{having a cell loss}|\ r \textit{ cells in the system}) \\ &= \frac{\sum_{i=1}^{\lfloor \frac{r}{m} \rfloor} \frac{i}{n} \binom{n}{i} \left(T_{n-i,r-i*m,m} - C_{n-i,r-i*m,m}\right)}{T_{n,r,m}}. \end{aligned}$$

Therefore, the probability of cell loss is given by

$$P_{loss} = \sum_{r=m}^{n*m} P_{loss}(r) P(r \textit{ cells in the system}).$$

4.2 Heuristic approach

The cell loss probability, P_{loss}, obtained in the last subsection is based on two assumptions. We first assume that given r cells in the system, each occupancy configuration has an equal probability of occuring. However, this is not the case in reality. Since the server skips the empty queues and only visits the queues which have cells present, long queues will less likely be formed. Consequently, the cell loss probability derived from this assumption will always overestimate and could serve as an upper bound. Secondly, we use the queue length distribution of a cell multiplexer to approximate the stationary probability distribution of cells in the polling system. This approximation will become exact when the buffer sizes grow to infinity.

Knowing that these two assumptions are not exact for most of the cases, we propose a heuristic approach in order to provide a better estimation for cell loss probability of the polling system. Given that the buffer capacity of the multiqueue system is m, the buffer size Q_m of the multiplexer model will be equal to $n \times m$, where n is number of queues in the multiqueue system. The second assumption mentioned above is not exact because the polling system has a cell loss condition different from the multiplexer. In our multiplexer model, cell loss does not occur until Q_m cells are queued up in the buffer. However, an arriving cell is subjected to loss as long as the number of cells in the polling system is equal to or more than m. Therefore, we should expect different queue length distributions for these two systems given that the queue sizes are finite.

In view of this discrepancy, we propose a heuristic method to modify the aggregate queue length distribution of the polling system. The algorithm of the heuristic modification is as follows:

```
factor = 0;
for ( r = m + 1; r <= Q_m; r++ )
{
   factor += P_loss(r - 1) P(r - 1 cells in the polling system);
   if (P(r cells in the polling system) == 0 || P(r cells in the polling system) - factor ≤ 0 )
      P(r cells in the polling system) = 0;
   else
      P(r cells in the polling system) -= factor;
}.
```

After this modification, the *P(r cells in the polling system)* should be normalized. The idea behind this arrangement is simple and straightforward. We know that a cell loss may occur when an arrival finds m or more cells queued up in the system. Suppose this arrival sees r cells present in the system and gets lost, then the queue distribution after r will all be affected. We consider that it is suitable to deduct the cell loss probability that this typical arrival incurs from the distribution of having $m+1$ or more cells. After we make this modification, we use the normalized distribution to compute the cell loss rate. As it is shown in the next section, this modification provides a much better estimate than the previous one, especially when the cell loss rate is medium or high.

5 NUMERICAL RESULTS

Figures 5 to 6 show the influence of arrival burstiness toward mean delays under different queue capacities. Here, we only present three kinds of burstiness to represent three different cases. When C^2 equals 1, we can regard this arrival process as Bernoulli. The burstiness of voice is represented by the case where C^2 equals 20. We use $C^2 = 200$ to show the burstiness of data. From the figures it is clear that big errors would be made if we assume only Bernoulli arrival processes under the ATM environment. The effect of burstiness toward mean delays becomes more obvious when the queue capacities increase. The analytic results always overestimate the simulation results and can serve as an upper bound. By comparing figures 5 to 6, we can see that the discrepancy between analytic and simulation results diminishes when the buffer size increases. Under the extreme case of the infinite buffer where cell loss does not occur, as we have mentioned in section 2, the analytic results will become exact. In fact, when the queue size is 32, every point of analytic results which we measure falls in the range of the confidence interval of simulation. This shows that the analytical results cannot only serve as an upper bound, but also provide an accurate estimation when the queue size increases.

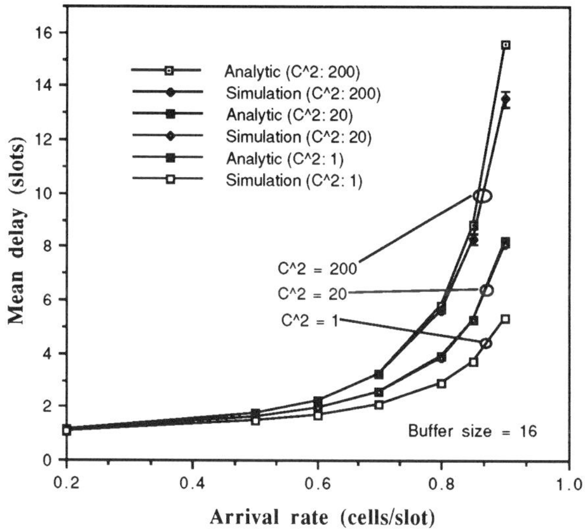

Figure 5: Mean delay w.r.t. different burstiness (queue capacity: 16).

The results regarding cell loss probability are shown in figures 7 to 9. In each figure we show both the upper and lower bounds as well as the results from the heuristic approach and the simulation model for comparison. The relationship among these results can be best illustrated in figure 8. The heuristic results follow simulation results closely and still give conservative estimation when the cell loss probabilities are from medium to high. For the rest of the cases, the effect is not so obvious because of low cell loss rate. However, they all follow the same trend.

6 CONCLUSION

In this paper, we presented the upper bounds for mean delay and cell loss probability that bursty arrivals incur in a finite capacity polling system. We first used a cell multiplexer with a finite buffer to obtain an upper bound for the mean delay and a lower bound for the cell loss rate of the polling system. It is shown that the upper bound of mean delay becomes an accurate estimation when the buffer size increases. For the cell loss probability, we developed a multiple urn model with uniform occupancy distribution which guaranteed the upper bound. A heuristic method was further proposed to provide

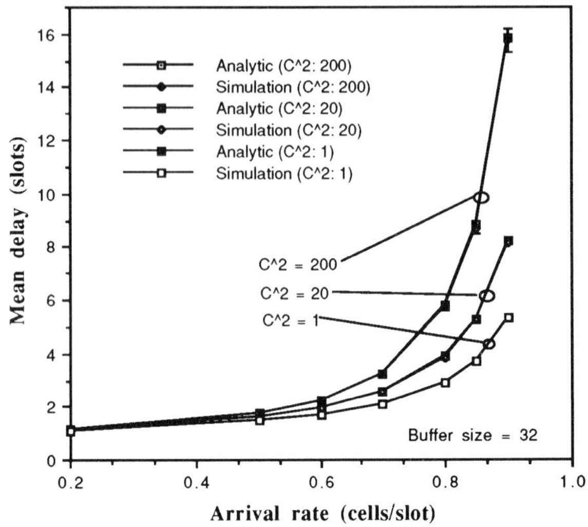

Figure 6: Mean delay w.r.t. different burstiness (queue capacity: 32).

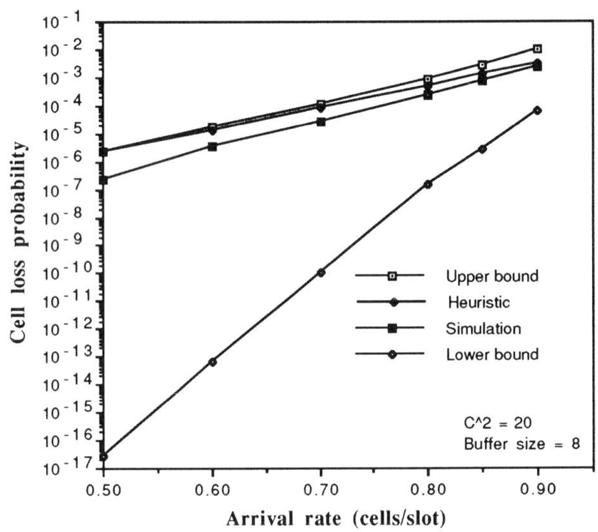

Figure 7: Cell loss probability when $C^2 = 20$, queue capacity $= 8$.

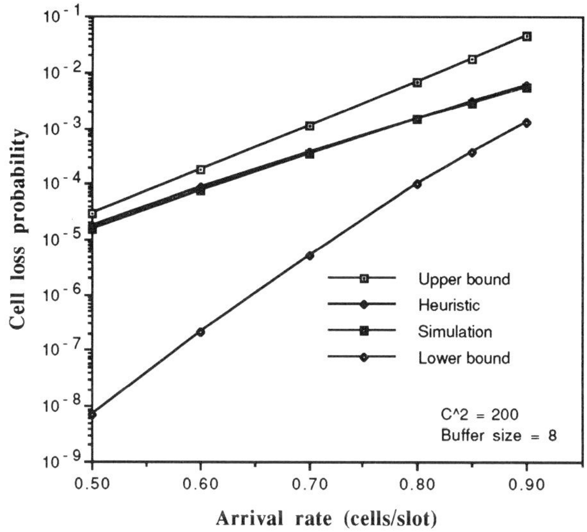

Figure 8: Cell loss probability when $C^2 = 200$, queue capacity = 8.

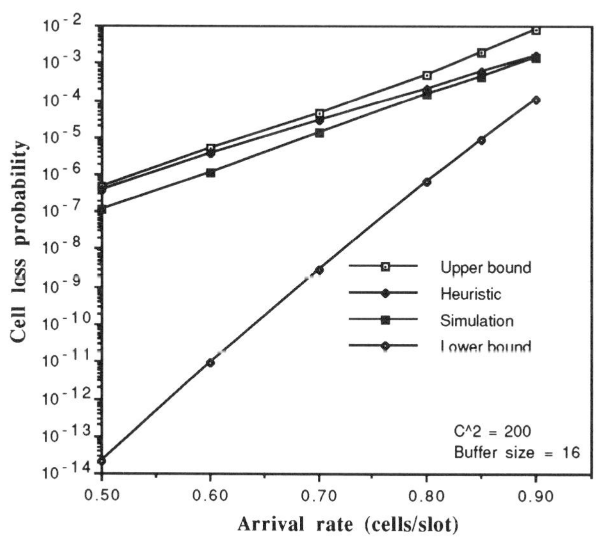

Figure 9: Cell loss probability when $C^2 = 200$, queue capacity = 16.

a better estimate of cell loss rate for cases which have medium to high cell loss rate. Our approach is validated through the comparison of these calculated bounds and heuristic results with the simulation results for a number of cases.

By comparing the results obtained from both the analytical and simulation models, we observe that the discrepancy between the mean delays from these two models is less than one percent as long as the cell loss rate is kept under 10^{-5}. In the context of the ATM environment, where the cell loss rate is required to be less than 10^{-9}, our approach is well applicable to finding the mean delay in the polling system. We conclude that our approach not only provides theoretical upper bounds but also gives accurate estimations under certain conditions. The significant advantage of this approach is such that the analytical difficulties which more realistic arrival processes introduce will not impose on the polling system directly. The complexity will be much smaller when we deal with a cell multiplexer.

7 REFERENCES

[1] K. Hajikano, K. Murakami, E. Iwabuchi, O. Isono, and T. Kobayashi, "Asynchronous transfer mode switching architecture for broadband ISDN - multistage self-routing switching," *IEEE International Conference on Communications*, vol. 2, pp. 911–915, 1988.

[2] Takagi, "Queueing analysis of polling models: an update," *Stochastic Analysis of Computer and Communication Systems*, pp. 267–318, H. Takagi (editor), Elsevier Science Publishers B.V. (North–Holland) Amsterdam, 1990.

[3] P. Tran-Gia and T. Raith, "Performance analysis of finite capacity polling systems with nonexhaustive service," *Performance Evaluation*, vol. 9, pp. 1–16, 1988.

[4] M. Lang and M. Bosch, "Performance analysis of finite capacity polling systems with limited-m service," *Proc. ITC-13*, vol. 14, pp. 731–735, 1991.

[5] P. J. Kuehn, "Multiqueue systems with nonexhaustive cyclic service," *B.S.T.J.*, vol. 58, no. 3, pp. 671–698, 1979.

[6] J. J. Hunter, *Mathematical Techniques of Applied Probability, Vol. 2, Ch. 9.* New York, NY: Academic Press, Inc., 1983.

[7] W. Feller, *An Introduction to Probability Theory and Its Applications, Vol. I, 3rd ed.* New York, NY: John Wiley & Sons, Inc, 1968.

Queueing Networks with Finite Capacity
R.O. Onvural and I.F. Akyildiz (Editors)

A DECOMPOSITION METHOD FOR THE ANALYSIS OF TANDEM QUEUEING NETWORKS WITH BLOCKING BEFORE SERVICE

A. Bouhchouch *, Y. Frein * and Y. Dallery **

* Laboratoire d'Automatique de Grenoble, UA 228 CNRS, BP 46, 38402 Saint Martin d'Hères, France.
** Laboratoire MASI (Méthodologie et Architecture des Systèmes Informatiques), UA 818 CNRS, Université Pierre et Marie Curie, 4 Place Jussieu, 75252 Paris Cedex 05, France.

Abstract

Queueing networks with finite buffers have received much attention in the past years. Different blocking mechanisms have been defined that are encountered in various applications such as manufacturing systems and communication networks. Closed form analytical solutions are not in general obtainable for queueing networks with finite buffers. As a result, much work has been devoted to deriving approximation solutions. In this paper, we deal with tandem queueing networks with finite buffers. Several different approximation methods have been proposed to analyze such networks. Most of them are based on decomposition and assume the so-called blocking-after-service mechanism. In this paper, we propose an approximation method for the analysis of tandem queueing networks with the so-called blocking-before-service mechanism. The service times are represented by phase type distributions. This method decomposes the original network into a set of subsystems consisting of two servers separated by a buffer. An iterative procedure is used to calculate the parameters of the different subsystems. The originality of this method lies is in the way it captures an essential feature of the blocking-before-service mechanism, namely that starvation and blocking may occur simultaneously. Numerical results show that this method provides very accurate results.

1. INTRODUCTION

Queueing network models in which buffers have finite capacity are often encountered in various applications such as manufacturing systems and communication networks. Different types of blocking mechanisms have been considered in the literature. Three main types of blocking mechanisms have been defined and referred to as blocking-after-service, blocking-before-service and repetitive blocking [Perros, 1989]. In the blocking-after-service mechanism, a server is blocked if the destination buffer of the customer is full after completion of the service of a customer. In blocking-before-service, the service of a customer is not allowed to start until there is a room available in its destination buffer. In repetitive blocking, a customer upon service completion attemps to join its destination buffer. If this buffer is full, the customer receives another service and this is repeated until space becomes available in the destination buffer.

Queueing networks with finite buffers have received much attention in the past years; see [Perros, 1989; Onvural, 1990; Dallery and Frein, 1992; Dallery and Gershwin, 1991] for surveys of results and lists of references. Exact closed form analytical solutions are not in general obtainable for queueing networks with finite buffers. As a result, much effort has been devoted to obtaining approximate solutions. Various methods have been proposed that can handle different classes of queueing networks with finite buffers. Among them, the class of open tandem queueing networks with blocking-after-service mechanism has received much attention. Some authors assume exponential service time distributions while others consider general service time distributions most often represented as phase type (PH) distributions [Neuts, 1981]. Also, some authors consider the case where customers arrive to the network according to a given arrival process while others assume that the first server is saturated, i.e., there is an infinite supply of customers in front of it. A model with an external arrival process can be equivalently transformed into a saturated model if the external arrival process is Poisson; see e.g. [Dallery and Frein, 1992].

Several different methods have been proposed for analyzing tandem queueing networks with blocking-after-service, e.g. [Altiok, 1982 and 1989; Altiok and Ranjan, 1987; Gün and Makowski, 1989; Jun and Perros, 1990; Perros and Altiok, 1986; Pollock, Birge and Alden, 1985; Takahashi, Miyahara, and Hasegawa, 1980]. All these methods are based on the decomposition of the original network into a set of subsystems. A comparison and unifying presentation of these methods can be found in [Dallery and Frein, 1989 and 1992]. It appears that there are three main approaches. Among them, the so-called *symmetrical* approach has the following interesting feature: starvation and blocking phenomena play a similar role in the method. A decomposition method based on this symmetrical approach has been proposed independently by Altiok and Ranjan [Altiok and Ranjan, 1987] and Gün and Makowski [Gün and Makowski, 1989]. They consider a saturated model with servers having PH distributions. This method has proved to perform very well and provide accurate results.

In this paper, we consider tandem queueing networks with finite buffers and blocking-before-service mechanism. Servers have general service time distributions represented as PH distributions. For this class of networks, Gün and Makowski [Gün and Makowski, 1989] have proposed a slightly different version of their method. However, the results obtained by this method are not as good as in the case of blocking-after-service. This is particularly true for networks with small buffer sizes. The reason is that their method does not capture in an adequate way an essential feature of the blocking-before-service mechanism, namely that starvation and blocking may occur simultaneously. The purpose of this paper is to present a new method for the analysis of tandem queueing networks with finite buffers and blocking-before-service mechanism. This method has similarities with that of Gün and Makowski: 1) it is also based on a decomposition of the original network into a set of subsystems consisting of two servers separated by a buffer; and 2) it uses the symmetrical approach. However, it differs in the way the blocking mechanism is captured in the decomposition. In particular, the behavior of the subsystems is fairly different. Numerical results reported in this paper show that the new method is significantly more accurate than that of Gün and Makowski.

The paper is organized as follows. In Section 2, the model considered in this paper as well as some relationships between the different blocking mechanisms are presented. A discussion pertaining to the behavior of the blocking-before-service mechanism is then provided. The new method proposed in this paper is presented in Section 3. A discussion pertaining to the comparison with the method of Gün and Makowski is also given. Numerical results are reported in Section 4.

2. TANDEM QUEUEING NETWORKS WITH FINITE BUFFERS

2.1. The tandem queueing network model

We consider a tandem queueing network consisting of a series of M+1 single servers separated by M buffers as illustrated in Figure 1. The servers will be denoted by S_i, for i = 0,..., M, and the buffers will be denoted by B_i, for i = 1,...., M. Let C_i denote the capacity of buffer B_i, including the server space in front of server S_i, and let $\mathbf{C} = (C_1, C_2,..., C_M)$ denote the buffer capacity vector. We assume that the successive service times at each server are independent and identically distributed (i.i.d.) random variables. We denote by Y_i the random variable representing the service time of server S_i. We assume that the first server is never starved and that the last server is never blocked. That is, we consider a saturated model.

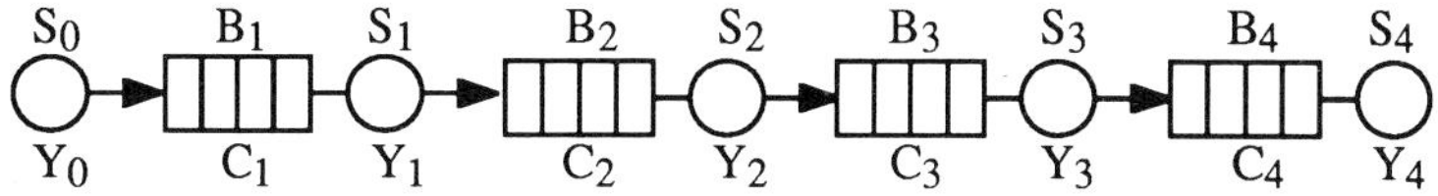

Fig. 1. The tandem queueing network model (illustrated for M=4)

Due to the finiteness of the buffers blocking may occur. Various blocking mechanisms have been considered in the literature. The most commonly used are blocking-after-service (BAS) and blocking-before-service (BBS). In blocking-after-service, upon completion of its service at server S_i a customer attemps to enter buffer B_{i+1}. If at that moment this buffer is full, the customer cannot be transfered until a space becomes available in buffer B_{i+1}. During this period of time server S_i is blocked, i.e., it can not serve any customer. In blocking-before-service, server S_i is allowed to start processing a customer only if there is a space available in buffer B_{i+1}. If buffer B_{i+1} is full, then server S_i is blocked and it will be allowed to start processing a customer when a space becomes available in buffer B_{i+1}. Blocking-before-service is further classified according to whether the service area may be occupied or not while the server is blocked. These two cases are referred as: BBS-SO (blocking-before-service with service area occupied while the server is blocked) and BBS-SNO (blocking-before-service with service area non-occupied while the server is blocked) [Perros, 1989].

Let us briefly discuss the relationships between the different blocking mechanisms. First, there is an equivalence between BAS and BBS-SNO mechanisms [Onvural and Perros, 1986; Dallery, Liu and Towsley, 1992]. Let $\mathbf{1}$ = (1, 1,..., 1) denote the unit vector. Consider a tandem queueing network with general service time distributions having buffer capacity vector $\mathbf{C}$ and operating under the BAS blocking mechanism. Consider the same tandem queueing network but with buffer capacity vector $\mathbf{C+1}$ and operating under the BBS-SNO blocking mechanism. Then, it was established in [Dallery, Liu and Towsley, 1992] that these two networks have exactly the same behavior (in terms of sample paths). The performance parameters of one can easily be derived from those of the other. In particular, they have the same throughput.

There is no such equivalence between the BAS (or BBS-SNO) and BBS-SO mechanisms except in the case of a two server one buffer system, i.e., M = 1. Indeed, in that case, BBS-SO and BBS-SNO mechanisms are equivalent. Therefore, a two server one buffer system with capacity C and BAS mechanism is equivalent to a two server one buffer system with capacity C+1 and BBS-SO mechanism. In the general case, i.e., M > 1, a relationship between the throughput of a tandem queueing network with BAS and BBS-SO mechanisms can be established. Let X(C, BAS) and X(C, BBS-SO) denote the throughputs of a tandem queueing network with general service time distributions and buffer capacity vector $\mathbf{C}$, operating under BAS and BBS-SO mechanisms, respectively. The following result is proved in [Dallery, Liu and Towsley, 1992]:

$$X(\mathbf{C}, \text{BBS-SO}) \le X(\mathbf{C}, \text{BAS}) \le X(\mathbf{C+1}, \text{BBS-SO}) \le X(\mathbf{C+1}, \text{BAS}) \qquad (1)$$

As a result of the above discussion, it appears that there is no need to develop specific methods for the case of BBS-SNO because of the equivalence with the BAS mechanism for which, as stated earlier, very efficient methods are available. As a result, we focus our attention on the case of BBS-SO mechanism. Since in the rest of the paper we only consider this blocking mechanism, we simply refer to it as blocking-before-service (BBS).

2.2. Behavior of the blocking-before-service mechanism

Before presenting the approximation method in Section 3, let us further investigate the behavior of tandem queueing networks with BBS mechanism. Let us first define starvation and blocking phenomena. Server S_i is said to be *starved* whenever buffer B_i is empty, that is there is no customers for the server to work on. Server S_i is said to be *blocked* whenever buffer B_{i+1} is full. According to the definition of the BBS mechanism, the transfer instant of a customer into buffer B_{i+1} is exactly the same as the processing completion instant of this customer by server S_i. Let us consider the time between the processing completion instants of two successive customers, say x and x+1, at server S_i. Equivalently, it is also the time between the transfer instants of customers x and x+1 into buffer B_{i+1}. After transfer of customer x into buffer B_{i+1}, server S_i may be in four different situations: 1) neither blocked nor starved; 2) blocked but not starved; 3) starved but not blocked; 4) simultaneously starved and blocked. Let us consider these four situations.

Situation 1. Server S_i is neither blocked nor starved at the processing completion instant of customer x. The server begins immediately the service of customer x+1. Therefore, the time between the transfer instants of customers x and x+1 is just the processing time of customer x+1 by server S_i, which is characterized by the random variable Y_i.

Situation 2. Server S_i is blocked but not starved at the processing completion instant of customer x. Customer x+1 is already present in buffer B_i waiting for receiving service. However, server S_i cannot start its processing because buffer B_{i+1} is full. This will last until there is a space available in buffer B_{i+1}. Thus, the time between the transfer instants of customers x and x+1 is composed of a blocking time followed by a processing time of server S_i. The blocking time corresponds to the time to the next processing completion of server S_{i+1} at the processing completion instant of customer x at server S_i. This time is just the residual processing time of server S_{i+1} if server S_{i+1} is not blocked at the processing completion instant of customer x at server S_i. Otherwise, it is the sum of the time for server S_{i+1} to become unblocked plus the time for server S_{i+1} to process one customer.

Situation 3. Server S_i is starved but not blocked at the processing completion instant of customer x. There is a space available in buffer B_{i+1}. However, customer x+1 is not yet present in buffer B_i and thus server S_i cannot start working. This will last until customer x+1 arrives into buffer B_i. Thus, the time between the transfer instants of customers x and x+1 is composed of a starvation time followed by a processing time of server S_i. The starvation time corresponds to the time to the next processing completion of server S_{i-1} at the processing completion instant of customer x at server S_i. This time is just the residual processing time of server S_{i-1} if server S_{i-1} is not blocked at the processing completion instant of customer x at server S_i. Otherwise, it is the sum of the time for server S_{i-1} to become unblocked plus the time for server S_{i-1} to process one customer.

Situation 4 (Figure 2). Server S_i is starved and blocked at the processing completion instant of customer x. That is, customer x+1 is not yet present in buffer B_i and buffer B_{i+1} is full. The processing of customer x+1 will begin when both the starvation and the blocking are over. Thus, the time between the transfer of customers x and x+1 is composed of the maximum between the starvation time and the blocking time followed by a service time of S_i. Two cases may happen. In the first case, the blocking time is larger than the starvation time (Fig. 2.a). In this case server S_i will still be blocked when customer x+1 arrives into buffer B_i. The processing of customer x+1 is then delayed by the residual blocking time at the end of starvation. In the second case, the starvation time is larger than the blocking time (Fig. 2.b). In this case the blocking of server S_i has not delayed the processing of customer x+1.

This fourth situation is an essential feature of the BBS mechanism: a server can be blocked and starved simultaneously. In BBS mechanism, the blocking of a server can affect the beginning of the processing of the next customer either totally (Situation 2), partially (Fig. 2.a), or not at all (Fig. 2.b). The aim of this paper is to propose an approximation that captures this feature.

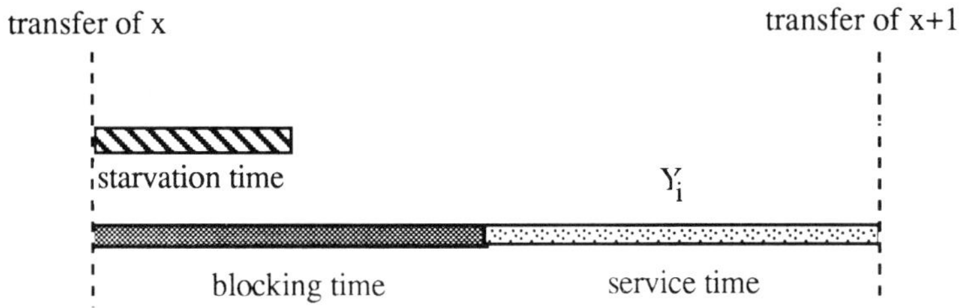

Fig. 2.a

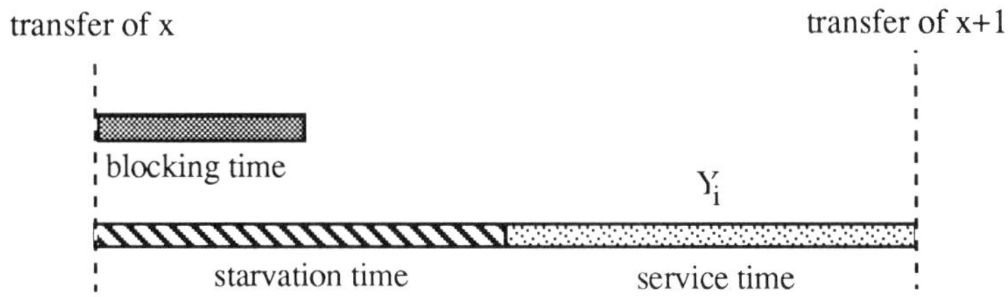

Fig. 2.b

Fig. 2. Time between transfer instants of customers x and x+1 (situation 4)

Finally, let us discuss some important properties of tandem queueing networks, namely reversibility and duality. The reverse of a tandem queueing network is the network obtained by reversing the flow of customers. The so-called reversibility property states that the reverse network has the same throughput as the original network. This property holds for the BAS mechanism [Muth, 1979] as well as for the BBS mechanism [Dallery, Liu and Towsley, 1990 and 1992]. However, in the case of BBS, a stronger property so-called duality also holds [Ammar and Gershwin, 1989; Dallery, Liu and Towsley, 1990]. It states that the behavior of the reverse network is equivalent (in terms of sample-paths) to that of the original network, provided that the customers of the reverse network are interpreted as the holes (empty space) of the original network, and vice-versa. As a result, there is a one to one correspondance between the performance parameters of the two networks. In particular, the probability of starvation (resp. blocking) of a server in the reverse network is equal to the probability of blocking (resp. starvation) of the corresponding server in the original network.

3. THE DECOMPOSITION METHOD

In this section, we present the new approximation method. The decomposition principle is presented in Section 3.1. In Section 3.2, we focus our attention to a special case, namely the case of a three server network with exponential service times. This will allow us to present the basic ideas. In Section 3.3, we show how this extends to the general case. The computational algorithm is presented Section 3.4. Finally, a discussion pertaining to the comparison of the proposed method with that of Gün and Makowski is presented in Section 3.5.

3.1. The decomposition principle

Consider the tandem queueing network model introduced in Section 2.1. As in [Altiok and Ranjan, 1987] and [Gün and Makowski, 1989], our method is based on the decomposition of the original network consisting of M+1 servers into M subsystems (see Figure 3). Each subsystem T(i), for i = 1,...., M is composed of an upstream server $S_u(i)$ and a downstream server $S_d(i)$ separated by a finite buffer B(i). The first server, $S_u(i)$, is never starved and the second server, $S_d(i)$, is never blocked. The service times of server $S_u(i)$ and $S_d(i)$ are assumed to be independent and identically distributed (i.i.d.) random variables denoted by $Y_u(i)$ and $Y_d(i)$, respectively. Subsystem T(i) approximates the flow of customers in buffer B_i of the original system. The intermediate buffer B(i) has the same capacity C_i as that of buffer B_i. Server $S_u(i)$ represents the part of the original model upstream of buffer B_i while server $S_d(i)$ represents the part of the original model downstream of buffer B_i. Each subsystem T(i) operates under the BBS blocking mechanism.

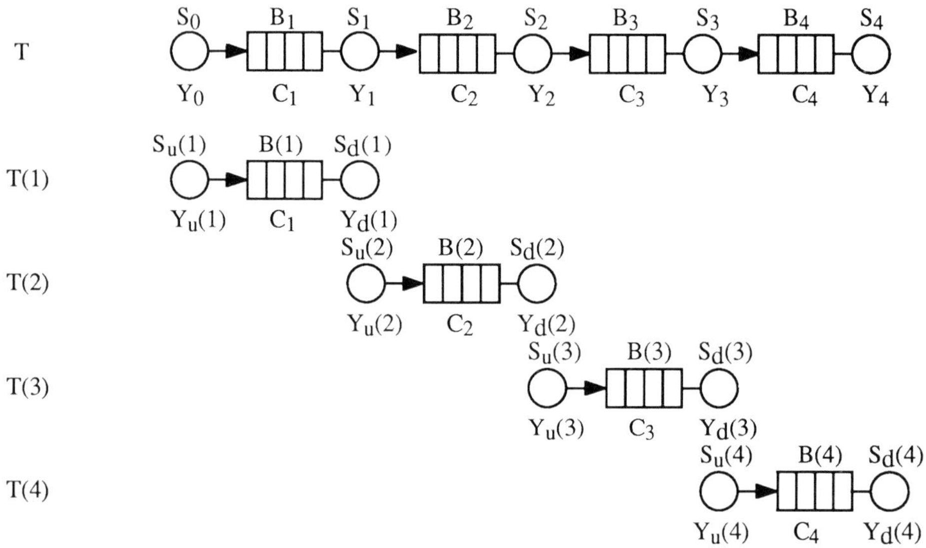

Fig. 3. Decomposition of the tandem queueing network

The goal of the method is to determine servers $S_u(i)$ and $S_d(i)$ in such a way that the flows into and out of buffer B(i) in subsystem T(i) closely match those of buffer B_i in system T, for all subsystems T(i), i = 1,...., M.

3.2. A three server network with exponential service times

Let us first focus on a special case, namely a tandem queueing network consisting of 3 servers (M = 2). Furthermore, we assume that the random variables Y_i, i = 0, 1, 2, are exponentially distributed. Let μ_i be the service rate of server S_i, i = 0, 1, 2. According to the general principle of our method presented above, this system is decomposed into two subsystems: T(1) and T(2). We need to characterize the behavior of each of these two subsystems and in particular characterize the service processes of the upstream and the downstream servers of each subsystem, that is characterize the random variables $Y_u(1)$, $Y_d(1)$, $Y_u(2)$ and $Y_d(2)$. First, as usual in decomposition methods, we have boundary conditions. Indeed, $S_u(1)$ represents the part of the original model upstream of buffer B_1 which consists only of server S_0. As a result, the service times of $S_u(1)$ should be identical to those of S_0 and therefore $Y_u(1) = Y_0$. Similarly, $S_d(2)$ represents the part of the original model downstream of buffer B_2 which consists only of server S_2. As a result, the service times of $S_d(2)$ should be identical to those of S_2 and therefore $Y_d(2) = Y_2$. In other words the service times of servers $S_u(1)$ and $S_d(2)$ are simply exponentially distributed with rates μ_0 and μ_2, respectively.

Let us first consider the characterization of the service process of the downstream server of subsystem T(1), that is $S_d(1)$. Server $S_d(1)$ represents (in an aggregate way) the part of the network downstream of buffer B_1, that is: the subnetwork consisting of server S_1, buffer B_2 and server S_2. Thus, the service process of server $S_d(1)$ must represent the effect of this subnetwork onto the behavior of the system. Therefore, it must reflect the service process of server S_1 as well as the possible blocking effect of server S_1 by server S_2. Ideally, the service process of server $S_d(1)$ should be such that the time between the departures of two successive customers from buffer B(1) (that is from subsystem T(1)), is the same as the time between the instants of transfer of two successive customers into buffer B_2 in the original system. Now, we have seen in Section 2.2 that this time consists of a service time of server S_1 possibly preceded by either a starvation time and/or a blocking time. This should be reflected in subsystem T(1). The starvation phenomena is captured in an explicit way in subsystem T(1) since the part of the system upstream of server S_1 is actually represented in T(1). Therefore what server $S_d(1)$ has to capture is the service time of server S_1 and the possible blocking time.

Thus, the service process of server $S_d(1)$ can be decomposed into two tasks that are performed successively. The first task represents the possible blocking time. It must be performed only if blocking occurs and its length must correspond to the time during which the server is blocked. The probability that this task need to be performed is approximated by the probability that buffer B(2) becomes full at the service completion instant of server $S_u(2)$ in subsystem T(2). Let $p_b(2)$ denote this probability. The blocking time is approximated by the residual service time of server $S_d(2)$ at the blocking instant of server $S_u(2)$ in subsystem T(2). Since the service time of $S_d(2)$ is exponentially distributed, this residual service time is exponentially distributed with the same rate, that is μ_2. The second task simply corresponds to the actual service process of a customer at server S_1. Therefore, its associated service time is characterized by the random variable Y_1, which in the special case considered here has an exponential distribution with rate μ_1. Thus, the service process of server $S_d(1)$ can be represented by a two stage PH distribution as illustrated in Figure 4.

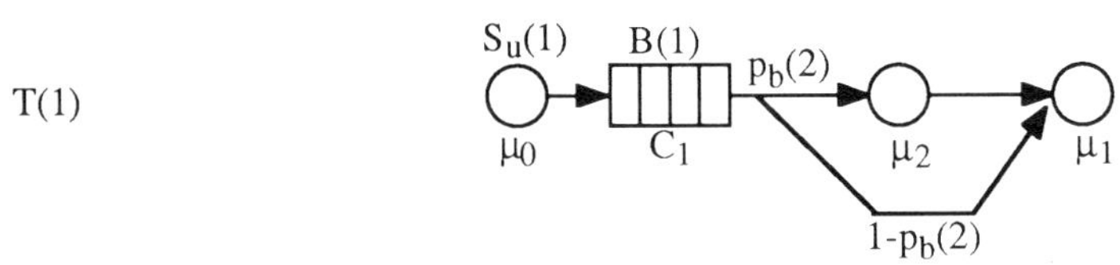

Fig. 4. Subsystem T(1)

Now, as stated in Section 2.2, an important feature of the BBS mechanism is that starvation and blocking may occur simultaneously. This is the fourth situation we considered. In this case, the next operation of the server can only begin when both starvation and blocking are over. The idle time of the server is then the maximum of the starvation time and the blocking time. Starvation is over as soon as a service completion occurs at server S_0. Blocking is over as soon as a service completion occurs at server S_2. What is important to notice is that these two activities, namely bringing a new customer into buffer B_1 and creating a new space into buffer B_2, are performed simultaneously and not one after the other. This should be reflected in subsystem T(1). This implies that the first part of the service process of server $S_d(1)$, i.e., the task of exponential length μ_2, must be performed even when buffer B(1) is empty, that is when server $S_d(1)$ is starved. This is an important feature of our decomposition method which was not encountered in decomposition methods dealing with BAS mechanisms.

We note that due to the above feature, namely that the first task of server $S_d(1)$ is performed even though server $S_d(1)$ is starved, the behavior of subsystem T(1) differs from that of the subsystems usually encountered in decomposition methods. In order to illustrate that, consider the case where buffer B_1 has capacity $C_1 = 3$. The markov chain associated to subsystem T(1) is given Figure 5. The markovian state is (n, j) where n is the number of customers in buffer B(1) (including that currently receiving service, if any) and j is the stage of the PH distribution of server $S_d(1)$, j =1, 2. The special feature of this Markov chain is that it can be in two different states when buffer B(1) is empty. State (0, 1) corresponds to the state where server $S_d(1)$ is performing the first task, although buffer B(1) is empty. State (0, 2) corresponds to the state where server $S_d(1)$ is waiting for a customer to arrive in buffer B(1) to begin to perform its second task.

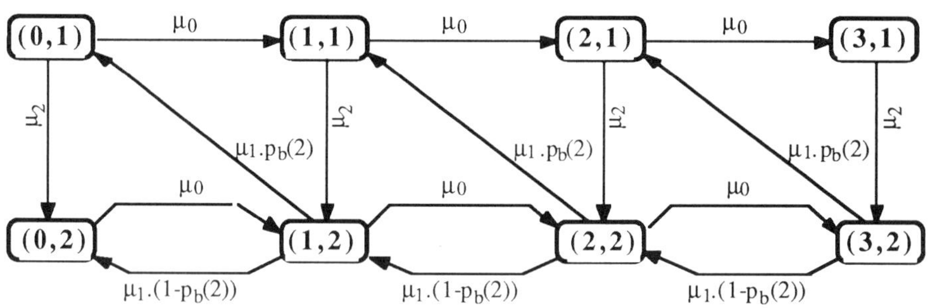

Fig. 5. Markov chain of subsystem T(1) ($C_1 = 3$)

Let us now consider the characterization of the service process of the upstream server of subsystem T(2), that is $S_u(2)$. As will appear, the following discussion is very similar to that pertaining to the characterization of server $S_d(1)$. As a result, it will not be as detailed. Server $S_u(2)$ represents (in an aggregate way) the part of the network upstream of buffer B_2, that is: the subnetwork consisting of server S_0, buffer B_1 and server S_1. Ideally, the service process of server $S_u(2)$ should be such that the time between the arrivals of two successive customers into buffer B(2) is the same as the time between the instants of transfer of two successive customers into buffer B_2 in the original system. Again, this time consists of a service time of server S_1 possibly preceded by either a starvation time and/or a blocking time. The blocking phenomena is captured in an explicit way in subsystem T(2) since the part of the system downstream of server S_1 is actually represented in T(2). Therefore what server $S_u(2)$ has to capture is the service time of server S_1 and the possible starvation time.

Thus, the service process of server $S_u(2)$ can be decomposed into two tasks that are performed successively. The first task represents the possible starvation time. The probability that this task need to be performed is approximated by the probability that buffer B(1) becomes empty at the service completion instant of server $S_d(1)$ in subsystem T(1). Let $p_s(1)$ denote this probability. The starvation time is approximated by the residual service time of server $S_u(1)$ at the starvation instant of server $S_d(1)$ in subsystem T(1). Since the service time of $S_u(1)$ is exponentially distributed, this residual service time is exponentially distributed with the same rate, that is μ_0. The second task simply corresponds to the actual service process of a customer at server S_1. Therefore, its associated service time is characterized by the random variable Y_1, which in the special case considered here has an exponential distribution with rate μ_1. Thus, the service process of server $S_u(2)$ can be represented by a two stage PH distribution.

Now, by using similar arguments as for the case of server $S_d(1)$, it appears that the first part of the service process of server $S_u(2)$, i.e., the task of exponential length μ_0, must be performed even when buffer B(2) is full, that is when server $S_u(2)$ is blocked. Let us again illustrate this by considering the case where buffer B_2 has capacity $C_2 = 3$. A markov chain can be associated to subsystem T(2). The markovian state is (n, j) where n is the number of customers in buffer B(2) (including that currently receiving service, if any) and j is the stage of the PH distribution of server $S_u(2)$, j =1, 2. The special feature of this Markov chain is that it can be in two different states when buffer B(2) is full. State (3, 1) corresponds to the state where server $S_u(2)$ is performing the first task, although buffer B(2) is full. State (3, 2) corresponds to the state where server $S_u(2)$ is waiting for a customer to arrive in buffer B(2) to begin to perform its second task.

3.3. The general case

Consider again the tandem queueing network consisting of M+1 servers and M buffers as defined in Section 2.1 and its decomposition into M subsystems as presented in Section 3.1. Consider a particular subsystem, say T(i). Recall that server $S_u(i)$ represents the part of the original model upstream of buffer B_i while server $S_d(i)$ represents the part of the original model downstream of buffer B_i. We need to generalize the analysis presented in Section 3.2 to handle the facts that: 1) service time distributions are of PH type; 2) except for subsystem T(1), there is more than one machine upstream of buffer B_i; except for subsystem T(M), there is more than one machine downstream of buffer B_i. First, we again have the following boundary conditions: the distribution of $Y_u(1)$ is identical to that of Y_0 (S_0 is never starved) and the distribution of $Y_d(M)$ is the same as that of Y_M (S_M is never blocked).

Let us consider the characterization of the service process of server $S_d(i)$, for any $i < M$. By generalizing the arguments given in Section 3.2, it appears that the service process of server $S_d(i)$ can be decomposed into two tasks that are performed successively. The first task represents the possible blocking time of server S_i. It must be performed only if blocking occurs and its length must correspond to the time during which the server is blocked. The probability that this task need to be performed is approximated by the probability that buffer B(i+1) becomes full at the service completion instant of server $S_u(i+1)$ in subsystem T(i+1). Let $p_b(i+1)$ denote this probability. The blocking time is approximated by the residual service time of server $S_d(i+1)$ at the blocking instant of server $S_u(i+1)$ in subsystem T(i+1). (This is because server $S_d(i+1)$ represents the part of the system downstream of buffer B_{i+1}.) Let $Y_d^r(i+1)$ denote the random variable associated with this residual service time. The second task simply corresponds to the actual service process of a customer at server S_i. Therefore, its associated service time is characterized by the random variable Y_i.

Consider now the characterization of the service process of server $S_u(i)$, for any $i > 1$. In a similar way, the service process of server $S_u(i)$ can be decomposed into two tasks that are performed successively. The first task represents the possible starvation time of server S_{i-1}. It must be performed only if starvation occurs and its length must correspond to the time during which the server is starved. The probability that this task need to be performed is approximated by the probability that buffer B(i-1) becomes empty at the service completion instant of server $S_d(i-1)$ in subsystem T(i-1). Let $p_s(i-1)$ denote this probability. The starvation time is approximated by the residual service time of server $S_u(i-1)$ at the starvation instant of server $S_d(i-1)$ in subsystem T(i-1). This is because server $S_u(i-1)$ represents the part of the system upstream of buffer B_{i-1}. Let $Y_u^r(i-1)$ denote the random variable associated with this residual service time. The second task simply corresponds to the actual service process of a customer at server S_{i-1}. The resulting subsystem T(i) is illustrated in Figure 6.

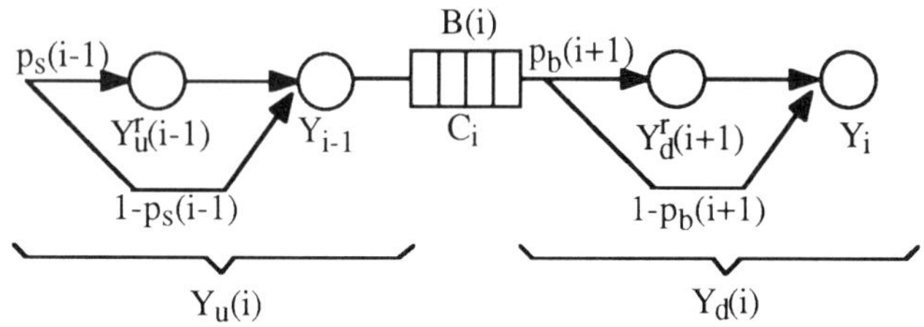

Fig. 6. subsystem T(i) (i = 2,....,M-1)

Again, the important feature that a server can be starved and blocked simultaneously (situation 4 described in Section 2.2) has to be captured in the decomposition. By generalizing the arguments of Section 3.2, it appears that the first task of server $S_u(i)$ should be performed (if required, accordingly to the probability $p_s(i-1)$), even when buffer B(i) is full. The second task however, cannot begin before there is a space available in buffer B(i). Similarly, the first task of server $S_d(i)$ should be performed (if required, accordingly to the probability $p_b(i+1)$), even when buffer B(i) is empty. The second task however, cannot begin before there is a customer in buffer B(i).

3.4. The computational procedure

In order to use the decomposition method presented above we need to be able: 1) to analyze each subsystem in isolation provided that the characterizations of its upstream and downstream servers are given; and 2) to determine the characterizations of the upstream and downstream servers of each subsystem.

Let us first address the first issue. Consider subsystem T(i). Suppose the characterizations of its upstream and downstream servers are given, i.e., the probabilities $p_s(i-1)$ and $p_b(i+1)$ as well as the distributions $Y_u^r(i-1)$ and $Y_d^r(i+1)$ given under a phase type representation. Then, subsystem T(i) can be analyzed by solving its underlying Markov chain. All performance parameters of interest can then be derived from the stationary probabilities of the Markov chain. The solution of the Markov chain can be obtained using any appropriate technique. However, it is important to recognize that the Markov chain has a special structure, namely a bloc tridiagonal structure. This was also the case in the method of Altiok and Ranjan [Altiok and Ranjan, 1987] and Gün and Makowski [Gün and Makowski, 1989]. The Markov chain however differs from that encoutered in their method because of the special feature pertaining to the simultaneous starvation and blocking phenomena (see Sections 3.2 and 3.3). Gün and Makowski [Gün and Makowski, 1987] used a matrix geometric technique [Neuts 1981]. An alternative technique that can be used in the case of bloc tridiagonal structures is the recursive technique [Herzog, Woo and Chandy, 1975]. It is shown in [Bouhchouch and Frein, 1991a and 1991b] how it can be efficiently used to solve the Markov chain associated with subsystem T(i).

Let us now address the second issue pertaining to the determination of the characterizations of the upstream and downstream servers of each subsystem. As it appears from Section 3.3, the characterization of the service process of server $S_u(i)$ depends on quantities pertaining to subsystem T(i-1). These quantities, namely the starvation probability $p_s(i\text{-}1)$ and the residual service time distribution $Y_u^r(i\text{-}1)$, given by its PH representation, can easily be derived from the stationary probabilities of the Markov chain associated with subsystem T(i-1). In a similar way, the characterization of the service process of server $S_d(i)$ depends on quantities pertaining to subsystem T(i+1). These quantities, namely the blocking probability $p_b(i+1)$ and the residual service time distribution $Y_d^r(i+1)$, given by its PH representation, can again be derived from the stationary probabilities of the Markov chain associated with subsystem T(i+1). Details can be found in [Bouhchouch, Frein and Dallery, 1992].

Therefore, as usual in decomposition methods [Dallery and Frein, 1992; Dallery and Gershwin, 1991], it turns out that the parameters of each subsystem T(i) (required in order to characterize its upstream and downstream servers) depend on quantities pertaining to its neighbors: subsystems T(i-1) and T(i+1). As a result, an iterative procedure must be used to determine the unknown parameters. This procedure is similar to that used in other decomposition methods based on a symmetrical approach [Altiok and Ranjan, 1987; Dallery and Frein, 1992; Dallery and Gershwin, 1991; Gün and Makowski, 1989]. The iterative algorithm consists of a forward pass (Step 1) and a backward pass (Step 2). The forward pass calculates new estimates of the service process of each upstream server $S_u(i)$, for i = 2,..., M, while the backward pass calculates new estimates of the service process of each downstream server $S_d(i)$, for i = M-1,..., 1. The iterative procedure is stopped when the parameters of all the servers have converged.

Computational algorithm

Step 0. Set $Y_u(1) = Y_0$ and $Y_d(M) = Y_M$.
Initialize $Y_d(i) = Y_i$, for i = 1, ..., M-1.

Step 1. For i = 1, 2,....., M-1:
- Step 1.1. Solve the Markov chain associated with subsystem T(i).
- Step 1.2.Obtain the starvation probability $p_s(i)$ and the PH distribution $Y_u^r(i)$.

Step 2. For i = M, M-1,..., 2:
- Step 2.1. Solve the Markov chain associated with subsystem T(i).
- Step 2.2. Obtain the blocking probability $p_b(i)$ and the PH distribution $Y_d^r(i)$.

Go to Step 1 until convergence.

An important property of tandem queueing networks is the so-called conservation of flow [Dallery and Gershwin, 1991]. It states that the throughputs of all the servers are identical. It is thus important that a decomposition method satisfies this property as well in the sense that all the subsystems should have the same throughput. As pointed out in [Dallery and Frein, 1992], the symmetrical approach does not involve an equation that prescribes in an explicit way that this is true. However, it was shown in [Dallery and Frein, 1989] that, in the case of tandem queueing networks with general service times and BAS mechanism, this property is satisfied by the symmetrical methods. We showed in [Bouchouch, Dallery and Frein, 1992] the following property.

Property 1. The proposed method satisfies the conservation of flow.

3.5. Discussion

The purpose of this section is to briefly discuss the relationships between our method and that proposed by Gün and Makowski [Gün and Makowski, 1989]. Gün and Makowski proposed two slightly different methods, one dealing with the BAS mechanism and the other with the BBS mechanism. We will refer to these two methods as GM/BAS method and GM/BBS method, respectively. Although these methods were originally presented in the context of networks having discrete service time distributions, they are readily applicable to the case of continuous distributions, which is what we are interested with in this paper. Note that the GM/BAS method is then the same as that independently proposed by Altiok and Ranjan [Altiok and Ranjan, 1987].

The GM/BBS method differs from ours because it does not capture the essential feature of the BBS mechanism, namely that a server may be simultaneously starved and blocked. The subsystem involved in the GM/BBS method has the same structure as in our method (Figure 6). However, its behavior is different. Indeed, the service process of server $S_u(i)$ is interrupted as soon as the buffer is full. In our method however, the first task associated with $Y_u^r(i-1)$ can still be performed, if required. Similarly, the service process of server $S_d(i)$ is interrupted as soon as the buffer is empty. In our method, the first task associated with $Y_d^r(i+1)$ can still be performed, if required. Thus, what is not captured by the GM/BBS method is the fact that when starvation and blocking of a server occur simultaneously in the original network, the two activities (bringing a new customer into its upstream buffer and creating a new space into its downstream buffer) are performed independently and simultaneously (i.e., concurrently) and not one after the other (see situation 4 in Section 2.2). As will appear in the next section, the fact that our method captures the above mentionned feature makes it significantly more accurate than the GM/BBS method. Moreover, it is worth pointing out that the two methods have the same computational complexity. Indeed, the computational burden of each method is due to the analysis of the subsystems in isolation. The Markov chains associated with each subsystem in both methods are very similar: they have the same number of states and same structure. Thus, the computational complexity of calculating the stationary probabilities is the same, provided that the subsystems are analyzed using the same technique (e.g., a recursive technique, a matrix-geometric technique,...) in both methods.

Another issue of interest pertaining to the GM/BBS method is the following. Consider again a tandem queueing network consisting of any number of servers having general service time distributions. Consider a first case where this network has buffer capacity vector **C** and is operated according to the BBS mechanism. Consider a second case where this network has buffer capacity vector **C-1** and is operated according to the BAS mechanism. Let us refer to these two networks as N_1 and N_2, respectively. As stated in Section 2.1, there is no equivalence between these two networks and as a result, they should have different performance parameters. In particular, they do not have the same throughput (see equation (1)). However, it turns out that the results obtained by applying the two methods proposed by Gün and Makowski are totally identical and in particular the throughputs are exactly the same. In other words, the GM/BBS method applied to a network with BBS mechanism and buffer capacity vector **C** is exactly identical to the GM/BAS method applied to a network with BAS mechanism and buffer capacity vector **C-1**. Let us briefly explain why.

The subsystem associated with buffer B_i of network N_2 in the GM/BAS method is shown in Figure 7. It has the same buffer capacity as that of B_i, that is C_i-1, and BAS mechanism. This subsystem can be transformed into an equivalent subsystem having buffer capacity C_i and BBS mechanism (see Section 2.1). Let us compare this subsystem with the corresponding subsystem involved in the GM/BBS method applied to network N_1. They have the same buffer capacity, same blocking mechanism and same service process for the upstream server. Although, the structure of the service process of the downstream process is different (the order in which the two tasks are performed is different), it is easy to see that the distribution of the random variable $Y_d^r(i)$ is the same. Therefore, the behavior of the two susbsystems is identical and as a result, the GM/BBS method applied to N_1 is exactly identical to the GM/BAS method applied to N_2. This result has been confirmed by numerical experiments. This helps explaining why the GM/BBS method does not perform as well as our method.

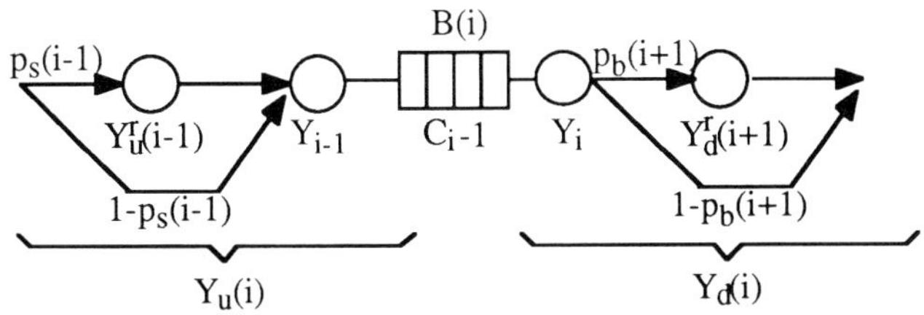

Fig. 7. subsystem T(i) associated with the GM/BAS method

A last interesting property of our decomposition method is that it is consistent not only with the reversibility property but also with the duality property of tandem queueing networks with BBS mechanism (see Section 2.2). This follows from the fact that the upstream and downstream servers of the subsystems play a totally symmetrical role in our decomposition method.

4. NUMERICAL EXAMPLES

Our method (referred to as new method) as well as the method of Gün and Makowski (the GM/BBS method simply referred to as GM method) have been implemented on a SUN 360 computer. The results are compared with those obtained by a discrete event simulation using QNAP2 [Veran and Potier, 1984]. The simulation time was equal to 10^6 times the largest mean service time of the network. We consider 4 examples consisting of balanced networks, i.e., networks in which all the servers have the same service time distribution and all the buffers have the same capacity, as well as unbalanced networks. For each example, the network is characterized by the number of servers, M+1, the buffer capacity vector, $C = (C_1, C_2,..., C_M)$ and the service time distributions of the different servers. For each example, we give the throughput calculated by each method and that obtained by simulation as well as the percentage of relative errors of each method with respect to simulation.

Examples 1 and 2, presented in Tables 1 and 2, pertain to balanced networks. All servers have the same service time distributions and all buffers have the same capacity C (including the server space). For these two examples, the results are provided for a capacity C varying from 1 to 4. The service time distribution for Example 1 is exponential with rate $\mu_i = 1$, for all i = 0,..., M. The service time distribution for Example 2 is a two-stage coxian distribution identical for all servers and denoted by $C_2(\lambda(1), \lambda(2), \alpha)$ where: $\lambda(1) = 1$ is the service rate of the first stage; $\lambda(2) = 0.1$ is the service rate of the second stage and $\alpha = 0.1$ is the branching probability. This distribution has mean $m_1 = 1$ and squared coefficient of variation c = 5. In these two examples we have 5 servers (M=4) (we have considered the cases M=2 and M=3 in [Bouhchouch, Frein and Dallery, 1992]). It appears that the new method performs significantly better than the GM method. This is in particular true for small buffer capacities. Our method has a very good acurracy, regardless of the buffer capacity, whereas the error associated with the GM method is very large for small buffer capacities. These results are not surprising since in the case of small buffer capacities, there is a significant probability of simultaneous starvation and blocking. Our method captures this phenomena whereas the GM method does not. On the other hand, for large buffers, the results obtained by the two methods are very close. The accuracy of our method is not affected by the choice of C while the GM method is. Also, it appears that the error of the GM method increases when the number of servers increases whereas the accuracy of our method remains as good.

Example 3 consists of a tandem network with 8 servers all having the same distribution which is the same as that pertaining to Example 2, namely the $C_2(1, 0.1, 0.1)$ distribution. However, the buffer capacities are different. The buffer capacity vector is **C** = (4, 3, 4, 1, 1, 1,

3). Example 4 has the same buffer capacity vector as that of Example 3. However, the servers have different service time distributions defined in Table 5. We consider distributions such that $0.517 \leq m_1 \leq 10$ and $0.25 \leq c \leq 4.0$. $C_2(\lambda(1), \lambda(2), \alpha)$ denotes a two stage coxian distribution with service rates $\lambda(1)$ and $\lambda(2)$, and branching probabilitie α; $E_k(\mu)$ denotes the k stage Erlang distribution with parameter μ (μ is the service rate of each stage), $H_k((a(1), ..., a(k)), (\lambda(1), ..., \lambda(k)))$ denotes the hyperexponential distribution with k stages where a(j) is the branching probability and λ the service rate of stage j, j = 1,..., k. The results pertaining to these two examples, shown in Tables 3 and 4 confirm that our method is significantly more accurate that the GM method. We tested other examples and the results obtained were similar.

We have also compared the average queue lengths. We give the average queue lengths calculated by each method and that obtained by simulation in the case of example 3 (table 6). These results show that our method provides very accurate results.

Buffer capacity	Simulation	New method	Relative error in %	GM method	Relative error in %
1	0.3322	0.3328	+ 0.2	0.2000	- 39.8
2	0.5112	0.5002	- 2.1	0.4649	- 9.1
3	0.6154	0.6049	- 1.7	0.5917	- 3.8
4	0.6834	0.6752	- 1.2	0.6700	- 2.0

Table. 1. Example 1: balanced network with 5 exponential servers (M = 4); *$\mu_i = \mu = 1$; $C_i = C$ with C = 1, ..., 4.*

Buffer capacity	Simulation	New method	Relative error (in %)	GM method	Relative error (in %)
1	0.1374	0.1385	+ 0.8	0.1000	- 27.2
2	0.1817	0.1815	- 0.1	0.1699	- 6.5
3	0.2102	0.2097	- 0.2	0.2029	- 3.5
4	0.2313	0.2313	0.0	0.2266	- 2.0

Table. 2. Example 2: balanced network with 5 Coxian-2 servers (M = 4); *service parameters: (1, 0.1, 0.1); $C_i = C$ with C = 1, ..., 4.*

Simulation	New method	Relative error (in %)	GM method	Relative error (in %)
0.1426	0.1425	- 0.1	0.1153	- 19.1

Table. 3. Example 3 network with 8 Coxian-2 servers (M = 7); *service parameters: (1, 0.1, 0.1); **C** = (4, 3, 4, 1, 1, 1, 3).*

Simulation	New method	Relative error (in %)	GM method	Relative error (in %)
0.0568	0.0573	+ 0.9	0.0433	- 23.8

Table. 4. Example 4: network with 8 servers (M = 7); *service parameters of Table 5: **C** = (4, 3, 4, 1, 1, 1, 3).*

Server	m_1	c	PH distribution
0	3.0	0.5	$E_2(0.667)$
1	2.0	1.5	$C_2(1, 0.333, 0.333)$
2	5.0	3.0	$C_2(0.4, 0.067, 0.167)$
3	3.0	0.75	$C_2(0.667,0.444,0.66)$
4	4.0	1.0	$C_2(0.5, 0.25, 0.5)$
5	10.0	4.0	$C_2(0.2, 0.025, 0.125)$
6	6.0	0.25	$E_4(0.667)$
7	0.517	1.477	$H_3((0.2, 0.3, 0.5), (1, 2, 3))$

Table.5. Service time distributions used in example 4

	Simulation	New method	Relative error (in %)	GM method	Relative error (in %)
Q_1	3.411	3.411	0	3.583	+ 5.0
Q_2	2.383	2.383	0	2.564	+ 7.6
Q_3	3.204	3.212	+ 0.2	3.472	+ 8.4
Q_4	0.6538	0.6553	+ 0.2	0.7319	+ 11.9
Q_5	0.4945	0.4962	+ 0.3	0.5013	+ 1.4
Q_6	0.3372	0.3368	- 0.1	0.2707	- 19.7
Q_7	0.4842	0.4909	+ 1.4	0.3520	- 27.3

*Table. 6. Example 3: network with 8 Coxian-2 servers (M = 7); service parameters: (1, 0.1, 0.1); **C** = (4, 3, 4, 1, 1, 1, 3).*

5. CONCLUSION

In this paper we have presented a new decomposition method for the analysis of tandem queueing networks with general service time distributions and BBS mechanism. The approach is similar to that proposed by Gün and Makowski [Gün and Makowski, 1989]. However, our method differs from their by the fact that it captures in an adequate way one essential feature of the BBS mechanism, namely the simultaneous occurence of starvation and blocking. Numerical results showed that the new method is significantly more accurate than the method of Gün and Makowski, 1989, especially for small buffer capacities. Since the computational complexity of the two methods is the same, this new method should be of interest.

6. REFERENCES

Altiok, T., 1982, "Approximate analysis of exponential tandem queues with blocking", Euro. J. Oper. Res. 11, pp. 390-398.

Altiok, T., 1989, "Approximate analysis of queues in series with phase-type service times and blocking", Oper. Res. 37, pp. 601-610.

Altiok, T., and Ranjan, R., 1987, "Analysis of production lines with general service times and finite buffers: a two-node decomposition approach", Industrial Engineering Dept., Rutgers University.

Ammar, M.H, and Gershwin, S.B., 1989, "Equivalence relations in queueing models of fork/join queueing networks with blocking", Performance Evaluation 10, pp. 233-245.
Bouhchouch, A., and Frein Y., 1991a, "Méthode récursive de résolution d'une PH/PH/1/N sans retour", Technical report LAG 91-28.
Bouhchouch, A., and Frein, Y., 1991b, "Analyse d'un réseau de files d'attente tandem à capacité limitée avec des distributions de service générales: cas du blocage avant service", Technical report LAG 91-100.
Bouhchouch, A., Frein, Y. and Dallery, 1992, "A decomposition method for the analysis of tandem queueing networks with blocking before service", Technical Report LAG 1992.
Dallery, Y., and Frein, Y., 1989, "On decomposition methods for tandem queueing networks with blocking", Tech. Rep. MASI n°293, Université Pierre et Marie Curie, July.
Dallery, Y., and Frein, Y., 1992, "On decomposition methods for tandem queueing networks with blocking", Operations Research, to appear.
Dallery, Y., and Gershwin, S.B., 1991, "Manufacturing flow line systems: a review of models and analytical results", Tech. Rep. MASI n° 91-18, Université Pierre et Marie Curie, April, to appear in Queueing Systems Theory and Applications.
Dallery, Y., Liu Z., and Towsley, D., 1990, "Equivalence, reversibility, symmetry and concavity in fork/join queueing networks with blocking", Tech. Rep. MASI n° 90-32, Université Pierre et Marie Curie, June.
Dallery, Y., Liu Z., and Towsley, D., 1992, "Properties of fork/join queueing networks with blocking under various operating mechanisms", Tech. Rep. COINS, University of Massachusetts at Amherst, May 1992.
Gün, L., and Makowski, A.M., 1987, "Matrix-geometric solution for finite queues with phase-type distribution", in: Courtois and Latouche, eds., Performance'87 (North-Holland, Amsterdam), pp. 269-282.
Gün, L., and Makowski, A.M., 1989, "An approximation method for general tandem queueing systems subject to blocking", in: Perros and Altiok, eds., Proc. First International Workshop on Queueing networks with blocking (North-Holland, Amsterdam), pp. 147-171.
Herzog, U., Woo, L., and Chandy, K.M., 1975, "Solution of queuing problems by a recursive technique", IBM J. Res. Dev. 19, pp. 295 - 300.
Jun, K.P., and Perros, H.G., 1990, "An approximate analysis of open tandem queueing networks with blocking and general service times", European Journal of Operational Research, 46, pp. 123-135.
Neuts, M.F., 1981, "Matrix-geometric solutions in stochastic models: an algorithmic approach", The John Hopkins Univ. Press, Baltimore.
Onvural, R.O., 1990, "Survey of closed queueing networks with blocking", ACM Computing Surveys, Vol. 22, N°2.
Onvural, R.O., and Perros, H.G., 1986, "On equivalencies of blocking mechanisms in queueing networks with blocking", Oper. Res. Letters 5, pp. 293 - 297.
Perros, H.G., 1989, "Open queueing networks with blocking", in H. Takagi, ed., Stochastic Analysis of Computer and Communication Systems (North-Holland, Amsterdam).
Perros, H.G., and Altiok, T., 1986, "An Approximate analysis of open networks of queues with blocking: tandem configurations", IEEE Trans. Soft. Eng. SE-12, pp. 450-461.
Pollock, S.M., Birge, J.R., and Alden, J.M., 1985, "Approximation analysis for open tandem queues with blocking: exponential and general distributions", IOE Rept. 85-30, Univ. of Michigan.
Takahashi, Y., Miyahara, H., and Hasegawa, T., 1980, "An approximation method for open restricted queueing networks", Oper. Res. 28, pp. 594-602.
Veran, M., and Potier, D., 1984, "A portable environment for queueing systems modelling", INRIA Report 314.

Queueing Networks with Finite Capacity
R.O. Onvural and I.F. Akyildiz (Editors)

Approximate Analysis Of Open M/M/C/K Queueing Networks

Youngnam Han and J. MacGregor Smith

Department of Industrial Engineering and Operations Research,
University of Massachusetts, Amherst, MA 01003

Abstract
In this paper, an approximation model of multi-server queueing networks with finite waiting room is developed and analyzed. Little work has been done on M/M/C/K compared to M/M/1/K systems. By revising the mean service time and arrival rate of nodes, we can obtain performance measures within reasonable error bounds. The methodology is based upon the idea of introducing a holding node and appropriate blocking parameters which affords a means to model the service time distribution of the node being blocked as a Coxian distribution. We focus on the simple topologies of series-parallel networks: i.e. tandem, merge, and split topologies with varying parameters of queue size, the number of servers, and arrival and service rates. Experimental results support our approximation.

1 Introduction

Queueing theory and queueing network models have matured and evolved in order to provide more sophisticated tools and techniques for analyzing and designing complex service systems. Except for certain restrictive assumptions on system parameters, exact product form solutions for system performance are rare. Most real world applications do not conform exactly to the assumptions of the underlying mathematical models, so much of the literature dealing with these intractable problems provides only approximate solutions. It is well known that the queueing systems with finite waiting room are often intractable, mainly because of the blocking property.

In this paper, we focus on the simplest topologies of series-parallel networks: i.e. series, split and merge which are comprised of two or three stations incorporating finite buffer spaces and single (and multiple) server(s). Some examples of such queueing situations are in manufacturing or assembly line processes in which customers (or, products) must proceed through a sequence of work stations, each station performing a given task. Work on multiple class or multiple server queues is emerging but there are only a few studies available to the best of our knowledge [7]. As mentioned, the finiteness of the buffer space introduces the blocking phenomena, and there are essentially two types of blocking mechanisms applied to various systems, which are modified forms of so-called production (Blocking After Service (BAS) and communication blocking (Blocking Before Service (BBS)[13]. Fundamentally, it is very difficult to analyze the exact nature of the blocking property of any real system, therefore most of the research points toward the use of analytical approximation techniques, simulation or their combination. Among those approximation methods successfully applied are the method of stages using Coxian Approximations [1, 2, 12, 14], one by a Markov Reward Process [4], the Isolation Method [10], and the Expansion Method [8, 9]. In his paper [12], R. Marie first introduced the idea of using the Coxian distribution for approximate solutions to load dependent closed queuing networks. By approximating the service time distribution via a Coxian distribution [1], Akyildiz and Sieber derived approximate expressions for the conditional throughput of load dependent closed queueing networks with general service time distributions. Later, Coxian approximations for open queueing networks with finite capacity queues for different topologies such as tandem [14], split and merge [2] were applied by Altiok and Perros. One specific difficulty in this method is that there is no explicit expression for the blocking probabilities at each stage of the approximation. Therefore those methods mainly rely on the robustness of the approximate or analytical expression of the blocking probabilities. Note that the above mentioned works were done for the case where the system consists of the stations with single servers. For other research done on cases of finite queue single server with exact result, refer to [3]. For queueing networks with multiple servers at each station, an analytic method using Markov Reward Process for the upper and lower bounds of the throughput of the tandem networks with finite queue capacities was introduced by Van Dijk and Van der Wal [4]. The isolation method introduced by Labetoulle and Pujolle [10] consists in subdividing the global system into N subsystems and studying them separately. Also, efforts to get the exact system performances instead of approximation have been made by several authors [11, 17]. An analysis of the distribution of waiting time in delay systems with two-stage finite tandem queues was presented by Langaris and Conolly [11]. In [17], Tsiotras et. al. provided the Markov Chain analysis of the blocking probabilities for multi-server finite capacity tandem queueing systems. When the system size gets larger and the topologies more complex, it seems almost impossible to achieve the exact system performance parameters, and this provides sound reasons for concentrating upon approximating algorithms. Also, an approximation method called Expansion Method by L. Kerbache and J. M. Smith [8, 9] and subsequent works with S. Daskalaki [16] have provided approximate solutions on the throughput of those systems with different series-parallel topologies of series, split and merge configurations. For those works done on systems with different blocking schemes and different topologies not mentioned here, see reference Gun's bibliography [5] and the volume by Perros and Altiok [15] for additional detailed references.

Section 2 is devoted to introducing the system under consideration with blocking properties, assumptions and notation used throughout the work. In section 3, we present the basic idea for approximating the effective service time of station one for the simplest tandem queues by using a Coxian distribution approximation and algorithms of estimating the throughput for each topology considered. We also present in section 3 the mean and variance of the effective service time and throughput, and compare the results with those by simulation, the Expansion Method and Markov Chain Analysis when possible. In section 4, we examine split and merge topologies. We conclude by making comments on our methodology and directions for further research in Section 5.

2 Problem Description

The class of queueing networks with blocking studied here consists of single (or, multiple) server queues embedded in tandem, split and merge topologies. It is assumed that arrival processes are Poisson and service time processes are exponentially distributed and all the arrival processes and service processes are independent of each other. Servers at each station are assumed to be homogeneous, that is, service time distribution for each server at the same station are independent and identical. We assume that servers of the departing (leaving) station never get blocked. We essentially consider production systems with finite buffer space which are modelled as delay systems, but our methodology can be successfully applied to loss systems.

In this paper, we only consider the simplest cases for each topology, however, our analysis can be extrapolated to more complex topologies. First, consider a queueing model in which service requires attention at two consecutive stations which can have multiple servers. Obviously, we can have more than one blocked server at station one, in which case servers are available on a first blocked, first unblocked basis. Secondly, consider the split topology (also known as Palm or Fork topology). Once the customer is served by one of N_{S_1} servers at station one, it either proceeds to station two with preassigned probability P_{SP} and to station three with probability $(1 - P_{SP})$. Thirdly, we also consider the merge topology (also known as a Join) which is comprised of three stations. Customers being served by station one or two compete for station three, where also service by station three is done on a first-blocked, first-unblocked basis.

3 Methodology

We introduce an approximation method for analyzing the tandem queueing system with finite buffer spaces by using a Coxian distribution for calculating the effective service time distribution of the customers at station one. By effective service time, we mean the time between being put into service and completion of service which consists of service time and holding time by blocking caused by the next station in line. Since the queue capacities of each stations are finite, there will be some customers blocked by station two after being processed by the server(s) at station one. To capture this, we introduce a holding node which can be considered as a server between those two stations (Figure 1). By the memoryless property of Poisson process, we can assume that the service time of the holding node introduced is exponentially distributed with the same rate as the next station, i.e. for series of queues, μ_2. Also, we denote P_{K_2} as the probability of being blocked by station two after being processed by the server at station one.

3.1 Effective Service Time Approximation

From Figure 1, we can easily calculate the Laplace transform of the distribution function of the effective service time, where each station has only one server as follows;

$$\begin{aligned}
\mathcal{L}[f(t)] &= \frac{\mu_1}{s+\mu_1}\left[(1 - P_{K_2}) + \frac{P_{K_2}\mu_2}{s+\mu_2}\right] \\
&= \frac{\mu_1(1-P_{K_2})}{s+\mu_1} + \frac{\mu_1 P_{K_2}\mu_2}{(s+\mu_1)(s+\mu_2)} \\
&= \frac{\mu_1 A}{(s+\mu_1)} + \frac{\mu_2 B}{(s+\mu_2)}, \qquad (1)
\end{aligned}$$

where

$$\begin{aligned}
A &= \frac{(\mu_1-\mu_2) - P_{K_2}\mu_1}{(\mu_1-\mu_2)} \\
B &= \frac{\mu_1 P_{K_2}}{(\mu_1-\mu_2)}.
\end{aligned}$$

The above expression is drawn from the assumption that customers served by station one can proceed without blocking to station two with probability $(1 - P_{K_2})$ and be blocked

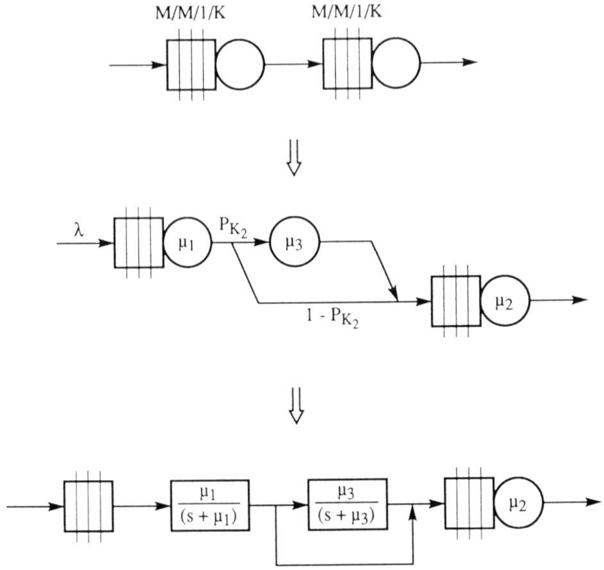

Figure 1: Coxian Approximation Model

with probability P_{K_2}. The customers blocked have to wait a certain amount of time whose distribution is assumed to be an exponential distribution with mean μ_2, identical with the server at station two. By the inverse Laplace transform, we get the probability density function of the service time of the customers at the first station as

$$f(\tilde{t}) = A\mu_1 e^{-\mu_1 \tilde{t}} + B\mu_2 e^{-\mu_2 \tilde{t}}.$$

Note that $f(\tilde{t})$ satisfies the condition that $\int_0^\infty f(t)dt = 1$, that is,

$$A + B = 1.$$

From the well-known properties of the generating function, we have

$$E(\tilde{T}) = \mu_1^{-1} + P_{K_2}\mu_2^{-1} \tag{2}$$

as a mean of effective service time and

$$E(\tilde{T}^2) = \frac{2(\mu_2^2 + P_{K_2}\mu_1\mu_2 + P_{K_2}\mu_1^2)}{(\mu_1\mu_2)^2} \tag{3}$$

as a second moment. Now, we have to determine P_{K_2} under the assumption that underlying processes are all independently and exponentially distributed. Even for such simple systems with single server, it is not an easy task to get the exact value of P_{K_2}. The computational effort, the running time and computer storage grows exponentially with system size. Thus, when the complexity of a system increases, exact solutions become more dubious. In what follows, we present a way of approximating P_{K_2} by adjusting the system parameters. We compare our results of the first and second moment of the effective service time distribution with those of simulation in Table 1.

Let's derive our methodology for approximating system throughput. The basic methodology can be characterized by three stages: *Decomposition, Compensation,* and *Adjustment.*

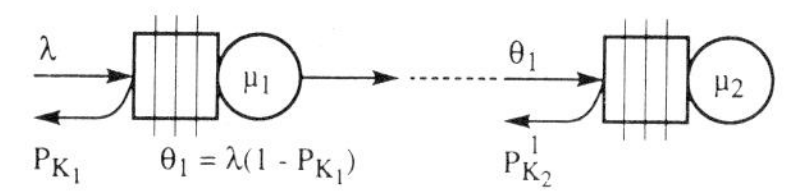

Stage 1.0: Decomposition

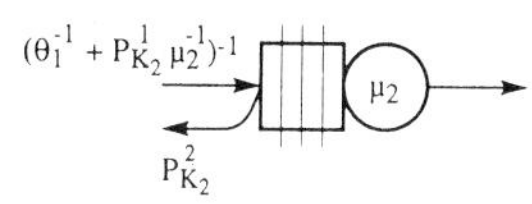

Stage 2.0: Compensation

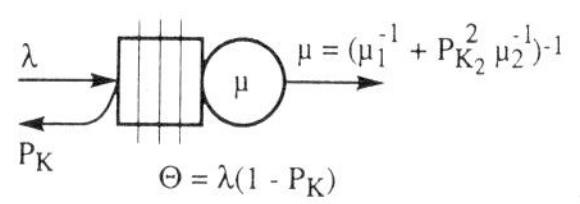

Stage 3.0: Adjustment

Figure 2: Series Decomposition

3.2 Series Throughput Approximation Methodology

- **Step 0.0 : (Initialization)**
  Calculate P_{K_1} of the first station from the analysis of $\mathrm{M}(\lambda)/\mathrm{M}(\mu_1)/N_{S_1}/K_1$.
- **Step 1.0 : (Decomposition)**
  - **Step 1.1 :** Decompose the stations into two.
  - **Step 1.2 :** Calculate the throughput of station one, $\theta_1 = \lambda\,(1 - P_{K_1})$.
  - **Step 1.3 :** Calculate the blocking probability $P^1_{K_2}$ from $\mathrm{M}(\theta_1)/\mathrm{M}(\mu_2)/N_{S_2}/K_2$.
- **Step 2.0 : (Compensation)**
  Adjust the blocking probability $P^2_{K_2}$ by analysing $\mathrm{M}((\theta_1^{-1} + P^1_{K_2}\mu_3^{-1})^{-1})/\mathrm{M}(\mu_2)/N_{S_2}/K_2$, where $\mu_3 = \mu_2 N_{S_2}$.
- **Step 3.0 : (Adjustment)**
  Let j be the number of customers in $\mathrm{M}/\mathrm{M}/N_{S_1}/K_1$ system in steady state. Approximate the blocking probability P_K of the system with arrival rate λ and service rate as follows:
  - **Step 3.1 :** if $j \le N_{S_1}$, then $\mu(j) = ((j\mu_1)^{-1} + P^2_K\mu_3^{-1})^{-1}$
  - **Step 3.2 :** else, then $\mu(j) = ((N_{S_1}\mu_1)^{-1} + P^2_K\mu_3^{-1})^{-1}$
- **Step 4.0 :** Approximate the system throughput Θ as $\lambda(1 - P_K)$

Simulation experiments to validate our approximation model and algorithms were performed using SIMAN Version 3.0 on VAX computer systems. Typically, while dealing with service rates of servers at node one ranging from 0.25 to 4.0 per unit time and(or) buffer spaces from 0 to 5, 10 runs of 10,000 time units each were generated, and the mean and standard deviation of effective service time and throughput for all 10 runs were computed. Thus, during each simulation run the number of entities getting through the system was anywhere between 2000 and 9000. Comparative experimental results for the effective service time from simulation and the algorithm are provided in Table 1 and 3. For system throughput, we also provide some results comparing different approximation methodologies for single-server two node series queueing system in Table 2.

Results of the first and second moments of effective service time distribution of customers at node one from simulation runs are compared with those from our approximation algorithm. Mean(AVG) and standard deviation(σ) results for the algorithm are obtained from equations (2) and (3) of our approximation. Note that effective service time is defined as the time period a customer enters service at station one until it leaves station one.

N_{Q_1}	N_{Q_1}	λ	μ_1^{-1}	μ_2^{-1}	AVG	σ (Sim)	C_v	AVG	σ (Algo)	C_v	AVG (% Error)	σ
0	0	1.0	0.5	1.0	1.0075	0.9151	0.9083	0.8805	0.9307	1.0570	12.6	1.7
			1.0	1.0	1.4090	1.1072	0.7858	1.3448	1.2533	0.9320	4.5	13.1
			2.0	1.0	2.2992	1.9067	0.8293	2.2752	2.1154	0.9298	1.0	10.9
			4.0	1.0	4.1854	3.8664	0.9238	4.1854	4.0418	0.9657	0.0	4.5
1	1		0.5	1.0	0.8963	0.8668	0.9671	0.7330	0.8134	1.1097	18.2	6.1
			1.0	1.0	1.2654	1.0953	0.8656	1.1975	1.1645	0.9724	5.2	6.3
			2.0	1.0	2.1265	1.9766	0.9295	2.1186	2.0550	0.9700	0.4	4.0
			4.0	1.0	4.0428	3.9923	0.9875	4.0458	4.0112	0.9914	0.1	0.5
2	2		0.5	1.0	0.8429	0.8365	0.9924	0.6883	0.7472	1.0856	18.3	10.7
			1.0	1.0	1.1888	1.0681	0.8985	1.1364	1.1199	0.9855	4.4	4.8
			2.0	1.0	2.0589	1.9810	0.9622	2.0578	2.0279	0.9855	0.1	4.2
			4.0	1.0	4.0114	3.9910	0.9949	4.0116	4.0029	0.9978	0.0	0.3

Table 1: Effective Service Time

As explained above, we examined the effects of varying service rates and(or) buffer sizes. It is observed from the simulation that as the service rate of a server increases, the chance of a customer being blocked by node two is higher which in turn results in longer holding times. Also, with the increase in buffer sizes, the effective service time decreases. In general, it is observed that with more customers experiencing blocking, the effective service time increases. It is observed that under the heavy traffic case, the results are more negative, whereas in light traffic, our approximation gives good results with reasonable error bounds. Also, since our approximation model adopts the method of stages as in Figure 1, it can be argued that in the case when the coefficient of variations $C_v(=\frac{\sigma}{AVG})$ is less than one(measured from both simulation results and algorithm), i.e. Erlangian distribution, our algorithm gives a more than reasonable approximation of effective service time distributions. The results are provided in Table 1, where we provide the % deviations of algorithmic mean and standard deviation from simulation results.

As expected, increasing the buffers, service rates and number of servers increase system throughput. Also, it can be observed that from the results of effective service time in Table 1, the larger the mean of effective service time is (i.e. the smaller the mean of service rate is), the higher is the system throughput. As seen in Table 1, the effect of increasing the buffer size, varying the service rate, or increasing the number of servers at each node is investigated. The results reveal that our approximation works well under any traffic conditions within reasonable error bounds.

Tandem single server two-node networks were also studied where arrival rates to the system and the service rate of a server at node two are fixed to 1 and the service rate of a server at node one ranges from 0.25 to 2.0. Buffer sizes were taken from 0 to 2 to gauge the effect of varying the waiting spaces.

In the next table reported in [3] with our method and Expansion Method [8] (Expan) added, mean arrival rate to the system and service rates of servers at both nodes are equal to be 1, whereas S_1 and S_2 represents the maximum number of customers allowed in node 1 and 2, respectively. Data from our algorithms is provided under the column of (H&S) with % deviation from the exact values in the last column(% Error). As we can see from Table 2, our algorithm gives very accepatable results of less than 2 % in these single server cases.

S_1	S_2	Taka	Expan	Exact	Hill	Boxma	H&S	% Error
1	1	.587	.527	.564	.570	.575	.565	0.2
2	2	.682	.635	.670	.671	.679	.679	1.3
3	3	.741	.703	.734	.734	.743	.746	1.6
4	4	.782	.750	.777	.777	.785	.790	1.7
5	5	.811	.784	.808	.808	.816	.821	0.1
7	7	.852	.830	.849	.849	.857	.862	1.5
10	10	.888	.872	.886	.886	.892	.897	1.2

Table 2: System Throughput

4 Split and Merge Topologies

In this penultimate section we provide our decomposition results of system throughput for splitting and merging topologies. We first present the splitting topology along with the experimental results, then the merging topology.

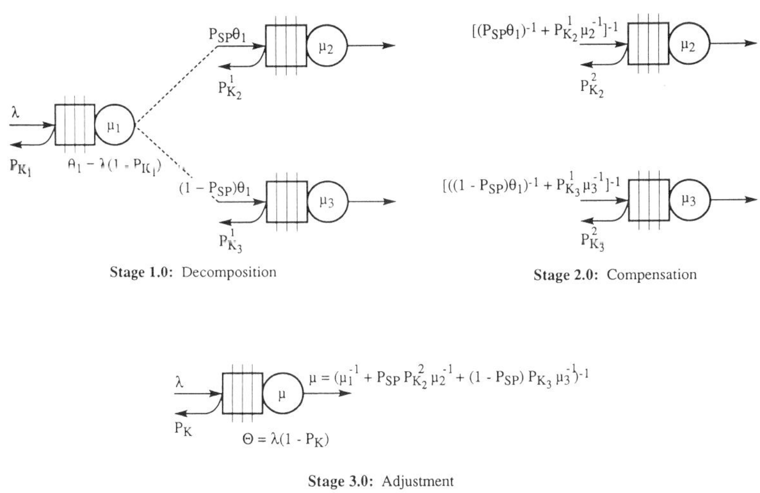

Figure 3: Splitting Decomposition

N_{Q_1}	N_{S_1}	N_{Q_2}	N_{S_2}	λ	μ_1	μ_2	Θ(Sim)	Θ(Algo)	% Error
0	1	0	1	1.0	4.0	1.0	0.7617	0.7953	4.0
				1.0	2.0	1.0	0.6809	0.7081	3.8
				1.0	1.0	1.0	0.5630	0.5646	0.3
				1.0	0.5	1.0	0.3967	0.3875	2.4
				1.0	0.25	1.0	0.2320	0.2284	1.6
1	1	1	1	1.0	1.0	1.0	0.6746	0.6789	0.6
2	1	2	1				0.7320	0.7459	1.9
3	1	3	1				0.7802	0.7898	1.2
4	1	4	1				0.8067	0.8207	1.7
5	1	5	1				0.8288	0.8438	1.8
0	1	0	2	1.0	0.5	1.0	0.4308	0.4238	1.6
				1.0	0.25	1.0	0.2368	0.2375	0.3
1	1	1	2	1.0	0.5	1.0	0.4686	0.4651	0.7
				1.0	0.25	1.0	0.2456	0.2470	0.6
2	1	2	2	1.0	0.5	1.0	0.4921	0.4834	0.9
				1.0	0.25	1.0	0.2498	0.2493	0.2
0	2	0	1	1.0	0.5	1.0	0.6172	0.6022	2.4
				1.0	0.25	1.0	0.4203	0.4030	4.1
1	2	1	1	1.0	0.5	1.0	0.7029	0.7019	0.1
				1.0	0.25	1.0	0.4631	0.4517	2.5
2	2	2	1	1.0	0.5	1.0	0.7580	0.7613	0.4
				1.0	0.25	1.0	0.4815	0.4752	1.3
1	2	1	2	1.0	4.0	1.0	0.9885	0.9990	1.0
				1.0	2.0	1.0	0.9830	0.9924	1.0
				1.0	1.0	1.0	0.9172	0.9471	3.3
				1.0	0.5	1.0	0.7676	0.7680	0.1
				1.0	0.25	1.0	0.4779	0.4737	0.9
0	3	0	3	1.0	0.25	0.25	0.5520	0.5466	1.0
1		1					0.5971	0.5996	0.4
2		2					0.6239	0.6331	1.5
0		0		1.0	0.5	1.0	0.8735	0.8715	0.2
1		1					0.9215	0.9226	0.1
2		2					0.9558	0.9514	0.5
0	5	0	5	1.0	0.125	0.125	0.5031	0.4989	0.3
1		1					0.5285	0.5302	0.3
2		2					0.5440	0.5501	1.1
0		0		1.0	0.25	1.0	0.8651	0.8625	0.3
1		1					0.9020	0.9010	0.1
2		2					0.9270	0.9266	0.04

Table 3: Experiments for Tandem Topology

4.1 Split Decomposition Algorithm

- **Step 0.0 : (Initialization)**
  Calculate P_{K_1} of station one from the analysis of M(λ)/M(μ_1)/N_{S_1}/K_1.

- **Step 1.0 : (Decomposition)**

  - **Step 1.1 :** Decompose the stations into two groups, station one and station two and three.
  - **Step 1.2 :** Calculate the throughput of station one, $\theta_1 = \lambda\,(1 - P_{K_1})$.

- **Step 1.3 :** Calculate the blocking probability $P^1_{K_2}$ with arrival rate $P_{SP} \cdot \theta_1$ for station two and $P^1_{K_3}$ with arrival rate $(1 - P_{SP}) \cdot \theta_1$ for station three.

- **Step 2.0 : (Compensation)**
  Adjust the blocking probabilities $P^2_{K_2}$ by analysing $\mathrm{M}((P_{SP}\theta_1)^{-1}+P^1_{K_2}\mu_4^{-1})^{-1})/\mathrm{M}(\mu_2)/N_{S_2}/K_2$, where $\mu_4 = \mu_2 N_{S_2}$ and $P^2_{K_3}$ by analysing $\mathrm{M}((P_{SP}\theta_1)^{-1}+P^1_{K_3}\mu_5^{-1})^{-1})/\mathrm{M}(\mu_3)/N_{S_3}/K_3$, where $\mu_5 = \mu_3 N_{S_3}$.

- **Step 3.0 : (Adjustment)**
  Let j be the number of customers in $\mathrm{M/M}/N_{S_1}/K_1$ system in steady state. Approximate the blocking probability P_K of the system with arrival rate λ and service rate as follows:

  - **Step 3.1 :** if $j \leq N_{S_1}$, then $\mu(j) = ((j\mu_1)^{-1}+P_{SP}P^2_{K_2}\mu_2^{-1}+(1-P_{SP})P^2_{K_3}\mu_3^{-1})^{-1}$

  - **Step 3.2 :** else, then $\mu(j) = ((N_{S_1}\mu_1)^{-1} + P_{SP}P^2_{K_2}\mu_2^{-1} + (1 - P_{SP})P^2_{K_3}\mu_3^{-1})^{-1}$

- **Step 4.0 :** Approximate the system throughput Θ as $\lambda(1 - P_K)$

Also, for the merge and split topologies, we provide the same sets of experiments under the variations of importance in Table 4 and Table 5. Our method is also capable of capturing the effects of the different topologies and the bottleneck effects due to them for both loss and delay systems, see Figure 5, and this displays its versatility [6]. Above all, the most striking feature about our algorithm is that it gives values of throughput within 5 % of those obtained from simulations in almost all the cases. This is extremely encouraging, given that a simulation analysis of any stochastic system of this nature would tend to have an inherent degree of variability, and it is very difficult to obtain exact values for most measure of performance.

Data from both simulation and algorithm is provided for different sets of parameters for each of the three separate topologies in order to show that our algorithm provides close approximation of the throughput. We provide comparisons with % errors of deviation from the simulation results at the last column of each table. In the experiment for split topology, we assume the splitting probability $P_{SP} = 0.5$. Observation reveals that for series and split topologies, our approximation works well with good error bounds of less than 4 % in almost all cases. In the merge topology however, it is observed that when there is a heavy blocking at the bottleneck node, our approximation ranges above 5 % in some cases, but never beyond 10 %. Generally our method works very well in light traffic systems which is understandable given the decomposition approach. Below, we provide the results of experiments we performed.

N_{Q_1}	N_{S_1}	N_{Q_2}	N_{S_2}	N_{Q_3}	N_{S_3}	λ	μ_1	μ_2	μ_3	Θ(Sim)	Θ(Algo)	% Error
0	1	0	1	0	1	1.0	1.0	1.0	1.0	0.6040	0.5943	1.6
						1.0	1.0	0.5	0.5	0.4719	0.4851	2.8
						1.0	0.5	1.0	1.0	0.4024	0.4022	0.05
						1.0	0.5	0.5	0.5	0.3540	0.3531	0.3
						1.0	0.5	0.25	0.25	0.2554	0.2672	4.5
1	1	1	1	1	1	1.0	1.0	1.0	1.0	0.7260	0.7176	1.2
						1.0	1.0	0.5	0.5	0.5922	0.6152	3.9
						1.0	0.5	0.5	0.5	0.4213	0.4240	0.6
0	2	0	2	0	2	1.0	1.0	1.0	1.0	0.9052	0.8982	0.8
						1.0	1.0	0.5	0.5	0.8502	0.8523	0.2
						1.0	0.5	1.0	1.0	0.7117	0.7066	0.7
1	2	1	2	1	2	1.0	1.0	1.0	1.0	0.9580	0.9773	2.0
						1.0	1.0	0.5	0.5	0.9379	0.9640	2.8
						1.0	0.5	1.0	1.0	0.7733	0.8159	5.5
0	2	0	1	0	1	2.0	1.0	1.0	1.0	1.0620	1.0258	3.4
1		1		1						1.2570	1.2656	0.7
2		2		2						1.3906	1.4121	1.5
0	2	0	1	0	1	2.0	1.0	2.0	2.0	1.3328	1.2683	4.8
1		1		1						1.5141	1.4853	1.9
2		2		2						1.6152	1.6039	0.7
0	2	0	2	0	2	2.0	1.0	1.0	1.0	1.3756	1.3466	2.1
1		1		1						1.5395	1.5166	1.5
2		2		2						1.6279	1.6176	0.6
0	2	0	1	0	2	2.0	1.0	2.0	1.0	1.3508	1.3067	3.3
1		1		1						1.5268	1.5009	1.7
2		2		2						1.6138	1.6107	0.2
0	1	0	2	0	2	2.0	1.0	1.0	1.0	.83692	.83820	0.6
1		1		1						.92546	.92772	0.2
2		2		2						.97038	.96594	0.5
0	1	0	2	0	2	2.0	2.0	1.0	1.0	1.2782	1.2618	1.3
1		1		1						1.4685	1.4648	0.3
2		2		2						1.5865	1.5827	0.2
0	1	0	1	0	2	2.0	2.0	2.0	1.0	1.2375	1.2244	1.0
1		1		1						1.4674	1.4499	1.2
2		2		2						1.5822	1.5762	0.4
0	3	0	3	0	3	1.0	1/4	1/4	1/4	0.6038	0.5900	2.3
1		1		1						0.6532	0.6467	1.0
2		2		2						0.6837	0.6810	0.4
0	5	0	5	0	5	1.0	1/8	1/8	1/8	0.5598	0.5454	2.6
1		1		1						0.5893	0.5779	2.0
2		2		2						0.6000	0.5969	0.5

Table 4: Experiments for Split Topology

4.2 Merge Decomposition Algorithm

- **Step 0.0 : (Initialization)**
  Calculate $P^1_{K_1}$ of station one from the analysis of M(λ)/M(μ_1)/N_{S_1}/K_1 and Calculate $P^1_{K_2}$ of station two from the analysis of M(λ)/M(μ_2)/N_{S_2}/K_2.

- **Step 1.0 : (Decomposition)**

- **Step 1.1 :** Decompose the stations into two groups, stations one and two and station three.
- **Step 1.2 :** Calculate the throughput of station one, $\theta_1 = \lambda\,(1 - P^1_{K_1})$ and the throughput of station two, $\theta_2 = \lambda\,(1 - P^1_{K_2})$.
- **Step 1.3 :** Calculate the blocking probability $P^1_{K_3}$ with arrival rate $(\theta_1 + \theta_2)$ for station three.

- **Step 2.0 : (Compensation)**
Adjust the blocking probability $P^2_{K_3}$ by analysing $\mathrm{M}(((\theta_1+\theta_2)^{-1}+P^1_{K_3}\mu_4^{-1})^{-1})/\mathrm{M}(\mu_3)/N_{S_3}/K_3$, where $\mu_4 = \mu_3 N_{S_3}$.

- **Step 3.0 : (Adjustment)**
Let j be the number of customers in $\mathrm{M}/\mathrm{M}/N_{S_1}/K_1$ system in steady state. Approximate the blocking probability $P^2_{K_1}$ of the system with arrival rate λ_1 and service rate as follows:

  - **Step 3.1 :** if $j \leq N_{S_1}$, then $\mu(j) = ((j\mu_1)^{-1} + \frac{\lambda_1}{\theta_1} P^2_{K_3}\mu_4^{-1})^{-1}$
  - **Step 3.2 :** else, then $\mu(j) = ((N_{S_1}\mu_1)^{-1} + \frac{\lambda_1}{\theta_1} P^2_{K_3}\mu_4^{-1})^{-1}$

  And the same for station two to get blocking probability of $P^2_{K_2}$,

  - **Step 3.3 :** if $j \leq N_{S_2}$, then $\mu(j) = ((j\mu_2)^{-1} + \frac{\lambda_2}{\theta_2} P^2_{K_3}\mu_5^{-1})^{-1}$
  - **Step 3.4 :** else, then $\mu(j) = ((N_{S_2}\mu_2)^{-1} + \frac{\lambda_2}{\theta_2} P^2_{K_3}\mu_5^{-1})^{-1}$

- **Step 4.0 :** Approximate the system throughput Θ as $\lambda_1(1 - P^2_{K_1}) + \lambda_2(1 - P^2_{K_2})$.

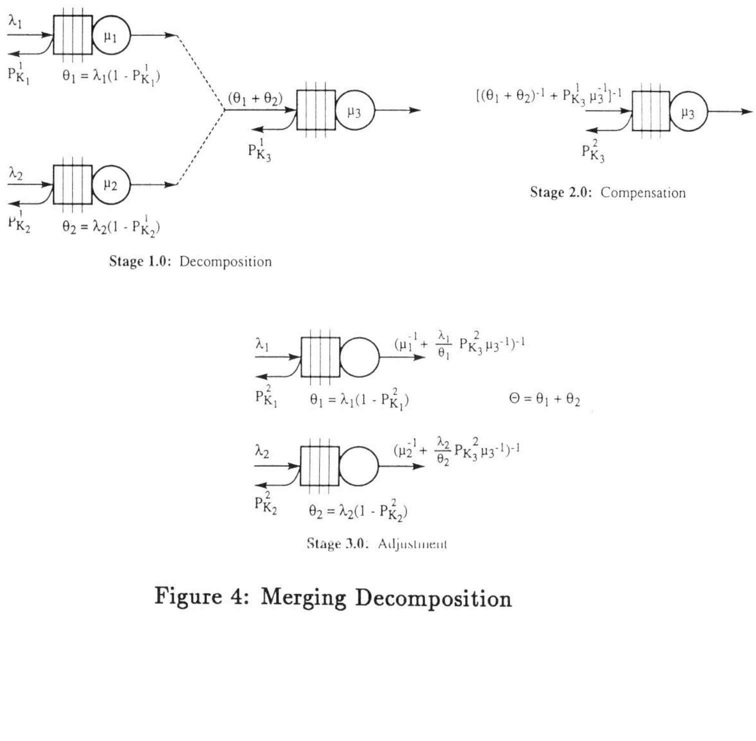

Figure 4: Merging Decomposition

N_{Q_1}	N_{S_1}	N_{Q_2}	N_{S_2}	N_{Q_3}	N_{S_3}	λ_1	λ_2	μ_1	μ_2	μ_3	Θ(Sim)	Θ(Algo)	% Error
0	2	0	2	0	1	1.0	1.0	1.0	1.0	2.0	1.5503	1.6658	7.4
1		1		1							1.7068	1.8457	8.1
2		2		2							1.7692	1.9252	8.8
0	2	0	2	0	1	1.0	1.0	1.0	1.0	4.0	1.7648	1.7645	0.1
1		1		1							1.8933	1.8974	0.2
2		2		2							1.9434	1.9526	0.5
0	1	0	1	0	2	1.0	1.0	1.0	1.0	1.0	1.2243	1.2250	0.1
1		1		1							1.3956	1.4709	5.4
2		2		2							1.5063	1.5441	2.5
0	2	0	2	0	2	1.0	1.0	1.0	1.0	1.0	1.5778	1.7092	8.3
1		1		1							1.7119	1.8594	8.6
2		2		2							1.7975	1.9303	7.4
0	2	0	1	0	1	1.0	1.0	1.0	1.0	2.0	1.3720	1.4196	3.5
1		1		1							1.5354	1.6225	5.7
2		2		2							1.6302	1.7258	5.9
0	3	0	3	0	3	1/2	1/2	1/8	1/8	1/4	0.5596	0.5610	0.3
1		1		1							0.6099	0.6155	0.9
2		2		2							0.6364	0.6490	2.0
0		0		0		1/2	1/2	1/4	1/4	1.0	0.8691	0.8737	0.5
1		1		1							0.9214	0.9233	0.2
2		2		2							0.9417	0.9516	1.1
0	5	0	5	0	5	1/2	1/2	1/8	1/8	1/8	0.8481	0.8417	0.8
1		1		1							0.8870	0.8873	0.1
2		2		2							0.9106	0.9173	0.7
0		0		0		1/2	1/2	1/8	1/8	1.0	0.8623	0.8626	0.03
1		1		1							0.8971	0.9010	0.4
2		2		2							0.9214	0.9266	0.6

Table 5: Experiments for Merge Topology

5 Summary and Conclusions

In this paper, we have presented an analytical approximation method for finite capacity multi-server queueing systems. The underlying methodology is based on a Coxian distribution approximation which provides the blocking probability of these systems and also a means to approximate the service time distribution of the station being blocked and the system throughput. Series, merge and split topologies were experimentally studied and the results clearly shows that the methodology is an effective tool for these finite multi-server system under light and medium traffic conditions. The compactness and simplicity of the algorithm are also noteworthy and should provide for its approximation to larger and more varied network topologies and their related optimization issues.

Acknowledgements This research was supported by National Science Foundation grants MSM-8715152 and MSS-9116666.

References

[1] I. F. Akyildiz and A. Sieber, "Approximate Analysis of Load Dependent General Queueing Networks." *IEEE Trans. on Soft. Eng.*, Vol. 14, No. 11, pp. 1357–1545, Nov. 1988.

[2] T. Altiok and H. G. Perros, "Approximate Analysis of Arbitrary Configurations of Open Queueing Networks with Blocking." *Annals. of Optns. Rsch.* Vol. 9, pp. 481-509, 1987.

[3] Peter C. Bell,"The Use of Decomposition Techniques for The Analysis of Open Restricted Queueing Networks."*Op. Res. let.* Vol. 1, pp. 230-235, 1982.

[4] N. M. van Dijk and J. van der Wal, "Simple Bounds and Monotonicity Results for Finite Multi-server Exponential Tandem Queues." *Queueing Systems*, Vol. 4 , No. 1, pp. 1–16, 1989.

[5] L. Gun, "Annotated Bibliography of Blocking Systems." *SRC Technical Report* 87-187, Systems Research Center, University of Maryland, College Park, MD, 1987

[6] Y. Han and J. MacGregor Smith, "Approximate Analysis of M/M/C/K Queueing Networks." paper presented at TIMS/ORSA Nashville meeting, May 12-15, 1991.

[7] S. Jain,"Optimization of Queueing Networks With Parallel Servers." M. S. Thesis. Dept. of Indusr. Engr. and Optns. Rsch., University of Massachusetts, 1989.

[8] L. Kerbache, "Analysis of Open Finite Queueing Networks." Ph.D. thesis, Department of Industrial and Operations Research, University of Massachusetts, Amherst, Mass. 1984.

[9] L. Kerbache and J. M. Smith, "The Generalized Expansion Method for Open Finite Queueing Networks." *Euro. J. Opnl. Res.*, Vol. 32, pp. 448-461, 1987.

[10] J. Labetoulle and G. Pujolle, "Isolation Method in a Network of Queues." *IEEE Trans. on Soft. Eng.*, Vol. SE- 6, No. 4, pp 373–381, July 1980.

[11] C. Langaris and B. Conolly, "On the Waiting Time of A Two-stage Queueing System with Blocking." *J. Appl. Prob.*, Vol. 21, pp. 628-638, 1984

[12] R. Marie,"An approximate Analytic Method for General Queueing Networks." *IEEE Trans. on Soft. Eng.*, Vol. SE- 5, No. 5, pp 530–538, Sept. 1979.

[13] Onvural, R.O. "Survey of Closed Queueing Network with Blocking," *ACM Computing Surveys, 22* (2), 83-121, 1990.

[14] H. G. Perros and T. Altiok, "Approximate Analysis of Open Networks of Queues with Blocking: Tandem Configuration." *IEEE Trans. on Soft. Eng.*, Vol. SE- 12, No. 3, pp. 450–461, March 1986.

[15] H. G. Perros and T. Altiok, "Queueing Networks with Blocking," North Holland Elsevier, 1989.

[16] J. M. Smith and S. Daskalaki, "Buffer Space Allocation in Automated Assembly Lines." *Oprn. Res.*, Vol. 36, No. 2, pp. 343–358, 1988.

[17] G. Tsiotras, H. Badr and E. Beltrami, "Blocking Probabilities for A Class of Two-station Tandem Queueing Models," to appear in *IIE Transactions.*

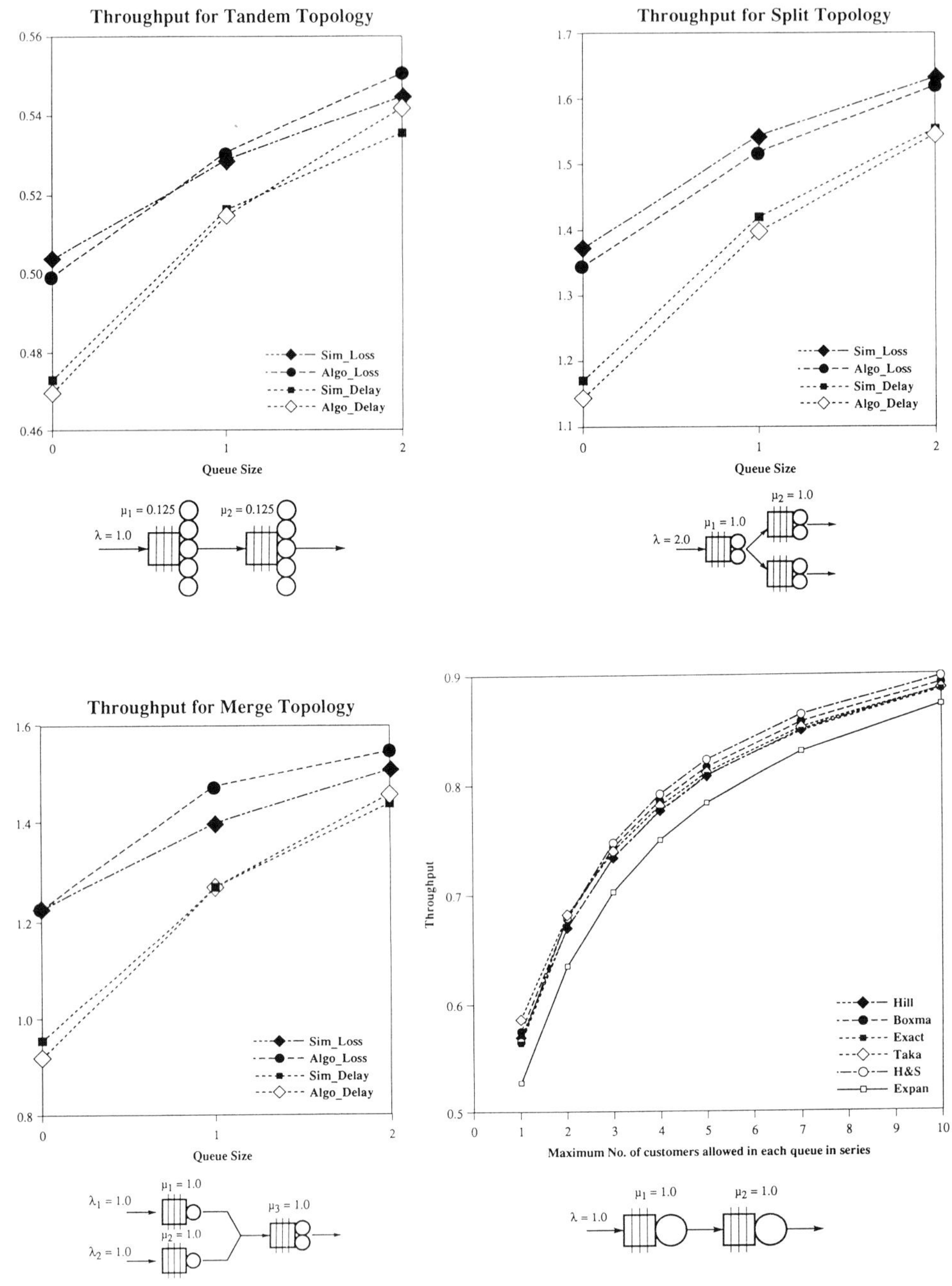

Figure 5: Sample Graphical Results

Queueing Networks with Finite Capacity
R.O. Onvural and I.F. Akyildiz (Editors)

ENTROPY MAXIMISED QUEUEING NETWORKS WITH BLOCKING AND MULTIPLE JOB CLASSES*

Demetres D Kouvatsos and Spiros G Denazis

Computer Systems Modelling Research Group
University of Bradford
Bradford, BD7 1DP
England

The method of entropy maximisation (MEM) is applied in a state space partitioning mode for the approximation of the joint stationary queue length distribution of an M/M/1/N queue with finite capacity, N(>1), multiple and distinct classes of jobs, R(>1), under a complete buffer sharing scheme and mixed service disciplines drawn from the first-come-first-served (FCFS), last-come-first-served with (LCFS-PR) or without (LCFS-NPR) preemption and processor sharing (PS) rules. The marginal and aggregate Maximum Entropy (ME) queue length distributions and the associated blocking probabilities per class are also determined. These ME results in conjunction with the first moments of the effective flows are used, as building blocks, in order to establish a new product-form approximation for arbitrary exponential open queueing networks with multiple classes of jobs under repetitive-service (RS) blocking with random destination (RD). The ME approximation captures the exact solution of open multiclass reversible queueing networks. Numerical experiments demonstrate a good accuracy level of the ME approximations in relation to simulation. Moreover, recent extentions of MEM for arbitrary GE-type queueing networks with RS-RD blocking and multiple classes of jobs are discussed.

1. INTRODUCTION

Queueing network models (QNMs) with finite capacity and multi-classes of jobs are widely recognised as powerful tools for analysing computer and

* This work is supported by the Science and Engineering Research Council (SERC), UK, under grant GR/F/29271.

communication systems and optimising their performance. Exact closed-form solutions for this type of QNMs are not generally attainable except for some special cases such as two-station cyclic queues and 'reversible' queueing networks, i.e., networks with finite capacity whose corresponding infinite capacity networks are reversible (c.f., Baskett et al [1], Kelly [2]). As a consequence, numerical techniques and analytic approximations are used for the study of general QNMs with finite capacity.

A useful bibliography on the subject has been compiled by Perros [3]. Moreover, two comprehensive surveys on open and closed QNMs under various blocking mechanisms has been carried out, respectively, by Perros [4] and Onvural [5]. However, not much work has been done so far on QNMs with multi-classes of jobs and finite capacity. Akyildiz and Von Brand [6 - 8] analysed open, closed and mixed QNMs with multi-classes and 'reversible' routing under repetitive-service (RS) blocking. Onvural [9] has shown that some product form stationary distributions for special single class networks hold also with multi-classes and Baskett et al type stations [1]. More recently, Onvural and Perros [10] developed a fairly accurate approximation procedure for open multi-class tandem queueing networks with finite capacity and Coxian-2 interarrival and service times under non-preemptive head-of-line priority rule.

In this paper earlier work by Kouvatsos and Xenios [11] and Kouvatsos and Denazis [12] on the maximum entropy method (MEM) for arbitrary single-class QNMs under repetitive-service (RS) blocking is extended to the multi-class case. A new maximum entropy (ME) product-from approximation is proposed for arbritrary exponential multi-class open QNMs with single-server stations of finite capacity under RS blocking with random destination (RD) and mixed service disciplines drawn from FCFS (first-come-first-served), LCFS-NPR (last-come-first-served non-preemptive), LCFS-PR (last-come-first-served with preemption) and PS (processor sharing) rules.

The principle of ME implies a decomposition of the open QNM into individual multi-class M/M/1/N queueing systems with finite capacity, N, under revised service-time distributions and censored interarrival processes. The multi-class M/M/1/N queueing system with distinct exponential interarrival and service times per class is solved via entropy maximisation and is used as a building block in the solution process.

The ME analysis of an M/M/1/N queue is presented in Sect.2. The ME product-form approximation for arbitrary exponential open QNMs and the associated MEM algorithm are given in Sect.3. Numerical validation experiments are carried out in Sect. 4. Extentions of the work to an approximate analysis of general GE-type QNMs are discussed in Sect.5. Concluding remarks follow in Sect.6.

2. THE MULTI-CLASS M/M/1/N QUEUE

Consider a stable M/M/1/N queue with finite capacity, N, a single server, R (R > 1) distinct job classes, exponential interarrival or service times and mixed scheduling disciplines drawn from FCFS, LCFS-PR, LCFS-NPR and PS rules. It is assumed that buffer management employs a complete sharing scheme, where buffers are allocated to jobs according to a given scheduling rule while no discrimination is made on the basis of job class. Moreover, the arrival process per class is considered to be "censored" i.e., arriving jobs of class i, i = 1,2,..., R are turned away when all N buffers are full.

2.1 Notation

With reference to the M/M/1/N queue, let at any given time $\underline{k} = (k_1, k_2, \ldots, k_R)$ be a system state, where $k_i (k_i \geq 0)$ is the number of jobs of class i, i = 1,2,...,R, S be a system state where jobs are positioned under an ordered arrangement, Q be the set of all feasible states S. For each state S and class i, i = 1,2,...,R let $n_i(S)$, be the number of class i jobs present in state S and $s_i(S)$, f(S) be auxiliary functions defined by $s_i(S) = 1$, if the job in service is of class i, or 0, otherwise, $f(S) = 1$, if $\Sigma_j n_j(S) = N$, or 0, otherwise. Moreover, let Λ_i be the rate of the overall arrival process, μ_i be the rate of the service process, π_i be the blocking probability that an arrival of class i finds the queue full (n.b., under exponential assumptions this is identical to the aggregate full state probabilities p(N)) and p(S), $p(k_1, k_2, \ldots, k_R)$ be the stationary state probabilities.

2.2 The Joint ME Solution

Motivated by earlier applications of the ME principle for the approximate analysis of single-class finite capacity queues and networks (c.f., Kouvatsos and Xenios [11], Kouvatsos [13]) and multi-class infinite capacity queues (c.f., Kouvatsos and Tabet-Aouel [14]), it is assumed that the following mean value constraints about the state probabilities {p(S)}, S ϵ Q, are known to exist:

(i) Normalisation, $\sum_{S \epsilon Q} p(S) = 1.$ (2.1)

(ii) Utilisation, $\sum_{S \epsilon Q} s_i(S)\, p(S) = \upsilon_i,\ 0 < \upsilon_i < 1,\ i=1,2,...,R.$ (2.2)

(iii) Mean queue length, $\sum_{S \epsilon Q} n_i(S)\, p(S) = L_i,\ \upsilon_i < L_i < N,\ i=1,2,...,R.$ (2.3)

(iv) Full buffer state probability, $\sum_{S \epsilon Q} f(S)\, p(S) = \varphi,\ 0 < \varphi < 1,$ (2.4)

satisfying the flow-balance conditions

$$\Lambda_i\,(1 - \pi_i) = \mu_i\ \upsilon_i\,,\ \ i=1,2,...,R. \tag{2.5}$$

The form of the state probability distribution, p(S), can be characterised by maximising the entropy functional $H(p) = -\ \Sigma_S\, p(S) \log p(S)$, subject to constraints (2.1) - (2.4). By employing Lagrange's method of undetermined multipliers the following solution is obtained

$$p(S) = \frac{1}{Z} \prod_{i=1}^{R} g_i^{\,s_i(S)}\, x_i^{\,n_i(S)}\, y^{f(S)}, \tag{2.6}$$

for all $S \epsilon Q$, where Z is the normalising constant and g_i, x_i, (i=1,2,...R) and y are the Langragian coefficients corresponding to constraints (2.2) - (2.4), respectively.

Defining the sets

$$S_0 = \{S/S \epsilon Q : s_i(S) = 0,\ i = 1,2,...,R\},\ Q_i = \{S/S \epsilon Q : s_i(S) = 1\},\ i = 1,2,...,R,$$

$$Q_{i;k_1,...,k_R} = \{S \epsilon Q_i : n_j(S) = k_j\ \&\ k_i \geq 1,\ j = 1,2,...,R\},\ i = 1,2,...,R,$$

it is implied, after some manipulation, that the ME joint state probability distribution is given by

$$P(S_0) = p(0,...,0) = \frac{1}{Z}, \tag{2.7}$$

$$p(k_1,\dots,k_R) = \sum_{i=1}^{R} \text{Prob}\,(Q_{i;k_1,\dots,k_R}) =$$

$$Z^{-1}\,\frac{\left(\sum_{j=1}^{R} k_j - 1\right)!}{\prod_{j=1}^{R} k_j\,!}\left[\prod_{j=1}^{R} x_j^{k_j}\right]\left[\sum_{i=1}^{R} k_i\,g_i\right] y^{\,\delta\left(\sum_{j=1}^{R} k_j\right)}, \qquad (2.8)$$

where $\delta\,(\Sigma_j\,k_j) = 1$, if $\Sigma_j\,k_j = N$, or 0, otherwise.

2.3 The Aggregate ME Solution

The aggregate state probability, $p(n)$, $n = 0,1,\dots,N$, can be completely specified by making use of (2.7) - (2.8), and is given by

$$p(0) = p(S_0) = Z^{-1}\,, \qquad (2.10)$$

$$p(n) = \sum_{\underline{k}\in A_n} p(\underline{k}) = Z^{-1}\,\frac{\rho(1 - \mathbf{X})}{1 - \rho}\,\mathbf{X}^{n-1},\ 1 \le n \le N - 1, \qquad (2.11)$$

where $A_n = \left\{\underline{k}{:}0 \le k_j \le N,\ j = 1,\dots,R\ \&\ \sum_{j=1}^{R} k_j = n\right\}$, $\rho = \sum_{j=1}^{R} \rho_j$, $\rho_j = \Lambda_j/\mu_j$, and

$$p(N) = Z^{-1}\,\frac{\rho(1 - \mathbf{X})}{1 - \rho}\,\mathbf{X}^{N-1}\,y\,, \qquad (2.12)$$

where $\mathbf{X} = \sum_{i=1}^{R} x_i$.

2.4 The Marginal ME Solution

The marginal state probabilities per class $p_i(k)$, $i = 1,2,\dots,R$, $k = 0,1,\dots,N$, can be determined by using their standard definition in terms of the joint state probabilities (2.7) - (2.8). To this end, using Pascal's triangle inequality and carrying out laborious but not complex operations, the following cost effective recursive expressions are obtained:

$$Z^{(1)} = 1 + \frac{\rho(\,1 - \mathbf{X})}{1 - \rho}\;y \qquad (2.13)$$

$$Z^{(N)} = \frac{1 - \mathbf{X}}{1 - \rho} + \mathbf{X}\, Z^{(N-1)}, \quad N \geq 2, \tag{2.14}$$

$$p^{(N)}(S_0) = 1 / Z^{(N)} \tag{2.15}$$

For i=1,2,...,R and N ≥ 2,

$$p_i^{(1)}(0) = 1 - \frac{1}{Z^{(1)}} \left[\frac{1 - \mathbf{X}}{1 - \rho} \right] \rho_i\, y, \tag{2.16}$$

$$p_i^{(N)}(0) = \frac{Z^{(N-1)}}{Z^{(N)}}\, \bar{x}_i\, p_i^{(N-1)}(0) + \frac{1}{Z^{(N)}}\, A_i, \tag{2.17}$$

$$p_i^{(1)}(1) = \frac{1}{Z^{(1)}}\, x_i\, g_i\, y, \tag{2.18}$$

$$p_i^{(N)}(1) = \frac{Z^{(N-1)}}{Z^{(N)}} \left[\bar{x}_i\, p_i^{(N-1)}(1) + x_i\, p_i^{(N-1)}(0) \right] - \frac{1}{Z^{(N)}}\, x_i (1 - g_i), \tag{2.19}$$

$$p_i^{(N)}(k) = \frac{Z^{(N-1)}}{Z^{(N)}} \left[\bar{x}_i\, p_i^{(N-1)}(k) + x_i\, p_i^{(N-1)}(k-1) \right], \quad k=2,\ldots,N, \tag{2.20}$$

(n.b., $p_i^{(N-1)}(N) = 0$) where $\bar{x}_i = \mathbf{X} - x_i$ and $A_i = (1 - \bar{x}_i) + \frac{1 - \mathbf{X}}{1 - \rho}(\rho - \rho_i)$.

Note that details of proofs for expressions (2.7) - (2.20) can be found in Kouvatsos & Georgatsos [15].

2.5 The Langrangian Coefficients $\{x_i\}$, $\{g_i\}$, $\{y\}$

As in the case of single class G/G/1/N queues (c.f., [11, 13]), the Lagrangian coefficients $\{x_i\}$ and $\{g_i\}$ can be approximated analytically by making asymptotic connections to an infinite capacity queue as $N \to +\infty$. Assuming (a) $\rho < 1$ and $\mathbf{X} < 1$ and (b) $\{x_i\}$, $\{g_i\}$ and $\{y\}$ are invariant to the buffer capacity size N, it can be established that

$$x_i = \frac{\bar{n}_i - \rho_i}{\bar{n}}, \quad g_i = \frac{(1 - \mathbf{X})\rho_i}{(1 - \rho)x_i}, \tag{2.21}$$

and - using flow balance condition (2.5) -

$$y = \frac{1 - \rho}{1 - \mathbf{X}}, \tag{2.22}$$

where $\bar{n} = \sum_{i=1}^{R} \bar{n}_i$ and $\bar{n}_i$ is the (asymptotic) mean queue length of an M/M/1 queue.

For a multi-class M/M/1 queue statistics $\{\bar{n}_i, i=1,2,...,R\}$ are determined by (c.f., [14], [17]).

$$\bar{n}_i = \begin{cases} \rho_i + \dfrac{\Lambda_i}{1-\rho} \displaystyle\sum_{j=1}^{R} \dfrac{\rho_j^2}{\Lambda_j}, & \text{if FCFS or LCFS-NP rule,} \\ \dfrac{\rho_i}{1-\rho}, & \text{if LCFS-PR rule,} \\ \rho_i \left\{ 1 + \dfrac{1}{1-\rho} \displaystyle\sum_{j=1}^{R} \dfrac{h_j}{h_i} \rho_j \right\}, & \text{if PS rule,} \end{cases}$$

where $\rho_i = \Lambda_i/\mu_i$, $\rho = \sum_{i=1}^{R} \rho_i$ and $\{h_i, i=1,2,...,R\}$ is a set of discriminatory weights that impose differential treatment to different classes of jobs (c.f., Kleinrock [16]). In the context of entropy maximisation the effect of different scheduling disciplines on the joint and marginal ME solutions is clearly captured by the set of asymptotic queue lengths $\{\bar{n}_i, i=1,2,...,R\}$. Expressions (2.21) - (2.22) are also adopted by analogy to a single-class G/G/1/N queue [11, 13] for the case $\rho > 1$, $X > 1$. Under the Baskett et al [1] assumptions the ME solutions reduce to the exact ones (c.f., [6]) with Lagrangian coefficients $g_i = y = 1$ and $x_i = \rho_i$ (via (2.21) - (2.22)). Finally, closed-form expressions for marginal statistics such as $\upsilon_i = P(Q_i)$ and L_i, i=1,2,...,R can be easily established in terms of the Lagrangian coefficients $\{g_i, x_i, y\}$ and finite capacity, N.

3. EXPONENTIAL OPEN MULTI-CLASS QUEUEING NETWORKS WITH RS-RD BLOCKING

Consider an exponential open multi-class QNM at equilibrium with arbitrary configuration containing M single-server queues of finite capacities N_ℓ, $\ell=1,\ldots,M$. The network operates under a complete buffer sharing scheme, RS-RD blocking and FCFS, LCFS-NPR, LCFS-PR and PS scheduling rules.

3.1 Notation

For each QNM the notation of Sect.2 applies with the incorporation of ℓ, m subscripts, as appropriate, indicating queueing stations "1,2,...M" or external world "o". In addition, for each class i(i=1,2,...,R) let $\Lambda_{0;\ell i}$ be the overall external arrival rate of class i to queue ℓ, $a_{\ell,i;m,j}$ be the transition probability that a class i completer from queue ℓ, attempts to join queue m as class j, m = 0,...,M, j=1,2,..,R, $\pi_\ell = p_\ell(N_\ell)$ be the blocking probability of an external arriver or a completer in the network is blocked by queue ℓ, $\pi_{c\ell,i}$ be the blocking probability that a class i completer from queue ℓ (in the first or subsequent attempts) is blocked.

3.2 The joint ME solution

The state of the network at any given time can be described by an integer valued vector $\underline{k}=(\underline{k}_1,\underline{k}_2,\ldots,\underline{k}_M)$ where $\underline{k}_\ell=(k_{\ell,1},\ldots,k_{\ell,R})$ and $k_{\ell,i}$, i = 1,2,...,R is the number of class i jobs at queue ℓ, ℓ = 1,2,...,M. Let $p(\underline{k})$ be the joint stationary probability that the queueing network is at state $\underline{k}$.

The form of the ME solution $p(\underline{k})$, subject to normalisation and the marginal constraints of the type (2.2) - (2.4), can be clearly established in terms of the product-form approximation

$$p(\underline{k}) = \prod_{\ell=1}^{M} p_\ell(\underline{k}_\ell), \tag{3.1}$$

where $p_\ell(\underline{k}_\ell)$ is the marginal ME solution expressed by (2.7), (2.8). The ME solution implies an $M/M/1/N_\ell$ queue-by-queue decomposition of the QNM with revised interarrival and service times. Assuming a Poisson arrival process at each queue, the modified MEM algorithm of Kouvatsos and Denazis [12] (applied to the single class case) is extended in Sect.3.3 to produce a MEM algorithm for open multi-class queueing networks with RS-RD blocking.

Notice that the MEM algorithm [11, 12], based on the concept of queue-by-queue decomposition for estimating the Lagrangian coefficients of (3.1), does not quite capture the exact solution for open 'reversible' networks with finite capacity. This may be attributed to the way that the first moment of the effective flow is calculated under renewality assumptions which do not strictly hold. However, the ME solution of these networks, subject to the normalisation and constraints (2.1) - (2.4), may be viewed as the truncated ME solution of the corresponding open reversible network with infinite capacity. In this context, it can be verified either by analogy or by substituting (3.1) into the truncated global balance equations, that the ME solution is exact with $g_{\ell,i} = y_{\ell} = 1$ and $x_{\ell,i} = \rho_{\ell,i} = \Lambda^*_{\ell,i}/\mu_{\ell,i}$, $\ell=1,2,...,M$; $i=1,2,...,R$ (c.f., [6, 7, 9]), where $\Lambda^*_{\ell,i}$ are the job flow rates of the infinite capacity reversible network. It clearly follows from ME solution (3.1) that the queue length distribution of each queueing station in the 'reversible' open network is exactly the same as that for the station being in isolation under a Poisson arrival process with parameters $\Lambda^*_{\ell,i}$ and the original exponential service times with parameters $\mu_{\ell,i}$, $\ell=1,2,...,M$; $i=1,2,...,R$.

3.3 The MEM Algorithm

The MEM algorithm for arbitrary multi-class open QNMs with single server-queues exponential external interarrival and service times, class switching, complete buffer sharing, RS-RD blocking and FCFS, LCFS-NPR, LCFS-PR and PS scheduling rules can be described as follows:

BEGIN

INPUTS

$\{M, R, N_{\ell}, \mu_{\ell i}, \Lambda_{0;\ell,i}; \ell = 1,2,...,M; i = 1,2,...,R$, FCFS, LCFS-NPR, LCFS-PR, PS scheduling rules, $\{a_{\ell,i;m,j}\}$, $\ell = 1,2,...,M$; $m = 0,1,...,M$; $i,j = 1,2,...,R\}$.

STEP 1 IF NETWORK IS 'REVERSIBLE' THEN

STEP 1.1 Calculate job flow rates:

$$\Lambda^*_{\ell,i} = \Lambda_{0;\ell,i} + \sum_{\ell \neq m=1}^{M} \sum_{j=1}^{R} \alpha_{m,j;\ell,i} \Lambda^*_{m,j}, \forall_{\ell,i,m,j};$$

STEP 1.2 Preserve original arrival and service rates:

$$\Lambda_{\ell,i} \leftarrow \Lambda^*_{\ell,i}, \forall \ell,i; \ \hat{\mu}_{\ell,i} \leftarrow \mu_{\ell,i}; \forall \ell,i;$$

STEP 1.3 {Go to STEP 5}

STEP 2 FEEDBACK CORRECTION: For each queue ℓ, $\ell=1,2,\dots,M$ and each class i, i=1,2,...,R (with $a_{\ell,i;\ell,i} > 0$) substitute:

$$\mu_{\ell,i} \leftarrow \mu_{\ell,i} (1 - a_{\ell,i;\ell,i});$$

$$a_{\ell,i;m,j} \leftarrow a_{\ell,i;m,j} / (1 - a_{\ell,i;\ell,i}), \text{(for all } m \neq \ell \ \& \ j \neq i); \quad a_{\ell,i;\ell,i} \leftarrow 0;$$

STEP 3 INITIALISATION: $\pi_\ell = p_\ell(N_\ell) \leftarrow 0.5$, $\ell=1,2,\dots,M$

STEP 4 SOLVE SYSTEM OF NON-LINEAR EQUATIONS $\{\pi_\ell\}$ USING NEWTON-RAPHSON

$$\pi_\ell = p_\ell(N_\ell) \leftarrow \frac{(1 - \rho_\ell)}{1 - \rho_\ell^2 \mathbf{X}_\ell^{N_\ell - 1}} \rho_\ell \mathbf{X}_\ell^{N_\ell - 1}, \ \ell = 1,2,\dots,M;$$

STEP 4.1 Calculate effective flow transition probabilities $\{\bar{a}_{\ell,i;m,j}\}$:

$$\bar{a}_{\ell,i;m,j} \leftarrow a_{\ell,i;m,j} * (1 - \pi_m) / (1 - \pi_{c\ell,i}), \ \forall \ell,i,m,j;$$

STEP 4.2 Calculate effective job flow rates:

$$\lambda_{\ell,i} \leftarrow \lambda_{0;\ell,i} + \sum_{\ell \neq m=1}^{M} \sum_{j=1}^{R} \bar{a}_{m,j;\ell,i} \lambda_{m,j}, \ \forall \ell,i,m,j;$$

$$\lambda_{0;\ell,i} \leftarrow \Lambda_{0;\ell,i} (1 - \pi_\ell), \ \forall \ell,i;$$

STEP 4.3 Calculate effective service rates : $\hat{\mu}_{\ell,i} \leftarrow \mu_{\ell,i}(1 - \pi_{c;\ell,i}), \forall_{\ell,i};$

STEP 4.4 Calculate blocking probabilities $\{\pi_{c;\ell,i}\}$:

$$\pi_{c;\ell,i} \leftarrow \sum_{\ell \neq m=1}^{M} \left(\sum_{j=1}^{R} a_{\ell,i;m,j} \right) \pi_m, \quad \forall \ell,i,m,j;$$

STEP 4.5 Calculate overall arrival rates $\Lambda_{\ell,i}$: $\Lambda_{\ell,i} \leftarrow \lambda_{\ell,i} / (1-\pi_\ell)$, $\forall \ell,i$;

STEP 5 OUTPUT

Solve each queueing station ℓ, $\ell = 1,2,...,M$ in isolation as a censored multi-class $M/M/1/N_\ell$ queue with parameters $(\Lambda_{\ell,i}, \hat{\mu}_{\ell,i})$ under a given schedulive rule;

END

For arbitrary open QNMs the main computational cost of the ME algorithm is of $O(k\, M^3)$, where k is the number of iterations in Step 4 and M^3 is the number of operations for inverting the associated Jacobian matrix. However, if a quasi-Newton numerical method is employed, this cost may be reduced to be of $O(k\, M^2)$. Note that the computational cost of STEP 4 is exactly the same as that of the single class case (c.f., [11]). This is due to the fact that the blocking probability $\pi_\ell = p_\ell(N_\ell)$, $\forall$ i=1,2,...,R. Moreover, the existence and unicity for the solution of the non-linear system of Step 4 cannot be shown analytically due to the complexity of the expressions of the blocking probabilities $\{\pi_k$, k = 1,2,...,M$\}$; nevertheless, numerical instabilities were never observed during extensive experimentations under any feasible set of initial values.

For 'reversible' QNMs the computational effort is mainly due to the solution of the linear system in Step 1.1. Moreover, notice that the recursive formulae used as building blocks in Step 5 (c.f., Sect. 2.4) for obtaining the performance measures of the network constitute simpler and more cost effective procedures than the ones advocated in [7].

4. NUMERICAL RESULTS

An indication of the credibility of the proposed MEM is demonstrated via a typical numerical experiment presented in the Appendix. The experiment refers to a two-class open central server QNM with three queues operating under LCFS-PR, FCFS and LCFS-NPR rules respectively. For each queue ℓ the following statistics are presented: the aggregate mean queue length, L_ℓ, response time, R_ℓ, idle and full probabilities $\{p_\ell(0)\}$, $p_\ell(N_\ell)\}$ (Table 1.1), marginal utilization, $\upsilon_{\ell,i}$ mean queue length, $L_{\ell,i}$ and idle probabilities $\{p_{\ell,i}(0)\}$ (Table 1.2).

Relative comparisons are carried out between analytic ME solutions (MEM) and simulation (SIM) results (obtained via QNAP-2 [18]). For validation purposes two kind of error indicators are used: Percentage differences (%D) for aggregate statistics and error tolerance (TOL) for marginal statistics (n.b., TOL is defined for mean queue lengths by $|SIM(L_{\ell,i}) - MEM(L_{\ell,i})| / \Sigma_i SIM(L_{\ell,i})$ and for both utilisations and idle probabilities by the absolute derivations between SIM and MEM results.

It can be observed from Tables 1.1 - 1.2 that the ME solutions are very comparable in accuracy with those obtained via simulation. In general, following an extensive comparative study all statistics are found to have error tolerance less than 0.06. More numerical experiments can be supplied on request by the authors. Note that occasional higher values of %D are due to very small (near zero) aggregate and/or marginal probabilities.

5. MEM EXTENTIONS TO GENERAL OPEN MULTI-CLASS QUEUEING NETWORKS WITH RS-RD BLOCKING

Consider a general open network at equilibrium with arbitrary configuration, M single-server queues, multiple job classes, class switching, RS-RD blocking, complete buffer sharing and mixed service disciplines drawn from FCFS, LCFS-NPR, LCFS-PR and PS rules. A product-form approximation for the state probability $p(\underline{k})$ can be clearly established as in (3.1), subject to normalisation and the marginal constraints of the type (2.2) - (2.4), where ϕ is replaced by $\phi_{k,i}$, the full buffer state probability of queue k, k=1,7,...,M, given that a job of class i, i=1,2,...,R is in service. This ME solution suggests a decomposition of the network into individual G/G/1/N censored queues with revised interarrival and service times reflecting the complete buffer sharing scheme, the RD-RS blocking mechanism and the service disciplines involved. The implementation of the MEM algorithm is broadly based on the algorithm of Section 3.3 and the GE-type algorithmic implementation of the single class QNMs under RS-RD blocking (c.f., [12]). In summary, the following main steps are involved:

Step 1: Calculate a GE-type service process per class i, i=1,2,...,R, at each queue k, k=1,...,M accounting for all possible delays due to blocking;

Step 2: Determine a GE-type revised arrival process per class i, i=1,2,...R at queue k, k=1,2,...,M so that the new process in isolation imitates the corresponding process within the network;

Step 3: Use the revised processes of Steps 1 and 2 to analyse approximately (using ME results of Section 5.1) each queue k, k=1,2,...,M as a censored GE/GE/1/N queue in isolation to obtain the statistics of the entire network.

Notice that it is no longer safe to make an invariability assumption (c.f., Section 2.5) about the Lagrangian coefficients $\{g_{k,i}\}$ and $\{x_{k,i}\}$, i=1,2,...,R with respect to N_k, k=1,2,...,M. The analytic determination of mean queue lengths $\{L_{k,i}\}$ and, subsequently, $\{g_{k,i}\}$ and $\{x_{k,i}\}$ has to be achieved directly via probabilistic arguments at the finite capacity level of the GE-type queue (c.f., [19]). However, the blocking probability of an arriver at queue k, k=1,2,...,M depends on the aggregate state probability (under a complete sharing scheme) and can be approximated by focusing on a single class GE/GE/1/N censored queue (c.f., [11]). Finally, the Lagrangian coefficients $\{y_{k,i}\}$ can be estimated by making use of flow-balance conditions of the type (2.5). It turns out that $\{y_{k,i}\}$ and subsequently $\{\upsilon_{k,i}\}$ depend only on aggregate coefficient $X_k = \Sigma_i x_{k,i}$ and finite capacity, N_k, k=1,2,...M. Consequently, $\{y_{k,i}\}$ and $\{\upsilon_{k,i}\}$ can be computed directly without the knowledge of the individual values of Lagrangian coefficients $\{x_{k,i}\}$, k=1,2,...,M; i=1,2,...,R.

Further details on MEM algorithm for general QNMs under RS-RD blocking and numerical results can be seen in [19].

6. CONCLUSIONS

A new ME approximation algorithm is proposed for obtaining the queue length distributions of arbitrary multiple class open exponential QNMs with finite capacity, complete buffer sharing, RS-RD blocking and mixed service disciplines drawn from the FCFS, LCFS-NPR, LCFS-PR and PS rules. A ME solution for a censored M/M/1/N queue with multiple classes of jobs plays the role of a building block towards a queue-by-queue decomposition analysis of the entire network under revised class arrival and service rates per queue. The ME algorithm captures the exact solution of open multiclass reversible queueing networks and has been found to have a very satisfactory error level in relation to simulation. Recent advances of MEM towards the approximate analysis of arbitrary GE-type QNMs with RS-RD blocking and multiple classes are discussed. The detailed description of the extended MEM algorithm together with numerical results appears in [19].

REFERENCES

[1] Baskett, F., Chandy, K.M., Muntz, R.R. and Palacios, F.G., "Open Closed and Mixed Networks with Different Classes of Customers, JACM 22, No. 2, pp. 248-260, 1975.

[2] Kelly, F.P., "Reversibility and Stochastic Networks", Wiley, New York, 1979.

[3] Perros, H.G., "A Bibliography of Papers on Queueing Networks, with Finite Capacity Queues", Perf. Eval. Vol. 10, pp. 255-260, 1989.

[4] Perros, H.G., "Approximation Algorithms for Open Queueing Networks with Blocking", in Stochastic Analysis of Computer and Communication Systems, Takagi (ed.), North-Holland, pp. 451-494, 1990.

[5] Onvural, R.O., "Survey of Closed Queueing Networks with Blocking", ACM Computing Surveys, Vol. 22, No. 2, pp. 83-121, 1990.

[6] Akyildiz, I.F. and Von Brand, H., "Exact Solutions for Open, Closed and Mixed Queueing Networks with Rejection Blocking", Thoret. Comput. Sci. 64, pp. 203-219, 1989.

[7] Akyildiz, I.F. and Von Brand, H., "Computational Algorithm for Networks of Queues with Rejection Blocking", Acta. Inf. 26, pp. 559-576, 1989.

[8] Akyildiz, I.F. and Von Brand, H., "Central Server Models with Multiple Job Classes, State Dependent Routing, and Rejection Blocking", IEEE Trans. Softw. Eng., to appear.

[9] Onvural, R.O., "A Note on the Product-Form Solutions of Multi-Class Closed Queueing Networks with Blocking", Perf. Eval., Vol. 10, pp. 247-253, 1989.

[10] Onvural, R.O. and Perros, H.G., "Approximate Analysis of Multi-Class Tandem Queueing Networks with Coxian Parameters and Finite Buffers", in Performance 1990, King, P.J.B., et al (eds.), North-Holland, p. 131-141, 1990.

[11] Kouvatsos, D.D and Xenios, N.P., "Maximum Entropy Analysis of General Queueing Networks with Blocking", in Queueing Networks with Blocking, Perros, H.G. and Altiok, T. (eds.), North-Holland, pp. 281-309, 1989.

[12] Kouvatsos, D.D. and Denazis, S.G., "Comments on and Tuning to: MEM for Arbitrary Queueing Networks with Repetitive-Service Blocking under Random Routing, Tech. Report CS-18-91, Bradford University, May 1991.

[13] Kouvatsos, D.D., "Maximum Entropy and the G/G/1/N Queue", Acta Informatical, 23, pp. 545-565, 1986.

[14] Kouvatsos, D.D. and Tabet-Aouel, N., "Product-From Approximations for an Extended Class of General Closed Queueing Networks", in Performance 1990, King, P.J.B., et al (eds.), North-Holland, pp. 301-315, 1990.

[15] Kouvatsos, D.D. and Georgatsos, P., "A Maximum Entropy Approximation for a Restricted G/G/1/N queue with Multi-Classes of Jobs under an Unrestricted Buffer Sharing Scheme", Res. Rep., DDK/PG, Bradford University, 1988.

[16] Kleinrock, L., "Queueing Systems, Vol. 2: Computer Applications", Wiley and Sons, New York, 1976.

[17] Kouvatsos, D.D., Georgatsos, P. and Tabet-Aouel, N., "A Universal Maximum Entropy Algorithm for General Multiple Class Open Networks with Mixed Service Disciplines", Modelling Techniques and Tools for Computer Performance Evaluation, Puigjauer, R., and Potier, D. (eds), Plenum Pub. Corp., pp 397-419, 1989.

[18] Veran, M. and Potier, D., "QNAP-2: A Portable Environment for Queueing Network Modelling", Proc. Int. Conf. on Modelling Techniques and Tools for Performance Analysis, INRIA, 1984.

[19] Kouvatsos, D.D. and Denazis, S., "Approximate Analysis of Multi-Class General Open Queueing Networks with RS-RD Blocking and Mixed Service Disciplines", Res. Rep., CS-32-92, Bradford University (in preparation).

APPENDIX

Experiment 1: Central Server Model
Raw Data: M = 3, R = 2

Queue 1: Service Discipline: LCFS-PR

$\bar{\lambda}_{o;1} = (1.6,2)$, $\bar{\mu}_1 = (3,4)$, $N_1 = 15$

$a_{11;21} = 1/6$, $a_{11;22} = 1/6$, $a_{11;31} = 2/9$, $a_{11;32} = 1/9$,

$a_{12;21} = 0.125$, $a_{12;22} = 0.125$, $a_{12;31} = 0.125$, $a_{12;32} = 0.125$,

$a_{11;o} = 1/3$, $a_{12;o} = 0.5$

Queue 2: Service Discipline: FCFS

$\bar{\mu}_2 = (5,2)$, $N_2 = 10$

$a_{21;11} = 0.5$, $a_{21;12} = 0.5$

$a_{22;11} = 2/3$, $a_{22;12} = 1/3$

Queue 3: Service Discipline: LCFS-NPR

$\bar{\mu}_3 = (2,2)$, $N_3 = 12$

$a_{31;11} = 1/5$, $a_{31;12} = 4/5$

$a_{32;11} = 1/5$, $a_{32;12} = 4/5$

Table 1.1: Aggregate Statistics

Queue	Measures	SIM	MEM	%D
1	L_1	14.39	14.399	0%
	R_1	4.680	4.771	1.94%
	$P_1(0)$	0	0	0%
	$P_1(15)$	0.623	0.625	0.3%
2	L_2	4.294	4.377	1.9%
	R_2	4.623	4.737	2.5%
	$P_2(0)$	0.136	0.138	1.5%
	$P_2(10)$	0.07	0.075	7.1%
3	L_3	9.334	9.306	0.3%
	R_3	12.53	12.508	0.2%
	$P_3(0)$	0.008	0.008	0%
	$P_3(12)$	0.256	0.255	0.4%

Table 1.2: Marginal Statistics

Queue	Class	Measures	SIM	MEM	TOL
1	1	$\upsilon_{1,1}$	0.648	0.65	0.002
		$L_{1,1}$	9.377	9.359	0.001
		$p_{1,1}(0)$	0	0	0
	2	$\upsilon_{1,2}$	0.352	0.35	0.002
		$L_{1,2}$	5.016	5.04	0.002
		$p_{1,2}(0)$	0.002	0.002	0
2	1	$\upsilon_{2,1}$	0.29	0.246	0.044
		$L_{2,1}$	2.007	2.004	0
		$p_{2,1}(0)$	0.272	0.286	0.014
	2	$\upsilon_{2,2}$	0.574	0.616	0.042
		$L_{2,2}$	2.287	2.373	0.008
		$p_{2,2}(0)$	0.217	0.197	0.02
3	1	υ_{31}	0.605	0.603	0.002
		$L_{3,1}$	5.684	5.661	0.001
		$p_{3,1}(0)$	0.0167	0.017	0
	2	$\upsilon_{3,2}$	0.387	0.388	0.001
		$L_{3,2}$	3.649	3.645	0
		$p_{3,2}(0)$	0.04	0.04	0

Queueing Networks with Finite Capacity
R.O. Onvural and I.F. Akyildiz (Editors)

EXACT ANALYSIS OF MULTI-JOB CLASS NETWORKS OF QUEUES WITH BLOCKING-AFTER-SERVICE*

Ian F. Akyildiz and Chueng-Chiu Huang

School of Electrical Engineering
Georgia Institute of Technology
Atlanta, GA 30332
U.S.A.

Abstract

Two models of closed queueing networks with blocking after service and multiple job classes are analyzed. The first model is a network with N stations and each station has either Type II or Type III. The second model is a star like queueing network also called as central server model in which the stations may have either Type I or Type IV, with the condition that the neighbors of these stations must be of Type II or Type III such that blocking will be caused only by this set of station types. Exact product form solutions are obtained for the equilibrium state probabilities in both models. Formulae for performance measures such as throughput and mean number of jobs are also derived.

Key Words: Performance Evaluation, Queueing Networks, Finite Buffer Capacity, Blocking, Deadlock, Equilibrium State Probabilities, Throughput, Mean Number of Jobs.

* This work was supported by National Science Foundation (NSF) under Grant No. CCR-90-11981.

1 Introduction

In recent years there has been an increased interest in the analysis of queueing networks with blocking. This is probably due to the realization that these queueing networks are useful in modelling computer systems, communication networks, and flexible manufacturing systems. The set of rules that dictate when a station becomes blocked and when it becomes unblocked is commonly referred to as the *blocking* mechanism. There are basically only a few blocking mechanisms that have been extensively studied in the literature [3,13]. We consider the so-called *blocking after service* [3,13] (in short form BAS) mechanism, i.e., if a job finishes service at a station and wants to enter a station which is full, it stays in the server of the source station, waiting for a space to be available in the destination station. This blocking policy is also known type 1 blocking, transfer blocking, manufacturing blocking, production blocking and non-immediate blocking in the literature [3,13]. Several papers consider this blocking policy [1,2,3,4,5,6,7,9,12,13,14]. Onvural and Perros [14] and Akyildiz and von Brand [4] consider closed queueing networks with BAS mechanism for limited number of jobs. Akyildiz and Liebeherr [5] determine necessary and sufficient conditions for deadlock-freedom. Akyildiz and von Brand [4] considered queueing networks with BAS mechanism. In the first model they analyze a two-station model with different types of stations and single class of jobs. They derive an exact product form solution for the equilibrium state probabilities. In [4] they also analyze a second model with N stations, multiple classes of jobs and the limited number of jobs $K = \min_{i \in N} \{B_i\} + 1$, where B_i is the buffer capacity of station i. If the stations that will cause blocking are type I station in $BCMP$ model, an exact product form solution is obtained.

In this paper we consider two models of closed queueing networks with BAS mechanism. In both models we assume that there are N stations and R classes of jobs. The first model is a general topology network with N stations where each station is either of Type II or Type III [8,10,11]. The second model is a star like queueing network (also known as a central server model) where the central server must have the station Type I or Type IV [8,10,11] and the neighbors must be either of Type II or Type III stations. In the second model we assume that the blocking will be caused only by the central server. For both models we obtain exact product form solution for equilibrium state probabilities. We also derive formulae for performance measures such as throughput and mean number of jobs.

The paper is organized as follows: In section 2 introduce common notations and definitions for both models. In section 3 we analyze the first model. In section 4 we investigate the central server model. In section 5 we conclude the paper.

2 Basic Model Description

Both models contain N stations, R job classes and K total number of jobs. The types of stations we consider in this paper are from $BCMP$ model [8,10,11]:

- Type I: The service discipline is first-come-first-served (FCFS); all job classes have the same service time distribution and the service rate can be state-dependent where $\mu(k)$ will denote the service rate with k jobs.
- Type II: There is a single server and the service discipline is processor sharing (PS). Each job class may have a distinct service time distribution.
- Type III: The number of servers is greater or equal to the maximum number of jobs which can be queued at the station; infinite servers (IS). Each job class may have a distinct service time distribution.
- Type IV: There is a single server and the queueing discipline is last-come-first-served pre-emptive-resume (LCFS-PR). Each job may have a distinct service time distribution.

Here we assume that all service times follow exponential distribution which depend on the station and on the class. Note that our results can easily be extended to general service time distributions for Types II, III and IV. For the sake of simplicity here we give results only for exponential cases.

The transition probabilities are denoted by $p_{ir,js}$, $1 \leq i, j \leq N$ and $1 \leq r, s \leq R$, where a class r job departs from station i and visits station j and becomes a class s job. We also assume that the routing matrix $\mathbf{P} = [p_{i,r,j,s}]$, is irreducible. We define

$$\alpha_{js} = \sum_{i=1}^{N} \sum_{r=1}^{R} \alpha_{ir} p_{ir,js}. \tag{1}$$

Let the vector $\mathbf{n} = (\mathbf{x}_1, \cdots, \mathbf{x}_N)$ denote the number of jobs in N stations and their positions in each station, i.e., $\mathbf{x}_i = (x_{i1}, \cdots, x_{in_i})$, where x_{il} denotes the class of the job in station i at the position l. Let n_i be the number of jobs in station i, and $n_i = n_{i1} + \cdots + n_{iR}$, where n_{ir} is the number of class r jobs in station i. We note that if station i is of Type II or III, then there is no requirement for the order of the jobs, thus, we denote $\mathbf{x}_i$ as $\mathbf{n}_i$, $\mathbf{n}_i = (n_{i1}, \cdots, n_{iR})$.

A product form solution for equilibrium state probabilities exists when there is no blocking in the network [8,10,11]:

$$p(\mathbf{n}) = C \prod_{i=1}^{N} f_i(\mathbf{x}_i), \tag{2}$$

where C is the normalization constant such that the sum of equilibrium state probabilities will be equal to one and $f_i(\mathbf{x}_i)$ is defined by the type of station i,

$$f_i(\mathbf{x}_i) = \begin{cases} \left(\frac{1}{\mu_i}\right)^{n_i} \prod_{l=1}^{n_i} \alpha_{ix_{il}}, & \text{if station } i \text{ is Type } I \\ n_i! \prod_{r=1}^{R} \frac{1}{n_{ir}!} \left(\frac{\alpha_{ir}}{\mu_{ir}}\right)^{n_{ir}}, & \text{if station } i \text{ is Type } II \\ \prod_{r=1}^{R} \frac{1}{n_{ir}!} \left(\frac{\alpha_{ir}}{\mu_{ir}}\right)^{n_{ir}}, & \text{if station } i \text{ is Type } III \\ \prod_{l=1}^{n_i} \left(\frac{\alpha_{ix_{il}}}{\mu_{ix_{il}}}\right). & \text{if station } i \text{ is Type } IV \end{cases} \tag{3}$$

We define T_1 as the set of Type I or IV stations, and T_2 as the set of Type II or III stations. Since we are considering the models with blocking, we need to define B_i as the capacity of station i (i.e., buffer and server capacity). In addition to avoid the self loop that cause *deadlock* we assume that

$$p_{ii} = 0, \text{ for } i = 1, \cdots, N.$$

For convenience we define the following quantities for each station type. For type II and III station, the service rate are dependent on the number of jobs in the station and on their classes but independent of their order. Thus, we let $\mathbf{n}_i = (n_{i1}, \cdots, n_{iR})$ for $i \in T_2$, where n_{ir} denotes the number of class r active jobs in station i. A job is said to be active if it is not blocked. Also we let $n_i = n_{i1} + \cdots + n_{iR}$ be the total number of active jobs in station i for $i \in T_2$.

Let b denote the number of blocked jobs. Thus, with the active jobs we have

$$K = \sum_{i=1}^{N} n_i + b, \tag{4}$$

with

$$0 \leq b \leq K - \min_{1 \leq i \leq N} \{B_i\}. \tag{5}$$

We assume that a job will choose its destination station and the class when it finishes its service. Considering the b blocked jobs we need to know their locations and their classes. We also assume that the blocked jobs will join that station, which caused blocking according to $First - Come - First - In$ $(FCFI)$ scheduling discipline. Thus, we let $\mathbf{y} = (\mathbf{y}_i)$ and $\mathbf{y}_i = (y_{i1}, \cdots, y_{ib_i})$ if $b_i \geq 1$ otherwise $\mathbf{y}_i = \emptyset$ where b_i denotes the number of jobs blocked by station i, and y_{il}, $1 \leq y_{il} \leq N$ denotes the location of the lth job blocked by station i. Similarly, $\mathbf{z} = (\mathbf{z}_i)$ and $\mathbf{z}_i = (z_{i1}, \cdots, z_{ib_i})$ where z_{il}, $1 \leq z_{il} \leq R$ denotes the class of the lth blocked job. To have the complete information about the system we define the state space $S = \{(\mathbf{n}, \mathbf{x}, \mathbf{y}, \mathbf{z})\}$, where $\mathbf{x} = (\mathbf{x}_i, i \in T_1)$ and $\mathbf{n} = (\mathbf{n}_i, i \in T_2)$. For a given a state $(\mathbf{n}, \mathbf{x}, \mathbf{y}, \mathbf{z})$ the service rate of station i for $i \in T_2$ is $\mu_i(\mathbf{n}_i) = \mu_{i1}(n_{i1}) + \cdots + \mu_{iR}(n_{iR})$ where $\mu_{ir}(n_{ir}) = n_{ir}\mu_{ir}$ if station i is of Type III, and if station i is of Type II, then $\mu_{ir}(n_{ir}) = \frac{n_{ir}}{n_i}\mu_{ir}$. For $i \in T_1$ then the service rate is: if station i is of type I, then $\mu_i(\mathbf{x}_i) = \mu_i$ for $n_i > 0$, otherwise it is equal to 0. If station i is of Type IV, then $\mu_i(\mathbf{x}_i) = \mu_{in_i}$ for $n_i \leq B_i$.

Let e_{ir} denote an NxR matrix with all zero elements except the element at the position of ith row and rth column with value 1.

3 General Topology Network with Type II and III Stations

In this section we consider a general topology network with only Type II and III stations. We assume the following constraint in the model:

$$K < \min_{1 \leq i,j \leq N; p_{ij} > 0} \{B_i + B_j\}, \tag{6}$$

This constraint says that only one of the N stations may cause blocking at any time. The reason of making this assumption is to avoid more than two jobs moving at one event. Thus, the state space is reduced to $S = (\mathbf{n}, \mathbf{y}, \mathbf{z})$, where $\mathbf{y}$ and $\mathbf{z}$ are b by 1 vectors, because if there are any blocked jobs, then they are all blocked by the same station.

Now we consider the transitions of the jobs for a nonblocking state $(\mathbf{n}, \mathbf{y}, \mathbf{z})$, where $\mathbf{y}$ and $\mathbf{z}$ are empty vectors $\emptyset$. In this case the system will reach the state $(\mathbf{n} - e_{ir} + e_{js}, \emptyset, \emptyset)$ with rate $\mu_{ir}(n_{ir}) p_{ir,js}$, if $n_j \neq B_j$. Otherwise, a class r job from station i will be blocked by station j and the transition will occur from state $(\mathbf{n}, \emptyset, \emptyset)$ to state $\left(\mathbf{n} - e_{ir}, \mathbf{y}', , \mathbf{z}'\right)$ where $\mathbf{y}' = (y_1)$ and $y_1 = i$. The same occurs for $\mathbf{z}' = (z_1)$ and $z_1 = s$. The rate to reach this state is the same as before, i.e., $\mu_{ir}(n_{ir}) p_{ir,js}$.

Now we consider the transitions from a blocking state. Suppose station k is full, i.e., $n_k = B_k$ and the current state is $(\mathbf{n}, \mathbf{y}, \mathbf{z})$, for $b \geq 0$. Then a class r job leaves station i and moves to station j, for $i \neq k, j \neq k$ and becomes a class s job and the system reaches the state $(\mathbf{n} - e_{ir} + e_{js}, \mathbf{y}, \mathbf{z})$ with rate $\mu_{ir}(n_{ir}) p_{ir,js}$. If a class r job moves from station i to station k and becomes a class s job, then the state $(\mathbf{n} - e_{ir}, \mathbf{y} + y_{b+1}, \mathbf{z} + z_{b+1})$ and $y_{b+1} = i$, $z_{b+1} = s$ will be reached. Note that in this case the transition rate is $\mu_{ir}(n_{ir}) p_{ir,js}$.

Let $\mathbf{y} = (y_1, \cdots, y_b)$ then the operator $\mathbf{y} + y_{b+1}$ will be $(y_1, \cdots, y_{b+1})$. Similar is also valid for the operator $\mathbf{z} + z_{b+1}$. When considering a class r job which moves from station k to station j and becomes a class s job with the transition rate $\mu_{kr}(n_{kr}) p_{kr,js}$, then one of the first blocked jobs will move to station k at the same time. Thus, the new state is $\left(\mathbf{n} - e_{kr} + e_{js} + e_{kz}, \mathbf{y}', \mathbf{z}'\right)$, with $\mathbf{y}' = \left(y_1', \cdots y_{b-1}'\right)$, $y_l' = y_{l+1}$ and $\mathbf{z}' = \left(z_1', \cdots z_{b-1}'\right)$, $z_l' = z_{l+1}$ for $1 \leq l \leq b-1$.

From these state transitions we obtain the global balance equation for a state $(\mathbf{n}, \mathbf{y}, \mathbf{z}) \in S$ if $n_i < B_i$ for $1 \leq i \leq N$, thus $\mathbf{y} = \emptyset$ and $\mathbf{z} = \emptyset$:

$$p(\mathbf{n}, \emptyset, \emptyset) \sum_{i=1}^{N} \sum_{r=1}^{R} \mu_{ir}(n_{ir}) = \sum_{i=1}^{N} \sum_{r=1}^{R} \sum_{j=1}^{N} \sum_{s=1}^{R} p(\mathbf{n} + e_{ir} - e_{js}, \emptyset, \emptyset) \mu_{ir}(n_{ir} + 1) p_{ir,js}, \tag{7}$$

and the global balance equation for a blocking state (suppose $n_k = B_k$):

$$
\begin{aligned}
& p(\mathbf{n},\mathbf{y},\mathbf{z}) \sum_{i=1}^{N}\sum_{r=1}^{R} \mu_{ir}(n_{ir}) \\
= & \sum_{i=1,i\neq k}^{N}\sum_{r=1}^{R}\sum_{j=1,j\neq k}^{N}\sum_{s=1}^{R} p(\mathbf{n}+e_{ir}-e_{js},\mathbf{y},\mathbf{z})\,\mu_{ir}(n_{ir}+1)\,p_{ir,js} \\
+ & \sum_{j\neq k}\sum_{s=1}^{R}\sum_{r=1}^{R}\sum_{y_0=1}^{N}\sum_{z_0=1}^{R} p(\mathbf{n}+e_{kr}-e_{js}-e_{kz_0},\mathbf{y}+y_0,\mathbf{z}+z_0)\,\mu_{kr}(n_{kr})\,p_{kr,js} \\
+ & \sum_{s=1}^{R} p(\mathbf{n}+e_{y_b s},\mathbf{y}-y_b,\mathbf{z}-z_b)\,\mu_{y_b s}(n_{y_b s}+1)\,p_{y_b s,kz_b}. \qquad (8)
\end{aligned}
$$

The first term on the right hand side of (8) is the total flow-in rate to the state $(\mathbf{n},\mathbf{y},\mathbf{z})$ from state $(\mathbf{n}+e_{ir}-e_{js},\mathbf{y},\mathbf{z})$. We note that the moving job is from station i to station j, $i\neq k, j\neq k$. The second term on the right-hand side of (8) is the flow-in rate to state $(\mathbf{n},\mathbf{y},\mathbf{z})$ by a class r job moving out from station k, and in the mean time another blocked job, which is class z_0, will move into station k from station y_0. As a consequence a transition occurs from the state $(\mathbf{n}+e_{kr}-e_{js}-e_{kz_0},\mathbf{y}+y_0,\mathbf{z}+z_0)$ to the state $(\mathbf{n},\mathbf{y},\mathbf{z})$.

For notational convenience we let y_0, z_0 denote the location and the class of the blocked job that will move into station k while a job leaves station k and joins station i. The third term in (8) is the flow-in rate to the state $(\mathbf{n},\mathbf{y},\mathbf{z})$ by a joining job at station k.

We note that a given state will be visited and the departure job is from a station i, $i\neq k$, then we know where the job comes from, i.e., $y_b = i$ and its class is switched from s to z_b. So that the transition is from the state $(\mathbf{n}+e_{y_b s},\mathbf{y}-y_b,\mathbf{z}-z_b)$ to the state $(\mathbf{n},\mathbf{y},\mathbf{z})$.

Theorem 1. The model has the following product form solution for equilibrium state probabilities

$$
p(\mathbf{n},\mathbf{y},\mathbf{z}) = C\prod_{i=1}^{N} f_i(\mathbf{n}_i)\left(\prod_{l=1}^{b}\frac{\sum_{t=1}^{R}\alpha_{y_l t}p_{y_l t,iz_l}}{\sum_{t=1}^{R}\mu_{it}(n_{it})}\right)^{I_{\{n_i=B_i\}}} \qquad (9)
$$

where $f_i(\mathbf{n}_i)$ is defined

$$
f_i(\mathbf{n}_i) = \left\{\begin{array}{ll} n_i!\prod_{r=1}^{R}\frac{1}{n_{ir}!}\left(\frac{\alpha_{ir}}{\mu_{ir}}\right)^{n_{ir}} & \text{if station } i \text{ is type } II \\ \prod_{r=1}^{R}\frac{1}{n_{ir}!}\left(\frac{\alpha_{ir}}{\mu_{ir}}\right)^{n_{ir}} & \text{if station } i \text{ is type } III \end{array}\right). \qquad (10)
$$

and C is a normalization constant

$$
C^{-1} = \sum_{(\mathbf{n},\mathbf{y},\mathbf{z})\in S}\prod_{i=1}^{N} f_i(\mathbf{n}_i)\left(\prod_{l=1}^{b}\frac{\sum_{t=1}^{R}\alpha_{y_l t}p_{y_l t,iz_l}}{\sum_{t=1}^{R}\mu_{it}(n_{it})}\right)^{I_{\{n_i=B_i\}}} \qquad (11)
$$

Proof. We escape the proof of the nonblocking states since it is exactly the same as in the *BCMP* model [8,10,11]. For the global balance equation of blocking states we assume that the state k has $n_k = B_k$ and the number of blocked jobs is b, then for (9) and (10) we have

$$p\left(\mathbf{n}+e_{ir}-e_{js},\mathbf{y},\mathbf{z}\right)=p\left(\mathbf{n},\mathbf{y},\mathbf{z}\right)\frac{\alpha_{ir}}{\mu_{ir}\left(n_{ir}+1\right)}\frac{\mu_{js}\left(n_{js}\right)}{\alpha_{js}} \tag{12}$$

if $i \neq k$ and $j \neq k$.

Next we consider the departure event of a job leaving station $i \neq k$ and visiting station k

$$\begin{aligned} & p\left(\mathbf{n}+e_{kr}-e_{js}-e_{kz_0},\mathbf{y}+y_0,\mathbf{z}+z_0\right) \\ & = p\left(\mathbf{n},\mathbf{y},\mathbf{z}\right)\frac{\alpha_{kr}}{\mu_{kr}(n_{kr}+1)}\frac{\mu_{js}(n_{js})}{\alpha_{js}}\frac{\mu_{kz_0}\left(n_{kz_0}\right)}{\alpha_{kz_0}}\frac{\sum_{t=1}^{R}\alpha_{y_0t}p_{y_0t,kz_0}}{\sum_{t=1}^{R}\mu_{kt}(n_{kt})}\end{aligned} \tag{13}$$

and for the departure event of a job leaving station k and visiting station i

$$p\left(\mathbf{n}+e_{y_bs},\mathbf{y}-y_b,\mathbf{z}-z_b\right)=p\left(\mathbf{n},\mathbf{y},\mathbf{z}\right)\frac{\alpha_{y_bs}}{\mu_{y_bs}\left(n_{y_bs}+1\right)}\frac{\sum_{t=1}^{R}\mu_{kt}\left(n_{kt}\right)}{\sum_{t=1}^{R}\alpha_{y_bt}p_{y_bt,kz_b}}. \tag{14}$$

By substituting (12) – (14) to the right-hand side of (8) separately and using (1), we obtain

$$\begin{aligned} & \sum_{i=1,i\neq k}^{N}\sum_{r=1}^{R}\sum_{j=1,j\neq k}^{N}\sum_{s=1}^{R}p\left(\mathbf{n}+e_{ir}-e_{js},\mathbf{y},\mathbf{z}\right)\mu_{ir}\left(n_{ir}+1\right)p_{ir,js} \\ & = p\left(\mathbf{n},\mathbf{y},\mathbf{z}\right)\sum_{j=1,j\neq k}^{N}\sum_{s=1}^{R}\frac{\mu_{js}(n_{js})}{\alpha_{js}}\left[\sum_{i=1,i\neq k}^{N}\sum_{r=1}^{R}\alpha_{ir}p_{ir,js}\right]\end{aligned} \tag{15}$$

$$\begin{aligned} & \sum_{j\neq k}\sum_{s=1}^{R}\sum_{r=1}^{R}\sum_{y_0=1}^{N}\sum_{z_0=1}^{R}p\left(\mathbf{n}+e_{kr}-e_{js}-e_{kz_0},\mathbf{y}+y_0,\mathbf{z}+z_0\right)\mu_{kr}\left(n_{kr}\right)p_{kr,js} \\ = {} & p\left(\mathbf{n},\mathbf{y},\mathbf{z}\right)\sum_{j\neq k}\sum_{s=1}^{R}\sum_{r=1}^{R}\sum_{z_0=1}^{R}\frac{\mu_{js}\left(n_{js}\right)}{\alpha_{js}}\frac{\alpha_{kr}}{\alpha_{kz_0}}\frac{\mu_{kz_0}\left(n_{kz_0}\right)}{\sum_{t=1}^{R}\mu_{kt}\left(n_{kt}\right)}p_{kr,js}\left[\sum_{y_0=1}^{N}\sum_{t=1}^{R}\alpha_{y_0t}p_{y_0t,kz_0}\right] \\ = {} & p\left(\mathbf{n},\mathbf{y},\mathbf{z}\right)\sum_{j\neq k}\sum_{s=1}^{R}\frac{\mu_{js}\left(n_{js}\right)}{\alpha_{js}}\sum_{r=1}^{R}\left[\alpha_{kr}p_{kr,js}\sum_{z_0=1}^{R}\left(\frac{\mu_{kz_0}\left(n_{kz_0}\right)}{\sum_{t=1}^{R}\mu_{kt}\left(n_{kt}\right)}\right)\right] \\ = {} & p\left(\mathbf{n},\mathbf{y},\mathbf{z}\right)\sum_{j\neq k}\sum_{s=1}^{R}\frac{\mu_{js}\left(n_{js}\right)}{\alpha_{js}}\left[\sum_{r=1}^{R}\alpha_{kr}p_{kr,js}\right]\end{aligned} \tag{16}$$

By combining (15) and (16) we have

$$\begin{aligned} (15)+(16) & = p\left(\mathbf{n},\mathbf{y},\mathbf{z}\right)\sum_{j\neq k}\sum_{s=1}^{R}\frac{\mu_{js}\left(n_{js}\right)}{\alpha_{js}}\left[\sum_{i=1,i\neq k}^{N}\sum_{r=1}^{R}\alpha_{ir}p_{ir,js}+\sum_{r=1}^{R}\alpha_{kr}p_{kr,js}\right] \\ & = p\left(\mathbf{n},\mathbf{y},\mathbf{z}\right)\sum_{j\neq k}\sum_{s=1}^{R}\frac{\mu_{js}\left(n_{js}\right)}{\alpha_{js}}\left[\sum_{i=1}^{N}\sum_{r=1}^{R}\alpha_{ir}p_{ir,js}\right] \\ & = p\left(\mathbf{n},\mathbf{y},\mathbf{z}\right)\sum_{j\neq k}\sum_{s=1}^{R}\mu_{js}\left(n_{js}\right).\end{aligned} \tag{17}$$

In (14) we substitute the third term of (8) and obtain

$$\begin{aligned}
&\sum_{s=1}^{R} p\left(\mathbf{n}+e_{y_b s}, \mathbf{y}-y_b, \mathbf{z}-z_b\right) \mu_{y_b s}\left(n_{y_b s}+1\right) p_{y_b s, k z_b} \\
&\quad = p(\mathbf{n}, \mathbf{y}, \mathbf{z}) \frac{\sum_{t=1}^{R} \mu_{kt}\left(n_{kt}\right)}{\sum_{t=1}^{R} \alpha_{y_b t} p_{y_b t, k z_b}} \sum_{s=1}^{R} \alpha_{y_b s} p_{y_b s, k z_b} \\
&\quad = p(\mathbf{n}, \mathbf{y}, \mathbf{z}) \sum_{t=1}^{R} \mu_{kt}\left(n_{kt}\right).
\end{aligned} \tag{18}$$

Thus, with (17) and (18) we have $p(\mathbf{n}, \mathbf{y}) \sum_{j=1}^{N} \sum_{s=1}^{R} \mu_{js}(n_{js})$ which is equal to the right-hand side of (8). **QED**

Now we derive the throughput of the system $\lambda = \sum_{i=1}^{N} \lambda_i$, where λ_i is the throughput of station i, and the mean number of jobs $\overline{k}_i$ in station i.

Theorem 2. The *throughput* of station i is computed from

$$\begin{aligned}
\lambda_i = \sum_{(\mathbf{n},\mathbf{y},\mathbf{z}) \in S} p(\mathbf{n}, \mathbf{y}, \mathbf{z}) \Bigg[& \sum_{r=1}^{R} \mu_{ir}(n_{ir}) \left(1 - \sum_{k=1}^{N} \sum_{s=1}^{R} p_{ir,ks} I_{\{n_k = B_k\}}\right) \\
& + \sum_{k=1}^{N} I_{\{n_k = B_k\}} I_{\{y_1 = i\}} \left(\sum_{r=1}^{R} \mu_{kr}(n_{kr})\right)\Bigg]
\end{aligned} \tag{19}$$

The *mean number of jobs* at station i is

$$\overline{k}_i = \sum_{(\mathbf{n},\mathbf{y},\mathbf{z}) \in S} p(\mathbf{n}, \mathbf{y}, \mathbf{z}) \left(\sum_{r=1}^{R} n_{ir} + \sum_{l=1}^{b} I_{\{y_l = i\}}\right). \tag{20}$$

Proof. For any given state $(\mathbf{n}, \mathbf{y}, \mathbf{z})$ we consider the throughput of station i. Suppose the number of class r active jobs is n_{ir}, thus, the service rate is $\mu_{ir}(n_{ir})$. However, a job which completed its service in station i can leave station i only by choosing a nonfull station j, i.e., $n_j < B_j$ with probability $\sum_{s=1}^{R} p_{ir,js}$. When considering the blocked jobs, if the first blocked job in station i is blocked by station k, i.e., $y_1 = i$, then this job has the same rate with which will leave station i and join the station k as the service rate of station k, i.e., $\sum_{r=1}^{R} \mu_{kr}(n_{kr})$. Thus, (19) follows.

Similarly, for given a state $(\mathbf{n}, \mathbf{y}, \mathbf{z})$ we have the number of jobs in station i which will be equal to the active jobs $\sum_{r=1}^{R} n_{ir}$ plus the blocked jobs in this station, i.e., $\sum_{l=1}^{b} I_{\{y_l = i\}}$, so that (20) follows. **QED**

4 Central Server Model with Different Types of Stations

In this section we analyze central server models (star networks) with BAS mechanism. We put the constraint that the central server must have station type either I or IV and all other stations may be of Type II or III. We further assume that only stations of Type I or IV will cause blocking, and the neighbors of such stations must be of Type II or III. The total number of jobs must satisfy the constraint $\min_{i\in T_1}\{B_i\} < K < \min_{i\in T_2}\{B_i\}$ and $p_{ir,js} = 0$, if $i, j \in T_1$.

We note that $p_{ir,js} = 0$ if $i, j \in T_1$ and for all r, s, then (1) will become:

$$\alpha_{js} = \sum_{i\in T_2}\sum_{r=1}^{R} \alpha_{ir} p_{ir,js} \qquad for \quad j \in T_1 \tag{21}$$

We denote T_A as the set of stations in T_1 if $n_i < B_i$, and T_B as the set of stations in T_1 with $n_i = B_i$. Note that $T_A \cup T_B = T_1$. In the following we give the global balance equation for a given state $(\mathbf{n}, \mathbf{x}, \mathbf{y}, \mathbf{z})$.

$$\begin{aligned} & p(\mathbf{n},\mathbf{x},\mathbf{y},\mathbf{z})\left(\sum_{i\in T_1}\mu_i(\mathbf{x}_i) + \sum_{i\in T_2}\sum_{r=1}^{R}\mu_{ir}(n_{ir})\right) \qquad (22)\\ = & \sum_{j\in T_A}\sum_{i\in T_2}\sum_{r=1}^{R} p\left(\mathbf{n}+e_{ir},\mathbf{x}-x_{jn_j},\mathbf{y},\mathbf{z}\right)\mu_{ir}(n_{ir}+1)\,p_{ir,jx_{jn_j}}\\ + & \sum_{j\in T_B}\sum_{r=1}^{R} p\left(\mathbf{n}+e_{y_{jb_j}r},\mathbf{x},\mathbf{y}-y_{jb_j},\mathbf{z}-z_{jb_j}\right)\mu_{y_{jb_j}r}\left(n_{y_{jb_j}r}+1\right)p_{y_{j,b_j}r,jz_{jb_j}}\\ + & \sum_{i\in T_2,j\in T_2}\sum_{s=1}^{R}\sum_{r=1}^{R} p(\mathbf{n}+e_{ir}-e_{js},\mathbf{x},\mathbf{y},\mathbf{z})\,\mu_{ir}(n_{ir}+1)\,p_{ir,js}\\ + & \sum_{i\in T_A,j\in T_2}\sum_{s=1}^{R}\sum_{x_{i0}=1}^{R} p(\mathbf{n}-e_{js},\mathbf{x}+x_{i0},\mathbf{y},\mathbf{z})\,\mu_i(\mathbf{x}_i+x_{i0})\,p_{ix_{i0},js}\\ + & \sum_{i\in T_B,j\in T_2}\sum_{s=1}^{R}\sum_{y_{i0}=1}^{N}\sum_{x_{i0}=1}^{R} p(\mathbf{n}-e_{js},\mathbf{x}+x_{i0}-x_{iB_i},\mathbf{y}+y_{i0},\mathbf{z}+z_{i0})\,\mu_i(\mathbf{x}_i+x_{i0}-x_{iB_i})\,p_{ix_{i0},js}. \end{aligned}$$

We consider a given state $(\mathbf{n}, \mathbf{x}, \mathbf{y}, \mathbf{z})$ then the total flow-out rate from this state is $p(\mathbf{n}, \mathbf{x}, \mathbf{y}, \mathbf{z})$ $\left(\sum_{i\in T_1}\mu_i(\mathbf{x}_i) + \sum_{i\in T_2}\sum_{r=1}^{R}\mu_{ir}(n_{ir})\right)$. In order to check the flow-in rates we consider the different events caused by these jobs: The first term on the right hand side of (22) denotes the departure event of a class r job from station $i \in T_2$ to station $j \in T_A$ to class x_{jn_j}. The second term denotes the events of a class r job in station $y_{jb_j} \in T_2$ which completed its service and is joining station $j \in T_B$. However, since station j is full, $n_j = B_j$, this job is blocked and becomes a class z_{jb_j} job and joins the list of blocked jobs waiting for station j. The third term denotes the event of moving of a class r job from station $i \in T_2$ to station $j \in T_2$ and becoming a class s job. The last two terms describing the events of moving of a class x_{i0} job from station $i \in T_A$, $i \in T_B$ to station $j \in T_2$ and becoming a class s job. We note that x_{i0} denotes the class in which the job leaves station i. y_{i0} and z_{i0} denote the station and the class of the blocked job at the first position of the list of blocked jobs which will move to station i. Note that $z_{i0} = x_{iB_i}$.

Theorem 3. The model has the following product form solution for equilibrium state probabilities:

$$p(\mathbf{n},\mathbf{x},\mathbf{y},\mathbf{z}) = C \prod_{i\in T_1} f_i(\mathbf{x}_i,\mathbf{y}_i,\mathbf{z}_i) \prod_{i\in T_2} f_i(\mathbf{n}_i), \tag{23}$$

where $f_i(\mathbf{x}_i,\mathbf{y}_i,\mathbf{z}_i)$ for $i \in T_1$ is defined as

$$f_i(\mathbf{x}_i,\mathbf{y}_i,\mathbf{z}_i) = \begin{cases} \left(\frac{1}{\mu_i}\right)^{n_i} \prod_{l=1}^{n_i} \alpha_{ix_{il}}, & \text{if } i \text{ is Type } I \text{ and } i \in T_A \\ \left(\frac{1}{\mu_i}\right)^{n_i} \prod_{l=1}^{B_i} \alpha_{ix_{il}} \prod_{l=1}^{b_i} \frac{\sum_{t=1}^{R} \alpha_{y_{il}t} p_{y_{il}t,iz_{il}}}{\mu_i}, & \text{if } i \text{ is Type } I \text{ and } i \in T_B \\ \prod_{l=1}^{n_i} \frac{\alpha_{ix_{il}}}{\mu_{ix_{il}}}, & \text{if } i \text{ is Type } IV \text{ and } i \in T_A \\ \prod_{l=1}^{B_i} \frac{\alpha_{ix_{il}}}{\mu_{ix_{il}}} \prod_{l=1}^{b_i} \frac{\sum_{t=1}^{R} \alpha_{y_{il}t} p_{y_{il}t,iz_{il}}}{\mu_{iz_{il}}}, & \text{if } i \text{ is Type } IV \text{ and } i \in T_B \end{cases} \tag{24}$$

and $f_i(\mathbf{n}_i)$ for $i \in T_2$, is defined as

$$f_i(\mathbf{n}_i) = \begin{cases} n_i! \prod_{r=1}^{R} \frac{1}{n_{ir}!} \left(\frac{\alpha_{ir}}{\mu_{ir}}\right)^{n_{ir}} & \text{if station } i \text{ is Type } II \\ \prod_{r=1}^{R} \frac{1}{n_{ir}!} \left(\frac{\alpha_{ir}}{\mu_{ir}}\right)^{n_{ir}} & \text{if station } i \text{ is Type } III \end{cases} \tag{25}$$

where C is a normalization constant

$$C^{-1} = \sum_{(\mathbf{n},\mathbf{x},\mathbf{y},\mathbf{z})\in S} \prod_{i\in T_1} f_i(\mathbf{x}_i,\mathbf{y}_i,\mathbf{z}_i) \prod_{i\in T_2} f_i(\mathbf{n}_i). \tag{26}$$

Proof. From (23) – (25) we have

$$p\left(\mathbf{n}+e_{ir},\mathbf{x}-x_{jn_j},\mathbf{y},\mathbf{z}\right) = p(\mathbf{n},\mathbf{x},\mathbf{y},\mathbf{z}) \frac{\mu_j(\mathbf{x}_j)}{\alpha_{jx_{jn_j}}} \frac{\alpha_{ir}}{\mu_{ir}(n_{ir}+1)} \tag{27}$$

$$\begin{aligned} & p\left(\mathbf{n}+e_{y_{jb_j}r},\mathbf{x}-x_{jn_j},\mathbf{y}-y_{jb_j},\mathbf{z}-z_{jb_j}\right) \\ &= p(\mathbf{n},\mathbf{x},\mathbf{y},\mathbf{z}) \frac{\mu_j(\mathbf{x}_j)}{\sum_{t=1}^{R} \alpha_{y_{jb_j}t} p_{y_{jb_j}t,jz_{jb_j}}} \frac{\alpha_{y_{jb_j}r}}{\mu_{y_{jb_j}r}\left(n_{y_{jb_j}r}+1\right)} \end{aligned} \tag{28}$$

$$p(\mathbf{n}+e_{ir}-e_{js},\mathbf{x},\mathbf{y},\mathbf{z}) = p(\mathbf{n},\mathbf{x},\mathbf{y},\mathbf{z}) \frac{\alpha_{ir}}{\mu_{ir(n_{ir}+1)}} \frac{\mu_{js(n_{js})}}{\alpha_{js}} \tag{29}$$

$$p(\mathbf{n}-e_{js},\mathbf{x}+x_{i0},\mathbf{y},\mathbf{z}) = p(\mathbf{n},\mathbf{x},\mathbf{y},\mathbf{z}) \frac{\alpha_{ix_{i0}}}{\mu_i(\mathbf{x}_i+x_{i0})} \frac{\mu_{js(n_{js})}}{\alpha_{js}} \tag{30}$$

$$\begin{aligned} & p(\mathbf{n}-e_{js},\mathbf{x}+x_{i0}-x_{iB_i},\mathbf{y}+y_{i0},\mathbf{z}+z_{i0}) \\ &= p(\mathbf{n},\mathbf{x},\mathbf{y},\mathbf{z}) \frac{\mu_{js(n_{js})}}{\alpha_{js}} \frac{\alpha_{ix_{i0}}}{\mu_i(\mathbf{x}_i+x_{i0}-x_{iB_i})} \frac{\sum_{t=1}^{R} \alpha_{y_{i0}t} p_{y_{i0}t,ix_{B_i}}}{\alpha_{ix_{iB_i}}} \frac{\mu_i(\mathbf{x}_i)}{\mu_i(\mathbf{z}_i+z_{i0})} \end{aligned} \tag{31}$$

Note that the last term on the right hand side of (31) is $\frac{\mu_i(\mathbf{x}_i)}{\mu_i(\mathbf{z}_i+z_{i0})} = 1$, because if the station i is of Type I, then the service rate is independent of the class. However, if the station is of Type

IV then $\mu_i(\mathbf{x}_i) = \mu_{ix_{iB_i}}$. Since we assume the blocked jobs are $FCFI$, then $\mu_i(\mathbf{z}_i + z_{i0}) = \mu_{iz_{i0}}$ and $x_{iB_i} = z_{i0}$. Now we substitute (27) – (31) on the right-hand side of (22) separately, and use (1) to obtain

$$\begin{aligned}
&\sum_{j\in T_A}\sum_{i\in T_2}\sum_{r=1}^{R} p\left(\mathbf{n}+e_{ir},\mathbf{x}-x_{jn_j},\mathbf{y},\mathbf{z}\right)\mu_{ir}(n_{ir}+1)\,p_{ir,jx_{jn_j}} \\
&\quad = p(\mathbf{n},\mathbf{x},\mathbf{y},\mathbf{z})\sum_{j\in T_A}\sum_{i\in T_2}\sum_{r=1}^{R}\frac{\mu_j(\mathbf{x}_j)}{\alpha_{jx_{jn_j}}}\alpha_{ir}p_{ir,jx_{jn_j}} \\
&\quad = p(\mathbf{n},\mathbf{x},\mathbf{y},\mathbf{z})\sum_{j\in T_A}\mu_j(\mathbf{x}_j) && (32)\\
& && (33)
\end{aligned}$$

$$\begin{aligned}
&\sum_{j\in T_B}\sum_{r=1}^{R} p\left(\mathbf{n}+e_{y_{jb_j}r},\mathbf{x},\mathbf{y}-y_{jb_j},\mathbf{z}-z_{jb_j}\right)\mu_{y_{jb_j}r}\left(n_{y_{jb_j}r}+1\right)p_{y_{j,b_j}r,jz_{jb_j}} \\
&\quad = p(\mathbf{n},\mathbf{x},\mathbf{y},\mathbf{z})\sum_{j\in T_B}\sum_{r=1}^{R}\frac{\mu_j(\mathbf{x}_j)}{\sum_{t=1}^{R}\alpha_{y_{jb_j}t}p_{y_{jb_j}t,jz_{jb_j}}}\alpha_{y_{jb_j}r}p_{y_{j,b_j}r,jz_{jb_j}} \\
&\quad = p(\mathbf{n},\mathbf{x},\mathbf{y},\mathbf{z})\sum_{j\in T_B}\mu_j(\mathbf{x}_j) && (34)\\
& && (35)
\end{aligned}$$

$$\begin{aligned}
&\sum_{i\in T_2, j\in T_2}\sum_{s=1}^{R}\sum_{r=1}^{R} p(\mathbf{n}+e_{ir}-e_{js},\mathbf{x},\mathbf{y},\mathbf{z})\mu_{ir}(n_{ir}+1)\,p_{ir,js} \\
&\quad = p(\mathbf{n},\mathbf{x},\mathbf{y},\mathbf{z})\sum_{j\in T_2}\sum_{s=1}^{R}\frac{\mu_{js(n_{js})}}{\alpha_{js}}\sum_{i\in T_2}\sum_{r=1}^{R}\alpha_{ir}p_{ir,js} && (36)\\
& && (37)
\end{aligned}$$

$$\begin{aligned}
&\sum_{i\in T_A, j\in T_2}\sum_{s=1}^{R}\sum_{x_{i0}=1}^{R} p(\mathbf{n}-e_{js},\mathbf{x}+x_{i0},\mathbf{y},\mathbf{z})\mu_i(\mathbf{x}_i+x_{i0})\,p_{ix_{i0},js} \\
&\quad = p(\mathbf{n},\mathbf{x},\mathbf{y},\mathbf{z})\sum_{j\in T_2}\sum_{s=1}^{R}\frac{\mu_{js(n_{js})}}{\alpha_{js}}\sum_{i\in T_A}\sum_{x_{i0}=1}^{R}\alpha_{ix_{i0}}p_{ix_{i0},js} && (38)\\
& && (39)
\end{aligned}$$

$$\sum_{i\in T_B, j\in T_2}\sum_{s=1}^{R}\sum_{y_{i0}=1}^{N}\sum_{x_{i0}=1}^{R} p(\mathbf{n}-e_{js},\mathbf{x}+x_{i0}-x_{iB_i},\mathbf{y}+y_{i0},\mathbf{z}+z_{i0})\mu_i(\mathbf{x}_i+x_{i0}-x_{iB_i})\,p_{ix_{i0},js}$$

$$= p(\mathbf{n},\mathbf{x},\mathbf{y},\mathbf{z}) \sum_{i\in T_B, j\in T_2} \sum_{s=1}^{R} \sum_{x_{i0}=1}^{R} \frac{\mu_{js(n_{js})}}{\alpha_{js}} \frac{\alpha_{ix_{i0}} p_{ix_{i0},js}}{\alpha_{ix_iB_i}} \sum_{y_{i0}=1}^{N} \sum_{t=1}^{R} \alpha_{y_{i0}t} p_{y_{i0}t,ix_iB_i}$$

$$= p(\mathbf{n},\mathbf{x},\mathbf{y},\mathbf{z}) \sum_{j\in T_2} \sum_{s=1}^{R} \frac{\mu_{js(n_{js})}}{\alpha_{js}} \left(\sum_{i\in T_B} \sum_{x_{i0}=1}^{R} \alpha_{ix_{i0}} p_{ix_{i0},js} \right) \tag{40}$$

By combining (36) – (40) we have

$$(36)+(38)+(40) = p(\mathbf{n},\mathbf{x},\mathbf{y},\mathbf{z}) \sum_{j\in T_2} \sum_{s=1}^{R} \frac{\mu_{js(n_{js})}}{\alpha_{js}} \left(\sum_{i=1}^{N} \sum_{r=1}^{R} \alpha_{ir} p_{ir,js} \right)$$

$$\mu_{js(n_{js})}. \tag{41}$$

The result follows by checking the flow-in rate equal to the flow-out rate. **QED**

Theorem 4.

The *throughput* of the ith station is computed from

$$\lambda_i = \sum_{(\mathbf{n},\mathbf{x},\mathbf{y},\mathbf{z})\in S} p(\mathbf{n},\mathbf{x},\mathbf{y},\mathbf{z}) \begin{cases} \mu_i(\mathbf{x}_i) & \text{for } i \in T_1 \\ \sum_{r=1}^{R} \mu_{ir}(r_{ir}) \left(1 - \sum_{j\in T_B} \sum_{s=1}^{R} p_{ir,js}\right) & \text{for } i \in T_2 \\ \quad + \sum_{j\in T_B} \mu_j(\mathbf{x}_j) I_{\{y_{j1}=i\}}. & \end{cases} \tag{42}$$

The *mean number of jobs* in the ith station is

$$\overline{k}_i = \sum_{(\mathbf{n},\mathbf{x},\mathbf{y},\mathbf{z})\in S} p(\mathbf{n},\mathbf{x},\mathbf{y},\mathbf{z}) \begin{cases} n_i & \text{for } i \in T_1 \\ \sum_{r=1}^{R} n_{ir} + \sum_{j\in T_B} \sum_{l=1}^{b_j} I_{\{y_{jl}=i\}} & \text{for } i \in T_2 \,. \end{cases} \tag{43}$$

Proof. Consider a state $(\mathbf{n},\mathbf{x},\mathbf{y},\mathbf{z})$. Since the jobs from stations of T_1 can never be blocked, thus the throughput of such station $i \in T_1$ is $\mu_i(\mathbf{x}_i)$. For stations in T_2 we have the throughput rate as the service rate $\sum_{r=1}^{R} \mu_{ir}(n_{ir})$. However, the job may choose the next station in T_B and a blocking may occur. If station $i \in T_2$ has blocked jobs which has the first priority to get into the destination station that caused their blocking, then the throughput rate of those jobs from station i is equal to the service rate of the station which caused blocking. Thus (42) follows.

The *mean number of jobs* is also obtained from given a state $(\mathbf{n},\mathbf{x},\mathbf{y},\mathbf{z})$. The number of jobs in station i for $i \in T_1$ is n_i. If $i \in T_2$, then the number of jobs in station i will include the active jobs of all classes $n_{i1} + \cdots + n_{iR}$ and the blocked jobs in that station. Thus, we take sum over the stations that cause blocking, $j \in T_B$, and check the vectors $\mathbf{y}_j$ if $y_{jl} = i$, then the lth job is located in station i, so that the result (42) follows. **QED**

5 Conclusions

We derived exact product form solution for the equilibrium state probabilities and computed throughput and mean number of jobs in two different models. We used a property of Type II and III stations that the locations of the blocked jobs has no effect on the service rates in those stations. Thus, we used this type of stations as kind of *optional* storage spaces for the blocking station. In other words, the system is treated as a virtual *nonblocking* system. By the well-defined state space, particularly the vector of the locations for the blocked jobs, we obtain the exact product form solutions. Another observation is that by the construction of the models we always have the constraint that no more than two jobs will move on a single departure event. The first model has a stronger constraint that is only one of the stations will cause blocking at a time. However, for the second model we allow more than one station to be of Type I or IV which may cause blocking at the same time. This is due to the constraint of the neighbors which will not cause blocking.

REFERENCES

[1] I. F. Akyildiz, "Exact Product Form Solution for Queueing Networks with Blocking", *IEEE Transactions on Computers*, Vol. C-36, No. 1, January 1987, pp. 122-125.

[2] I. F. Akyildiz, "On the Exact and Approximate Throughput Analysis of Closed Queueing Networks with Blocking", *IEEE Transactions on Software Engineering*, Vol. SE-14, No. 1, January 1988, pp. 62-71.

[3] I. F. Akyildiz and H. G. Perros, "Special Issue on Queueing Networks with Finite Capacity Queues: Introduction", *Performance Evaluation*, Vol. 10, No. 3, Dec. 1989.

[4] I. F. Akyildiz and H. von Brand, "Exact Solutions for Networks of Queues with Blocking-After-Service", *Technical Report, Georgia Tech, College of Computing, April 1991, GIT-COC-91-26.*

[5] I. F. Akyildiz and J. Liebeherr, "Optimal Deadlock Free Buffer Allocation in Multiple Chain Blocking Networks of Queues", *Proc. of the Int. Conference on the Performance of Distributed Systems and Integrated Communication Networks*, September 10-12, 1991.

[6] S. Balsamo, V. De Nitto Persone and G. Iazeoalla, "Identity and Reducibility Properties of Some Blocking and Non-blocking Mechanisms in Congested Networks", *Flow Control of Congested Networks*, NATO ASI Series, Springer Verlag, 1987, pp. 243-254.

[7] S. Balsamo and L. Donatiello, "On the Cycle Time Distribution in a Two-Stage Cyclic Network with Blocking", *IEEE Transactions on Software Engineering*, Vol. 15, No. 10, October 1989.

[8] F. Baskett, K. M. Chandy, R. R. Muntz and F. G. Palacios, "Open, Closed and Mixed Networks of Queues with Different Classes of Customers", *Journal of the ACM*, Vol. 22, Vol. 15, pp. 248-260, April 1975.

[9] P. P. Bocharov, "On the Two-Node Queueing Networks with Finite Capacity", *Proc. of the First Int. Workshop on Queueing Networks with Blocking*, North Holland, Dec. 1989, pp. 105-125.

[10] F. P. Kelly, "Networks of Queues with Customers of Different Types", *Journal of Applied Probability*, Vol. 12, pp. 542-554, 1975.

[11] F. P. Kelly, "Networks of Queues", *Advances in Applied Probability*, Vol. 8, No. 2, pp. 416-432, June 1976.

[12] R. O. Onvural, "A Note on the Product Form Solutions of Multiclass Closed Queueing Networks with Blocking", *Performance Evaluation*, Vol. 10, No. 3, December 1989, pp. 247-255.

[13] R. O. Onvural, "A Survey of Closed Queueing Networks with Finite Buffers", *ACM Computing Surveys*, Vol. 22, No. 2, June 1990, pp. 83-121.

[14] R. O. Onvural and H. G. Perros, "Some Equivalencies between Closed Queueing Networks with Blocking", *Performance Evaluation Journal*, Vol. 9, Dec. 1988, pp. 111-118.

Queueing Networks with Finite Capacity
R.O. Onvural and I.F. Akyildiz (Editors)

Sparse L-matrix computations : a new approach

J. Pellaumail

I.N.S.A. 20, avenue des Buttes de Coësmes - 35043 RENNES Cédex - France

Abstract

This paper is a survey of a graph theory approach for studying and computing solution y of linear system y m = x where m is some L-matrix. Two complementary ways are emphasized. The first one is the close relation between some techniques used in probability theory (imbedded process, taboo probabilities, ...) and some classical methods in numerical analysis (splitting, decomposition of M-matrices, ...). The second one rests upon "fluvial networks theorem" that gives an explicit solution q of some systems q m = 0. Message loss in asynchronous transfer mode (ATM) networks is especially studied.

A. BASIC THEOREMS

A.1. Introduction

Let m be a square matrix and x a unirow vector. Computing a solution y of the linear system y m = x is a fundamental step in a very large number of problems arising in physical science, engineering and economy. When m is large, this computation is far from being obvious. We focus our attention on such linear systems when m is some L-matrix, that is, the diagonal components m(i,i) of m are nonnegative and off-diagonal components m(i,j), $i \neq j$, of m are nonpositive.

Actually, we will consider only diagonally dominant matrices m, that is, (m s) is a nonnegative matrix where s is the unicolumn vector all the entries of which are equal to 1 (s is the vector associated with summations).

Such a linear system y m = x corresponds to an equilibrium between y(i) m(i,i), that may be considered as the outcoming stream from state i, and $x(i) - \sum_{j \neq i} y(j)\, m(j, i)$ that may be considered as the incoming stream in state i (let us recall that m(j,i) is nonpositive for $i \neq j$).

There are several areas where such linear systems play a prominent role : economy, electric networks, partial linear differential equations, Markov processes, ... For example, it is clear that there is an equilibrium between some economic streams. Such economic balances have been extensively studied by Leontief. They lead to linear systems y m = x where m is usually a μ-matrix as defined in section A.5 below. Actually several possible Leontief formulations may be stated for the same economic setting (for more details see [Sen]).

Electrical network theory is formulated in terms of two variables, called current and voltage, associated with each network element called a network branch. Let i be the unirow vector associated with the currents of the branches and v be the unirow vector associated with the voltages of the branches. These matrices satisfy linear systems associated with Kirchoff's current law, Kirchoff's voltage law and generalized Ohm's law. More precisely we have $i\ b = 0$, $v\ c = 0$ and Ohm's relations where b and c are π-matrices as defined in section A.5 below. Graph theory is a classical approach for solving such linear systems ([Che]).

Almost all physical problems lead to partial differential equations and a lot of work has been done to solve such equations. Some of which is devoted to the solution of linear partial differential equations by means of finite difference methods ; in such methods, one step is to solve a linear system $y\ m = x$ where m is a μ-matrix as defined in section A.5 below. It is for this reason that the notions of L-matrix, M-matrix, splitting, ... were introduced in numerical analysis.

A.2. Markov process

Let S be a finite set. An S-valued Markov process X is a process such that, for any time t, the evolution of X after time t depends on the past of X only through state X(t) of X at time t. This is not a "physical hypothesis" but a mathematical choice of the model and, especially, of state set S.

When, for some state set, the evolution is not markovian because some transition delays from state i to state j are not "exponentially distributed" it is possible to obtain a Markov model by introducing fictitious states. Several such models are now classical like Cox "positive" distribution, PH distribution, Cumani distribution, ... Actually, all these models are theoretically equivalent (see section C.2 below).

Scientific engineering abounds with examples for which Markov processes are successful models. In telephone theory and computer modelling, S has frequently to be chosen finite but large. Moreover, in such contexts, the evolution is very fast and the fundamental problem is to study and to compute the steady-state probabilities.

More precisely, let us consider an S-valued Markov process X(t) with a continuous-time parameter t. Let us define $a(i,i) := 0$ and, for $i \neq j$:

$$a(i,j) := \lim_{h \downarrow 0} \left\{ \frac{1}{h} \text{Prob}[\, X_{t+h} = j \mid X_t = i \,] \right\} \tag{1}$$

and let us assume that matrix a does not depend on t and is irreducible (see A.5 below). Then we have the following theorem :

Theorem : There exists a unique unirow vector p such that the following two properties are fulfilled :

(i) $\sum_{i \in S} p_i = 1$

(ii) for every element i of S : $p(i) \sum_{j \in S} a(i,j) = \sum_{j \in S} p(j)\, a(j,i)$

Two different direct proofs of this basic theorem are given in section A.7 and in section C.3 below.

The family p(i) where i belongs to S is the steady-state probability of process X and corresponds to an equilibrium between the first member in relation (ii), which is the outcoming stream of state i, and the second member in relation (ii) which is the incoming stream in state i.

Let us recall that s is the unicolumn vector the components of which are all equal to 1 and let us introduce square diagonal matrix d defined by :

$$d(i,i) := \sum_{j \in S} a(i,j) \tag{2}$$

Relations (i) and (ii) may be written respectively p s = 1 and p(d–a) = 0. Matrix (d–a) is an L-matrix such that (d–a) s = 0 : in fact, usually (d–a) is a π-matrix as defined in section A.5 below.

A.3. Irrigation and draining : definitions

Let m be a square matrix and let us introduce some definitions. We shall say that u m-irrigates v or, equivalently, that v m-drains u iff u = v or if there exists a sequence $\{ w(i), 1 \leq i \leq n \}$ such that u = w(1), v = w(n) and, for every i, $m(w(i), w(i+1)) \neq 0$. Let us consider the directed graph associated with m, that is, there is an edge from u to v iff $m(u,v) \neq 0$.Then v m-drains u iff it is possible to go from u to v along this directed graph.

We shall say that v m-drains F iff, for every element u of F, v m-drains u. We shall say that G m-drains F iff, for every element u of F, there exists an element v of G that m-drains u.

Let us recall that m is irreducible iff, for every pair (u,v), v m-drains u.

All the above definitions do not depend on the diagonal of m.

A.4. Fluvial networks theorem

Let a real function a, defined on $(S \times S)$, be a square matrix. Let us assume that, for every element v of S, a(v,v) = 0. For every element u of S, let T(u) be the set of all the "fluvial networks with terminal point u", i.e. all the mappings t from S to S such that t(u) = u and $v \neq u$ implies $t^k(v) \neq v$.

Let us define d as in section A.2 and let :

$$q(u) := \sum_{t \in T(u)} \prod_{v \in S, v \neq u} a(v, t(v)) \tag{3}$$

Then we have : q(d–a) = 0

Moreover $q(u) \neq 0$ iff u a-drains the whole space S.

A.5. μ-matrix and π-matrix : definitions

Let us recall that a square matrix m is an L-matrix iff, for every pair (u,v) with $u \neq v$, we have $m(u,u) \geq 0$ and $m(u,v) \leq 0$.

When ms ≥ 0, m is called diagonally dominant. For every diagonally dominant L-matrix m, let us call O(m) the set of the elements u such that :

$$\sum_{v \in S} m(u,v) > 0$$

We shall say that m is a μ-matrix if m is a diagonally dominant L-matrix such that O(m) m-drains the whole space S. In other words, m is a nonsingular diagonally dominant M-matrix (see [GoM], [GoV] or [BPl]. We shall say that m is a π-matrix if m is an irreducible L-matrix such that m s = 0. Particularly, when the positive matrix a is irreducible, (d–a) is a π-matrix where d is defined as in section A.2.

Verifying that certain matrices satisfy the above properties is easily checked by computers. We shall construct in the sequel some submatrices that theoretically satisfy the above properties. It is important that it is easy to check that these properties hold in spite of the roundoff errors.

A.6. Inversion of a μ-matrix

Theorem : Let a real mapping m, defined on $(S \times S)$, be a μ-matrix. Then m is nonsingular and the inverse of m is a nonnegative matrix. Moreover $m^{-1}(u,v)$ is positive when v m-drains u.

This theorem may be stated via theorem A.4.

A.7. Existence and uniqueness of p

Let us go back to the theorem given in section A.2, that is, there exists a unique unirow vector p such that p s = 1 and p(d–a) = 0 when a is an irreducible nonnegative matrix and d is the diagonal matrix associated with a such that (d–a) s = 0 ; moreover this vector p is positive.

This theorem is the first basic theorem on stationary finite set valued Markov processes. The irreducibility of a implies that all the entries of vector q constructed in section A.4 (via fluvial networks theorem) are positive. Thus, by normalization, we obtain the existence of p and we have only to prove its uniqueness.

That is easy to check by introducing a μ-matrix.

Remarks

a) Section A.6 shows that for studying a μ-matrix m one may introduce a matrix m' such that m's' = 0. Conversely, in section A.7 a μ-matrix is introduced for studying a π-matrix : both methods have a graph interpretation.
b) Fluvial networks theorem in section A.4 gives the existence and the uniqueness of p.

B. DECOMPOSITION THEOREMS

B.1. Introduction

Let X be an S-valued Markov process ; let a and d be the associated matrices as defined in section A.2. In general, and in particular for modelling computer or telephone systems, set S is very large and it is not possible to compute directly vector p such that p s = 1 and p(d–a) = 0.

Theoretically, for every nonnegative vector p' such that p' s = 1, we have :

$$p = \lim_{n \to \infty} p'(a\,d^{-1})^n \tag{4}$$

Perron-Frobenius theory and Birkhoff's contraction coefficients are two ways for proving the convergence of these series (see section C.3 below) ; actually, usually this convergence is very slow and relation (4) is not tractable for computing p. That implies that the Gauss-Seidel method and other related methods are also inefficient.

Moreover, computing p by simulation is closely related to relation (4) : namely, relation (4) tractable means that simulation is efficient and one has to choose simulation which is manageable and needs only a small memory. It is just why steady state probabilities are very often computed by simulation but the remark above holds for the other previous examples. From the theoretical point of view there is no difference between simulation and the Monte-Carlo method.

Sometimes, direct simulation is not usable, especially when some crucial states have a low probability. Usually, matrix a is very large but it is sparse : so almost all efficient methods rest upon a decomposition of the initial problem (or matrix) into smaller ones. The oldest and most classical decomposition is so-called splitting in numerical analysis and imbedded process in Markov process theory.

B.2. Splitting and imbedded process

Theorem : Let a, b, c, d be some square matrices with a = b+c and (d–c) nonsingular. Let x be a unirow vector. Let us define :

$r := (d-c)^{-1} b$ and $z := y(d-c)$

Then we have $y(d-a) = x$ iff $z = z\,r + x$.

Comments

a) In numerical analysis the decomposition a = b+c is called splitting.

b) A slightly more manageable theorem is given in section C.4 below.

B.3. Taboo probabilities

Theorem : Let $x = (x_1, \ldots, x_n)$ and

$$g = \begin{pmatrix} d(1) & & & \\ & (-w'(k)) & & \\ & & & \\ & d(k) & & \\ & -v(k) & & \\ & & & \\ & (0) & & d(n) \end{pmatrix}$$

meaning that g is a block Hessenberg matrix.

Let us define : $r'_2 := x_1\, d'_1$; $t'_2 := v_1\, d'_1$ and, for $2 \le k \le n$: $h_k = d_k - t'_k\, w'_k$.

Let u_k be the identity matrix ; let us assume that matrices h_k are nonsingular and let us define :

$r'_{k+1} := (r'_k\,,0) + (x_k + r_k\, w'_k)\, h_k^{-1}(t'_k\,, u_k)$

$t'_{k+1} := v_k\, h_k^{-1}(t'_k\,, u_k)$; $y := r'_{n+1}$

Then we have : $y\,g = x$

Moreover, if g is a μ-matrix, matrices h_k are μ-matrices too.

Comments : This theorem is proved in [Pel-4] ; let us give some comments. Usually, matrices h(k) are block Hessenberg matrices and the algorithm may be used again. In this algorithm, k increases from 1 to n and, for k = n, the solution y is computed ; on the other hand, for the proof, k has to be considered decreasing from n to 1.

Actually, when g is an Hessenberg matrix, this algorithm is close to the classical gaussian elimination without pivoting : then, what is the contribution of this theorem ?

At first, at each step, the heart of the algorithm is more transparent. In particular, matrices h(k) are μ-matrices and it is possible to check this property to be sure that the roundoff errors are not too large (see the end of section A.5). Moreover, this study and the convex method (see section B.4) below) show that the elimination order plays a prominent role with respect to the stability of the algorithm ; this study just gives us a way to choose a "good" elimination order and this is not a trivial problem in queueing networks according to a lot of possible block Hessenberg matrices asociated with a given problem.

This algorithm is closely related to the ones used in [Neu], [LJG] and [Leb] but, in these works, g is a μ-matrix. Let us recall that in the probabilist context, at the first step, g is a μ-matrix but, usually, by decomposition, at the second step, the inversion of a μ-matrix is needed.

Actually, in the probabilist context, the heart of this way has been studied - from a theoretical point of view - by a lot of mathematicians 25 years ago. The key to such studies rests upon the notion of taboo probabilities (see [Rev]). In the context considered here , g is not a π-matrix but a μ-matrix : thus it is not possible to use the classical probabilist arguments (imbedded process, taboo probabilities). In other words, in our context, the global way is analoguous to the one used by Neuts but the tools needed, namely the properties of the μ-matrices, are quite different.

B.4. Convex method

Convex method (see [Ast], [BeP], [Pel-1], [Sag], etc...) is a decomposition method (as above) which uses the convexity of the set of some solutions of some sub-system of the initial linear system. Let us recall (see [OPS]) a family of hypotheses for which this method holds.

Hypotheses : Let a nonnegative mapping a defined on $(S \times S)$ be an irreducible matrix. Let us assume that

$S = (H \times K)$ where $K := \{ k : 0 \leq k \leq n \}$

Let (H', H") be a partition of H such that card(H') / card(H") is small (in [BeP], H' has only one element).

Let us assume that the following two properties hold :

(i) $h \in H, h' \in H, 1 < j \leq k \leq n \Rightarrow a((h, k), (h', k-j)) = 0$
(a is a block Hessenberg matrix)

(ii) $h \in H, h'' \in H'', 0 < k \leq n \Rightarrow a((h, k), (h'', k-1)) = 0$

Algorithm : Let d be the diagonal matrix as defined in section A.2 and p be the unirow vector such that p s = 1 and p(d–a) = 0. Let x_k be the unirow vector associated with p(. ,k) ; in other words : $x_k(h) := p(h,k)$

Let us define :

$$z_k(h) := \sum_{i \in H} p(i, k+1)\, a[(i, k+1), (h,k)]$$

$x_k = (x'_k , x''_k)$ and $z_k = (z'_k , z''_k)$: block notation associated with H= (H',H").

For every initialization of x'_0, one may compute z'_n ; that gives $z'_n = x'_0\, \omega$; ω is not an L-matrix but ω is nonsingular and monotone.

Here is the heart of the algorithm : generally ω is ill-conditioned and x'_0 is almost equal to any row of w^{-1} ; namely x'_0 belongs to a "small" convex.

The theoretical proof of this last property is rather difficult and needs fluvial networks theorem (see [Pel-1]).

The particular case for which H'_k has only one element (for every integer k) is extensively studied in [BeP]. It is correlated to Neuts' works (see [Neu]) and gives an efficient way for computing the steady state probabilities of PH/M/1 and M/PH/1 queueing systems and the single queue with buffer.

C - MORE ON MATHEMATICS

C.1. Fluvial networks theorem

This theorem is expressed in section A.4. Let us sketch the proof (for more details see [Pel-3]).

For every element u of E, let R(u) be the set of all the "fluvial networks r the sea of which is a ring including u", i.e. of all the mappings r from E into E such that :

(i) $r(u) \neq u$

(ii) $\exists\, k$ with $r^k(u) = u$

(iii) $r^i(v) = v$ implies $\exists\, j$ such that $r^j(v) = u$

We have :

$$q(u) \sum_{v \neq u} a(u,v) = \sum_{r \in R(u)} \prod_{v \in E} a(v, r(v))$$

$$= \sum_{v \neq u} q(v)\, a(v, u)$$

The first equality is obtained by "cutting" the rings just "after u" ; the second equality is obtained by "cutting" the rings just "before u".

C.2. PH and positive Coxian distribution are equivalent

Let us recall (see [Neu]) that a continuous probability distribution F(.) on $[0,\infty]$ is of phase type (PH-distribution) if it is the distribution of the time until absorption in an S-valued Markov process X with a single absorbing state, where S is a finite set. In [Cum] F is said to be representable by a Multistate Homogeneous Markov Model (MHMM). Let us define :

$$a(i,j) := \lim_{h \downarrow 0} \frac{1}{h} \text{Proba}[\, X_{t+h} = j \mid X_t = i \,]$$

$$d(i) := \sum_{j \in S} a(i,j) \; ; \quad q(i) := \text{Prob}[\, X_0 = i \,]$$

Y(i) is a random variable associated with the time until absorption when the process is in state i at the initial time : in other words, the distribution of Y(i) is the distribution of $(\tau - t)$ where τ is the following stopping time :

$$\tau := \inf. \{\, t' : t' > t \,,\, X_{t'} \text{ is absorbed } \mid X_t = i \,\}$$

D(i) is a random variable the distribution of which is exponentially distributed with parameter d(i).

For every element i of S we have :

$$Y(i) = D(i) + \sum_j a(i,j) \frac{Y(j)}{d(i)} \tag{6}$$

Let u be the identity matrix and r be the square matrix defined by :

$$r(i,i) := 0 \text{ and, for } i \neq j \,,\; r(i,j) := \frac{a(i,j)}{d(i)}$$

Let q be the unirow vector associated with the family q(i), i in S.
Relations (6) may be written

$$Y = D + r\,Y$$

where Y and D are considered as unicolumn vectors ; that implies

$$(u-r)\,Y = D$$

(u–r) is a μ-matrix and is nonsingular ; that implies :

$$Y = (u-r)^{-1} D$$

where $(u-r)^{-1}$ is a nonnegative matrix. Then F is the distribution of the following random variable :

$$Z = \sum_{i \in S} q_i \; Y(i) = q(u-r)^{-1} D$$

Namely, Z is a positive combination of exponentially distributed random variables D(i), i in S. Such a random variable Z is a special case of the one representable by a Triangular MHMM and [Cum] (see lemma 1) shows that such a random variable Z has a "positive" Coxian distribution (for a proof with a formulation more mathematical than Cumani's one see [Pel-3]).

C.3. Contraction coefficient

Theorem : For every pair (x,y) of nonnegative unirow vectors let us define :

$$\delta(x,y) := \sup_{i,j}. \{\, |x_i\, y_j - x_j\, y_i| \,/\, |x_i\, y_j + x_j\, y_i| \,\}$$

Let S be a finite set. Let M be the set of all nonnegative mappings m defined on $(S \times S)$ and considered as square matrices.

For every element m of M let us define :

$$\gamma(m) := \operatorname*{Sup.}_{i,j,u,v} \{|m_{i,u}\, m_{j,v} - m_{i,v}\, m_{j,u}| \,/\, |m_{i,u}\, m_{j,v} + m_{i,v}\, m_{j,u}|\}$$

Then γ is a contraction coefficient, that is, the four following properties are fulfilled :

(i) γ is a [0,1] -valued mapping defined on M
(ii) a and b elements of M imply $\gamma(a\,b) \le \gamma(a)\,\gamma(b)$
(iii) x and y nonnegative vectors and a element of M imply
$\delta(x\,a, y\,a) \le \delta(x, y)\,\gamma(a)$
(iv) a positive matrix implies $\gamma(a) < 1$.

The proof of this theorem is stated in [Pel-3] .
This section C.3 gives a way for proving theorem in section A.2 different from the one given in section A.7. For more details see [Pel-3].

C.4. More on splitting

Let $f = \begin{pmatrix} r & -u \\ -v & w \end{pmatrix}$.

Let us assume that w is nonsingular and let :

$g := r - u\,w^{-1}\,v$; if g is nonsingular, we have :

$$f^{-1} = \begin{pmatrix} g^{-1} & g^{-1}\,u\,w^{-1} \\ w^{-1}\,v\,g^{-1} & w^{-1} + w^{-1}\,v\,g^{-1}\,u\,w^{-1} \end{pmatrix}$$

If f is a μ-matrix, r, w and g are μ-matrices too.
There is an analoguous result when f is a π-matrix.
This "theorem" is easy to check. In general for studying Markov processes it is more tractable than theorem given in section B.2. It is needed in section B.4.

C.5. On the ergodicity

Proposition : Let $a = \begin{pmatrix} a(1) & a(2) \\ a(3) & a(4) \end{pmatrix}$, $c = \begin{pmatrix} c(1) \\ c(2) \end{pmatrix}$, $s = \begin{pmatrix} s(1) \\ 1 \end{pmatrix}$

and $p = (p(1), p(2))$

where a(4), c(2) and p(2) are real numbers and a(1) is a square matrix. Let u be the identity matrix associated with a(1). Let us define $g := u(1) - a(1)$ and let assume that g is nonsingular, $p\,a = p$ and $p\,s = 1$. Then we have :

$p\,c = [a(3)\,g^{-1}\,c(1) + c(2)] \,/\, [a(3)\,g^{-1}\,s(1) + 1]$

and $p(2) = 1/\rho$ where

$\rho := a(3)\,g^{-1}\,s(1) + 1$.

C.6. Generalized imbedded process decomposition

Let G be a finite set ; let (F', F, F") be a partition of G. Let us consider nine matrices r', u', w', v, r, w, w", u" and r" indexed respectively by $F' \times F'$, $F' \times F$, $F' \times F''$, $F \times F'$, $F \times F$, $F \times F''$, $F'' \times F'$, $F'' \times F$ and $F'' \times F''$. Let :

$$g := \begin{pmatrix} r' & -u' & -w' \\ -v & r & -w \\ -v'' & -u'' & r'' \end{pmatrix}$$

Let y', y and y" be three unirow vectors indexed respectively by F', F and F".
<Let S be the following linear system :

$$S := (y', y, y'')\, g = 0$$

Let us assume that g is a μ-matrix ; then r is a π-matrix ; let us denote by m the inverse of r. Matrices $(r'-u'\ m\ v)$ and $(r''-u''\ m\ w)$ are μ-matrices and let us call m' and m" their inverse matrices respectively. Let us define :

$$h' := m'(w'+u'\ m\ w) \quad \text{and} \quad h'' := m''(v''+u''\ m\ v)$$

h"h' is an irreducible stochastic matrix (associated to some kind of generalized imbedded process) ; let z' be a nonnull unirow vector such that $z' = z'h''h'$. Let us define :

$$y' := z'\ h''\ m'' \ , \quad y'' := z'\ m'' \quad \text{and} \quad y := (y'\ u' + y''\ u'')\ m$$

Then (y', y, y") is a solution of system S.
The proof of this theorem is stated in Pel-5.

D. CELLS LOSS AND BUFFER

D.1. Introduction

In several transmission networks the incoming messages are stocked in some buffer until they may cross some multiplexer. When the buffer is full the new incoming messages are lost and the following fundamental question arises: how to calculate the buffer size for a given maximum loss ? In general this is done by simulation which is not efficient when the authorized loss is very small. For example, in the asynchronous transfer mode (ATM) the maximum loss is of the order of 10^{-9} and simulation is not available.

[Leb] is a masterly survey on ATM and especially on mathematical studies in this area. Let us recall some ATM properties needed in the sequel. All kinds of informations (voice, T.V., data transfers, ...) are sliced in short fixed-length packets called cells and cells of a same message are not carried together across the network.

Each channel has a framing structure : a typical size for one frame is 125 μs. Each frame is divided into a number of slots : a slot is the time needed to transmit a cell (as defined above). The slots' number n per trame is not fixed yet and depends on the technology. At each trame, n cells may transit via the considered channel. Let us remark that [Leb] does not take into account this framing structure.

Elementary computations show that, for a random incoming stream, a loss rate less than 10^{-8} for all cells imposes a traffic rate less than 1/2 (actually far less) which is unsatisfactory. Thus, to obtain a better traffic rate the loss rate less than 10^{-9} is to be reserved to some priority cells. In the sequel, all cells (having or not priority) are assumed to be carried on common paths.

From this viewpoint, a new question arises : how to compute the loss rate of the cells of low-priority with respect to the traffic rate for an imposed loss rate of the cells of high-priority. That is just the problem studied in [Leb]. Actually there are other mathematical works in this area but, in our opinion, [Leb] introduces the most efficient way.

Namely [Leb] assumes that the input cell stream is a Markov process modulated by a Markov chain : our model is very similar. The mathematical hypotheses of this model are given in section D.2 below and their "physical meaning" is given in section D.3.

D.2. Mathematical model

This mathematical model is an S-valued homogeneous Markov chain X, that is, a discrete time process. Let n, n' and n" be three positive integers. Set S is the set of all pairs (j,h) of nonnegative integers with $j \leq n'$ and $h \leq n''$.

Let J be the set of the nonnegative integers j with $j \leq n'$. Let b and b' be two real functions defined on J with :

$b'(0) = 0$, $b(n') = 0$, b' and b positive in all other cases and $b(j) + b'(j) < 1$ in all cases.

Let f be the function defined on S as follows :

$$f(j,h) := \inf. \{ n'', \sup. \{0,(h+j-n)\} \}$$

Let a be the function defined on S×S by :

$$\begin{aligned} a((j,h), (j',h')) &= 1 - b(j) - b'(j) \quad \text{if} \quad j' = j \quad \text{and} \quad h' = f(j,h) \\ &= b(j) \quad \text{if} \quad j' = j+1 \quad \text{and} \quad h' = f(j,h) \\ &= b'(j) \quad \text{if} \quad j' = j-1 \quad \text{and} \quad h' = f(j,h) \\ &= 0 \quad \text{in all other cases.} \end{aligned}$$

The matrix a is stochastic and drives the evolution of X, that is :

$$a(e,e') = \text{Proba} [X_{t+1} = e' \mid X_t = e]$$

D.3. Experimental model

Let us give the relation betwen the mathematical model stated in section D.2 above and the framing structure explained in section D.1. The delay between time t and time (t+1) corresponds to one frame : actually, if the slots' number per frame is very large (more than 500), this delay would correspond to several frames, but for the simplicity of the exposition, we will assume that the unit of time is one frame.

For a state (j,h) (an element of S), j is the number of active sources and h is the number of cells in the buffer ; n" is the buffer size ; n' is such that one may neglect the case where $j > n'$; n is the slots' number per frame. We have $n' > n$ and n small with respect to n". The state is observed at the beginning of a frame and we assume that the number of active sources may change only just before this observation : that is not a restricting hypothesis. Thus, if the state is (j,h) at time t (where t is the beginning of a frame), it is (j', f (j,h)) at time (t+1) because n cells may be sold during a frame.

Now we have to explain the evolution of the "modulator" j which is a Markov chain Y such that :

$$\mathrm{Prob}\,[Y_{t+1} = j+1 \mid Y_t = j] = b(j)$$

$$\mathrm{Prob}\,[Y_{t+1} = j-1 \mid Y_t = j] = b'(j)$$

$$\mathrm{Prob}\,[Y_{t+1} = j \mid Y_t = j] = 1 - b(j) - b'(j)$$

$$\mathrm{Prob}\,[Y_{t+1} = j' \mid Y_t = j] = 0 \quad \text{for } |j-j'| > 1 \;.$$

Namely b(j) and b'(j) are the probabilities to have one active source more or less respectively during a frame knowing that the number of active sources is j at the beginning of the frame. The probability to have more than one change in the number of active sources during a frame is assumed to be negligible which is not a restricting hypothesis : indeed b(j) and b'(j) are small. Examples of functions b(j) and b'(j) will be given in section D.6.

D.4. Equilibrium probability and loss

Let p be the positive unirow vector indexed by S the components of which are the steady-state probabilities of process X. We have p(u–a) = 0 and (u–a) is a π-matrix where u is the identity matrix associated with a.

Let γ be the nonnegative function defined on S as follows :

$$\gamma(n+j, n'') = j \quad \text{for } j \text{ positive}$$

$$\gamma(j,h) = 0 \quad \text{in all other cases.}$$

The problem is to compute the following quantity :

$$\nu := \sum_{e \in S} p(e)\,\gamma(e)$$

ν is the expectation of the number of lost cells per frame when the system is in equilibrium.

To solve the linear system p = p a and p s = 1 - which gives ν - one may use the classical Gauss elimination without pivoting (see section B. 3 above) but that needs a lot of computation. An alternative way is to use theorems in section C.5 and C.6 : the associated algorithm is very fast, needs a few computations but is somewhat mathematically difficult to justify ([Pel-5]).

D.5. Modulator evolution

The precise mechanisms used in ATM networks are not fixed yet. Several models and priority mechanisms have been suggested and studied in previous papers (for a detailed bibliography on this subject , see [Leb]). The algorithm explained above is mathematical and available for almost all these mechanisms, possibly with some modifications.

Let us give an example of a mechanism which seems to be realistic and for which the previous algorithm is efficient. Actually, we have only to explain the evolution of the modulator.

Let n(1) and n(2) be two integers with $0 \le n(1) \le n(2)$. Let λ, λ' and λ'' be three positive real numbers with $\lambda' < \lambda$. Let us assume that :

(i) for every j, $b'(j) = j\,\lambda''$

(ii) for $j < n+n(1)$, $b(j) = \lambda' + \lambda\,[n+n(2) - j]$

(iii) for $j \ge n+n(1)$, $b(j) = \lambda'$

Condition (i) is clear. Let us explain conditions (ii) and (iii) ; n+n(2) is the number of sources which are authorized to use the considered channel ; among them, j are active ; λ is the probability for one authorized nonactive source to become active during one frame.

Moreover, some sources have priority : λ' is the probability that one priority source becomes active during a frame.

At last, at the beginning of each frame, if $j \ge n+n(1)$, a mechanism prevents the authorized nonactive sources from becoming active, except for sources having priority.

D.6. Application

Let us give an example of a family of parameters : n = 40, n(1) = 3, n(2) = 12, $\lambda = \lambda'' = 5.10^{-3}$, $\lambda' = \lambda/10$. Let us define

$$q(j) := \sum_h p(j,h)$$

q(j) is the probability to have j sources active

$$\text{flow} := \sum_j j\, q(j)$$

flow is the expectation of the number of cells carried per frame and flow/n is the nominal capacity.

$$\text{asp} := \sum_{j \ge n+n(1)} q(j)\,[n+n(2) - j]\,\lambda$$

asp is the expectation per frame of the number of Authorized Sources which are Prevented from becoming active : asp/flow is the rate of such a loss.

For the previous family of parameters, the nominal capacity is 65 % and asp/flow is equal to $1{,}56.10^{-9}$. These quantities do not depend on the size n'' of the buffer.

Now, let us define

$$\text{nlc} := \sum_{j,h} p(j,h)\,(n''+n-j-h)^+$$

where $x^+ = x$ if $x \ge 0$ and $x^+ = 0$ if $x \le 0$.

nlc is the Number of Lost Cells per frame ; ρ := nlc/flow is the rate of such a loss.

The following table gives some values of this rate ρ close to 10^{-9} for some values of the size n" of the buffer :

n"	154	156	157	158
ρ	$1.1*10^{-9}$	$1.01*10^{-9}$	$9.7*10^{-10}$	$9.3*10^{-10}$

[BeP] **R. Bellamine et J. Pellaumail,** "Opérateurs positifs et probabilités stationnaires", RAIRO, Recherche Opérationnelle, n° 3, 1989.

[BoM] **R. Bott et J.P. Mayberry,** "Matrices and trees", Economic activity analysis", Wiley, 1954.

[BPl] **A. Berman and R.J. Plemmons,** "Nonnegative matrices in the mathematical sciences", Academic Press, 1979.

[Che] **W.K. Chen,** "Applied graph theory", North-Holland, 1971 and 1976.

[Chu] **K.L. Chung,** "Markov chains with stationary transition probabilities", Springer-Verlag, 1967.

[Cum] **A. Cumani,** "On the canonical representation of homogeneous Markov processes modelling failure-time distributions", Microelectron. Reliab. Vol. 22, n° 3, pp. 583-602, 1982.

[GoM] **G.H. Golub et G.A. Meurant,** "Résolution numérique des grands systèmes linéaires", Eyrolles, 1983.

[GoV] **G.H. Golub et C.F. Van Loan,** "Matrix computations", The Johns Hopkins University Press, 1983.

[Leb] **J.Y. Le Boudec,** "An efficient solution method for Markov models of ATM links with loss priorities", IEE J. Com. vol. 9, n° 3, 1991.

[LJG] **G. Latouche, P.A. Jacobs, D.P. Gaver,** "Finite Markov chains skip free in one direction", Naval Research Logistics Quarterly, vol. 31, pp. 571-588, 1984.

[Neu] **M.F. Neuts,**
1 - "Matrix-geometric solutions in stochastic models", The Johns Hopkins University Press, London, 1981.
2 - "Structured stochastic matrices of M/G/1 type and their applications", Marcel Dekker, 1989.

[OPS] **B. Oumehdi, J. Pellaumail et H. Saggou,** "Monotone ill-conditioned matrices and stationary probabilities", Congrès d'Analyse Numérique, juin 1990.

[Pel] **J. Pellaumail,**
1 - "Graphes et probabilités stationnaires", Ann. Inst. Henri Poincaré, vol. 26, n° 1, 1990.
2 - "Décompositions de L-matrices et probabilités stationnaires", C.R.A.S., T. 312, série I, pp. 459-462, 1991.
3 - "Systèmes markoviens discrets et applications", Rapport INRIA n° 1447, 270 pages, 1991.
4 - "Block-Hessenberg M-matrices and taboo probabilities", Computer and Information Sciences VI, Elsevier, pp. 469-477, 1991.
5 - "Décomposition de M-matrice et buffer d'un multiplexeur ATM", preprint.

[Rev] **D. Revuz,** "Markov chains" , North-Holland, 1975.

[Sag] **H. Saggou,** "Etude d'une file d'attente avec priorités et lois non exponentielles" Thèse de magister, Alger, 1989.

[Sen] **E. Seneta,** "Non-negative matrices and Markov chains", Springer-Verlag, 1981.

Queueing Networks with Finite Capacity
R.O. Onvural and I.F. Akyildiz (Editors)
1993 Elsevier Science Publishers B.V.

A Numerical Solution Method for PH/D/1/K Queueing Systems

R. German* and C. Lindemann**

Technische Universität Berlin, Institut für Technische Informatik, Fachgebiet Prozeßdatenverarbeitung und Robotik (Real-Time Systems and Robotics), Franklinstr. 28/29, 1000 Berlin 10, F. R. Germany

Abstract

In this paper we present an efficient numerical solution algorithm for PH/D/1/K queueing systems in case the phase-type arrival process can be specified by an acyclic Markov chain. The described approach is based on the embedded Markov chain technique as employed in the numerical solution process for Deterministic and Stochastic Petri Nets. Since the phase-type distribution is acyclic, this distribution can be transformed to the canonical representation of a Coxian distribution with real coefficients. A closed-form solution for the distribution of customers of the $E_r/D/1/K$ queueing system is presented. To illustrate the feasibility of the described solution technique some performance figures of the $E_r/D/1/K$ and the $H_2/D/1/K$ queueing systems are presented.

1. INTRODUCTION

Since actual computer and communication systems have finite capacity resources, queueing systems with limited waiting room have been proven as a useful modeling tool for their performance analysis. For the same reason this class of queueing systems is also well suited for modeling flexible manufacturing systems. For a recent survey paper and a list of bibliography we refer to [12] and [13], respectively. Communication networks and manufacturing systems typically contain features which have a deterministic delay (e.g. the transfer time of a packet or the service time of a manufacturing cell). In most work introduced in the literature exact solutions have only been derived for queueing systems and networks with blocking in which all delays are exponentially distributed (see e.g. [12]).

If the service time is represented by a deterministic distribution, approximate numerical solution techniques have been proposed. Yao and Buzacott investigated the approximation of a deterministic distribution by a two-phase Coxian distribution [14]. In case of two-dimensional generalizations of elementary M/G/1 queueing systems matrix-geometric techniques introduced

* Reinhard German was supported by a doctoral fellowship from the German Research Council (DFG) under grant Ho 1257/1-2.

** Christoph Lindemann was supported by the Federal Ministry for Research and Technology of Germany (BMFT) and by the German Research Council under grants ITR9003 and Ho 1257/2-1, respectively.

by Neuts [11] provide a feasible tool to iteratively calculate an exact solution. However, this technique is not suitable for queueing systems with finite capacity. Moreover, the automation of the solution process for a wide gamut of problems incorporates some difficulties because the repetitive structure in the probability matrix of the Markov chain is particular to each problem.

Lucantoni, Meier-Hellstern, and Neuts derived computational formulas for the distribution of the waiting time and queue length of a single-server queue with general Markovian arrival process, general distributed service time with vacations, and unlimited waiting room [8]. Blondia developed an approach for calculating the steady-state solution of a single-server queueing system with finite capacity where the arrival process is Neuts´ versatile Markovian point process and the service time may be arbitrary distributed [2].

In this paper we present an efficient numerical solution algorithm for single-server queues with finite capacity where the arrival process is an acyclic phase-type process and the service time is deterministic. This type of queueing system is referred to as PH/D/1/K queue and is included in the class of queues studied in [2]. We show that the steady-state distribution of customers at an arbitrary instant of time in PH/D/1/K queues can be calculated in a substantially simpler way than described in [2]. The described approach is based on the embedded Markov chain technique as employed in the numerical solution process for Deterministic and Stochastic Petri Nets [1]. Since the phase-type distribution is acyclic, this distribution can be transformed to a Coxian distribution with real coefficients [4]. We define a continuous-time Markov chain subordinated to the deterministic service process by all feasible completions of exponential phases which may occur concurrently to a service period. The entries of the probability matrix of the embedded Markov chain are determined by the transient state probabilities of this subordinated Markov chain at the instant of time equal to the service time. The steady-state probability distribution of customers in the queue is derived from the steady-state solution of the embedded Markov chain by means of a conversion matrix which takes into account sojourn times in states of the continuous-time behavior of the queue which are not visited in the discrete-time behavior of the embedded Markov chain. A closed-form solution for the distribution of customers of the E_r/D/1/K queueing system is derived. To illustrate the feasibility of the described solution technique various performance figures of the E_r/D/1/K and the H_2/D/1/K queueing systems are presented.

This paper is organized as follows. In section 2 a general description of the numerical solution method for PH/D/1/K queueing systems is introduced. In section 3 an efficient computational method for deriving the state transition probability matrix and the conversion matrix of the embedded Markov chain are introduced. Section 4 presents a closed-form solution for the distribution of customers of the E_r/D/1/K queueing system. Performance figures of the E_r/D/1/K and the H_2/D/1/K queueing systems are presented in section 5. Finally, concluding remarks are given.

2. GENERAL DESCRIPTION OF THE SOLUTION TECHNIQUE

The *phase-type distribution* (PH-distribution) can be defined in terms of a continuous-time Markov chain (CTMC) with r transient states 0,1,2,...,r-1 and one absorbing state r. The initial state of this CTMC is specified by a probability vector of dimension r on the transient states of the CTMC. The distribution of time spend in such a CTMC until the absorbing state r is reached specifies the PH-distribution [11]. As shown by Cumani [4] each acyclic PH-distribution can be uniquely transformed to the canonical representation of a Coxian distributions [3] with real rates λ_i and real transition probabilities α_i ($1 \leq i \leq r$). The Laplace transform of a PH distribution can be derived by summing up the product terms of all feasible paths to the absorbing state for all feasible initial states. In this way a sum of fractions with real zeros is built. Such a sum can be transformed to a single fraction. Subsequently, this fraction is transformed in the expression of the canonical form of a Coxian distribution. The transition probabilities α_i and the rates λ_i ($1 \leq i \leq r$) of this canonical form corresponding to a given acyclic PH distribution can be derived by comparison of coefficients.

Single-server queueing systems with PH distributed interarrival time, a deterministic service and finite capacity are denoted by PH/D/1/K queueing systems. This section describes a numerical method for deriving the steady-state solution of PH/D/1/K queueing systems in case the PH-distribution can be specified by an acyclic Markov chain. Since in each state at most one non-exponential state transition is feasible, the underlying stochastic process is included in the class of semi-Markov processes. The semi-Markov process can be obtained by employing the technique of the embedded Markov chain (EMC). Informally, the proposed numerical solution technique for PH/D/1/K queueing systems consists of the following steps:

1) The entries of the state transition probability matrix P of the embedded Markov chain have to be determined by calculating transient state transition probabilities of the CTMC of the arrival process. Since this Markov chain is subordinated to the deterministic service process, it is subsequently referred to as the *subordinated Markov chain.*
2) The steady-state solution of the embedded Markov chain is calculated by solving the linear system of its global balance equations.
3) The entries of the conversion matrix C of the embedded Markov chain have to be determined by calculating the mean sojourn times in the states of the subordinated Markov chain during a service period. The steady-state probability vector of the discrete-time embedded Markov chain is multiplied with this conversion matrix C in order to keep track of sojourn times in states of the subordinated Markov chain which are visited between two consecutive observation points.
4) The converted vector is normalized to yield the steady-state probability vector of the stochastic process underlying a PH/D/1/K queueing system. From this probability vector all performance measures of this queueing system can be easily derived.

A similar numerical approach has been proposed by Marsan and Chiola for deriving steady-state solutions of Deterministic and Stochastic Petri Nets (DSPNs [1]). A DSPN is a Petri Net in which a transition firing can be augmented with either an exponential or a deterministic delay. Marsan and Chiola proposed to employ an adaptive matrix exponentiation algorithm for calculating the entries of the state probability matrix P and the conversion matrix C [1]. Recently, we have introduced an improved numerical solution algorithm for DSPNs which is based on a refined randomization technique [7]. We showed that this solution algorithm is substantially more efficient than the original solution algorithm of Marsan and Chiola [1].

Each state $S_{n,i}$ of the underlying stochastic process is specified by the number of customers in service and by the number of phases already completed by the next arriving customer. The Coxian arrival process is specified by r pairs (α_i, λ_i). It is assumed that the first phase of the canonical representation of the Coxian distribution has always to be completed (i.e. $\alpha_0 = 1$) which builds only a minor restriction. The deterministic service time is specified by its delay τ. The canonical representation of the Coxian interarrival time distribution has r exponential phases, after each phase either the next phase is entered with probability α_i or death can occur with probability $1-\alpha_i$ and the customer arrival process is completed. As a consequence, the structure of the subordinated Markov chain is repeated every r states. This symmetry is exploited for efficiently determining the state transition probability of the embedded Markov chain.

Observing the stochastic behavior only at departure points of customers defines a discrete-time Markov chain which builds the embedded Markov chain (EMC). In case of states $S_{0,i}$ $(0 \leq i \leq r-1)$ the process is sampled at the instant of the next state transition. In case of states $S_{n,i}$ $(n \geq 1)$ the process is sampled at the departure point of a customer. To derive the steady-state solution of this EMC the matrix P of the state transition probabilities must be calculated.

Figure 1 shows the Markov chain subordinated to the deterministic service process. Note, the states $S_{0,i}$ $(0 \leq i \leq r-1)$ are not included in this Markov chain because in these states no customer is inside the system.

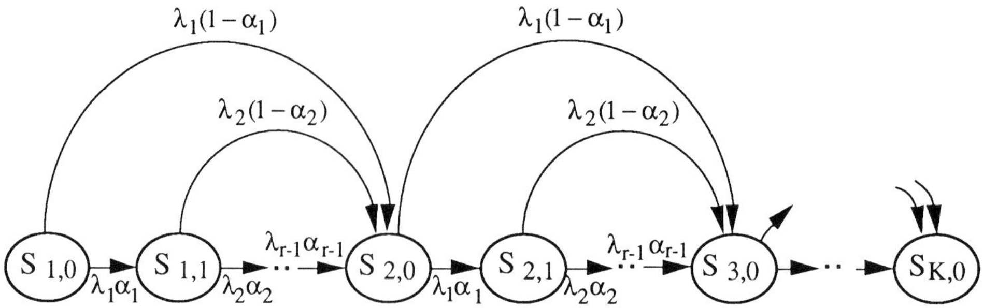

Figure 1: State transition diagram of the Markov chain subordinated to the service process

For each combination of an initial state $S_{N,I}$ and a current state $S_{n,i}$ $(N \geq 1,\ n \geq 1)$ the transient state transition probability $\pi_{(N,I)(n,i)}(\tau)$ is defined:

$$\pi_{(N,I)(n,i)}(\tau) = \text{Prob}(S_{n,i} \text{ at time } t = \tau \mid S_{N,I} \text{ at time } t = 0) \tag{1}$$

A matrix $P^{(PH)}(\tau)$ is defined by the transient state transition probabilities $\pi_{(N,I)(n,i)}(\tau)$ for all initial states $S_{N,I}$ with $1 \leq N \leq K-2$, $0 \leq I \leq r-1$ and with $N = K-1$, $I = 0$ and for all current states $S_{n,i}$ with $1 \leq n \leq K-1$, $0 \leq i \leq r-1$ and with $n = K$, $i = 0$. Thus $P^{(PH)}(\tau)$ has dimension $((K-2)r+1) \times ((K-1)r+1)$. The transient state transition probabilities for initial states $S_{N,I}$ with $N = K-1$, $1 \leq I \leq r-1$ and with $N = K$, $I = 0$ are neglected in $P^{(PH)}(\tau)$ because these states never can be reached by the EMC. The structure of $P^{(PH)}(\tau)$ depends on the CTMC which specifies the particular PH distribution of the arrival process. The efficient determination of this matrix represents an important step of the solution technique. In section 3.1 we present an efficient numerical algorithm for this purpose. Figure 2 shows the state transition diagram of the EMC underlying a PH/D/1/K queueing system.

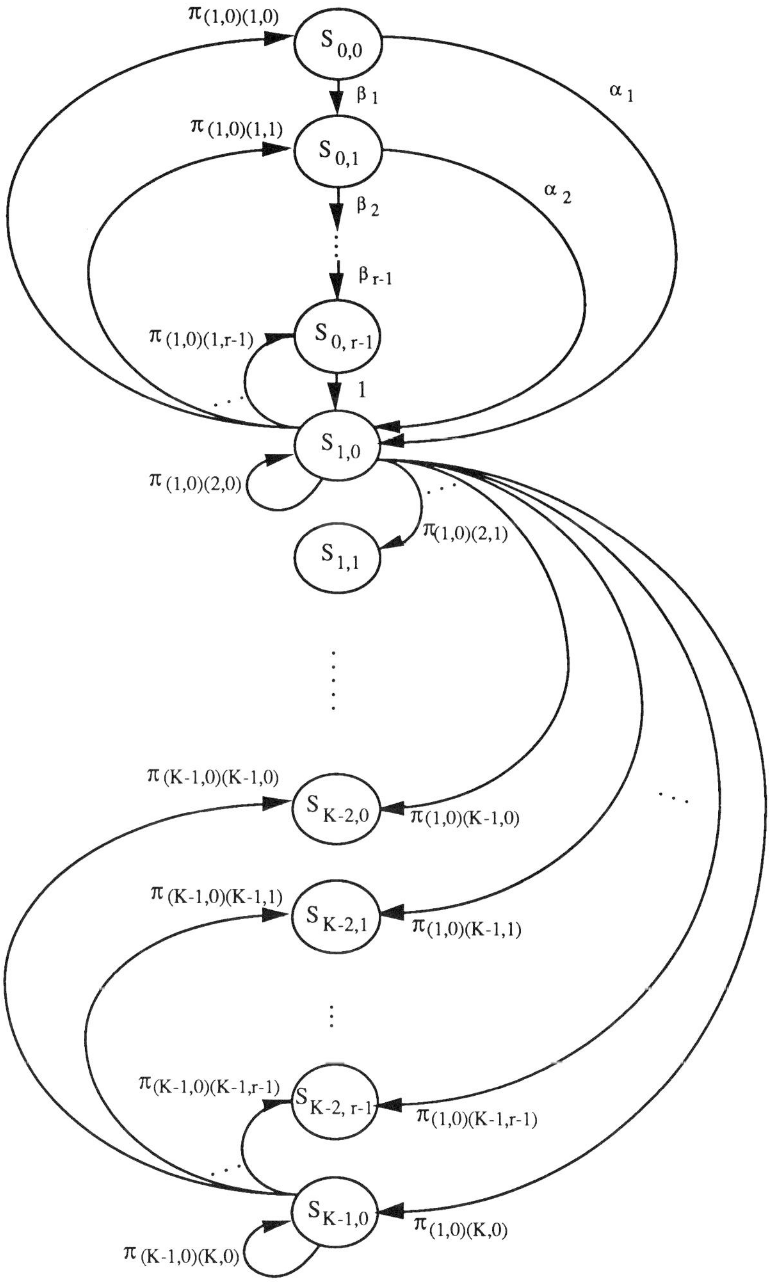

Figure 2: State transition diagram of the embedded Markov chain

If the matrix $P^{(PH)}(\tau)$ is known, the probability matrix P of the EMC is given by:

$$
\begin{array}{c|cccccccccc|}
 & 0,0 & 0,1 & 0,2 & \cdots & 0,r-1 & 1,0 & 1,1 & \cdots & \cdots & K-1,0 \\
\hline
0,0 & 0 & \alpha_1 & 0 & \cdots & 0 & 1-\alpha_1 & 0 & \cdots & \cdots & 0 \\
0,1 & 0 & 0 & \alpha_2 & \cdots & 0 & 1-\alpha_2 & 0 & \cdots & \cdots & 0 \\
\vdots & \vdots & & & \ddots & \vdots & \vdots & \vdots & & & \vdots \\
0,r-1 & 0 & \cdots & \cdots & \cdots & 0 & 1 & 0 & \cdots & \cdots & 0 \\
\hline
1,0 & & & & & & & & & & \\
\vdots & & & & & & P^{(PH)}(\tau) & & & & \\
\vdots & & & & & & & & & & \\
K-1,0 & & & & & & & & & & \\
\hline
\end{array}
$$

P is a square matrix of dimension (K-1)r+1 and consists of two submatrices. The contents of P is surrounded by a box and on the outside of this box the indices of the corresponding states of the EMC are given. The states $S_{0,i}$ $(0 \leq i \leq r-1)$ have possible state transitions to the next state or to state $S_{1,0}$. A state $S_{n,i}$ $(n > 0)$ has possible state transitions to the states $S_{n-1,i} \ldots S_{K-1,0}$. Note that only for the states $S_{1,0}$ and $S_{K-1,0}$ of this class all feasible state transitions are included in Figure 2. The steady-state solution of the EMC is determined by the solution of the linear system of global balance equations. The steady-state solution is a vector of dimension (K-1)r+1 and denoted by p.

$$p \cdot (P-I) = 0, \qquad \sum_{n=0}^{K-2} \sum_{i=0}^{r-1} p_{n,i} + p_{K-1,0} = 1 \tag{2}$$

Entries of the conversion matrix C are given by the mean sojourn times in states of the semi-Markov process up to the instant of time, when the process is sampled. In general, this conversion matrix C has the structure.

$$
\begin{array}{c|ccccc|cccccccc|}
 & 0,0 & 0,1 & 0,2 & \cdots & 0,r-1 & 1,0 & 1,1 & \cdots & \cdots & K-1,0 & K-1,1 & \cdots & K,0 \\
\hline
0,0 & 1/\lambda_1 & 0 & \cdots & \cdots & 0 & 0 & \cdots & \cdots & \cdots & \cdots & \cdots & \cdots & 0 \\
0,1 & 0 & 1/\lambda_2 & 0 & \cdots & 0 & \vdots & & & & & & & \vdots \\
\vdots & \vdots & & & & \vdots & \vdots & & & & & & & \vdots \\
0,r-1 & 0 & \cdots & \cdots & \cdots & 1/\lambda_r & 0 & \cdots & \cdots & \cdots & \cdots & \cdots & \cdots & 0 \\
\hline
1,0 & 0 & \cdots & \cdots & \cdots & 0 & & & & & & & & \\
\vdots & \vdots & & & & \vdots & & & & & C^{(PH)}(\tau) & & & \\
\vdots & \vdots & & & & \vdots & & & & & & & & \\
K,0 & 0 & \cdots & \cdots & \cdots & 0 & & & & & & & & \\
\hline
\end{array}
$$

C is a square matrix of dimension Kr+1 and consists of four submatrices. In section 3.2 it is shown how the submatrix $C^{(PH)}(\tau)$ can be calculated efficiently. The vector of the steady-state-solution of the EMC is extended by the vector of zeros $\mathbf{0}$ of the dimension r:

$$p' = (p, \mathbf{0}) \tag{3}$$

The steady-state solution of the stochastic process underlying a PH/D/1/K queueing system is derived by converting and normalizing the solution of the extended steady-state solution of the EMC.

$$q = p' \cdot C \tag{4}$$

$$q' = \frac{q}{\sum_{n=0}^{K-1} \sum_{i=0}^{r-1} q_{n,i} + q_{K,0}} \tag{5}$$

The steady-state distribution of the number of customers in the system is then derived by summing up on the number of phases for each feasible number of customers in the system.

$$P_n = \text{Prob}\{\text{n customers are in the system}\} = \begin{cases} \sum_{i=0}^{r-1} q'_{n,i} & 0 \le n \le K-1 \\ q'_{K,0} & n = K \end{cases} \tag{6}$$

3. EFFICIENT CALCULATION OF THE PROBABILITY MATRICES

3.1. Calculation of the Submatrix $P^{(PH)}(\tau)$ of the Probability Matrix P

This section introduces an efficient calculation method of the transient state probabilities of the subordinated Markov chain depicted in Figure 1. The entries of the matrix $P^{(PH)}(\tau)$ are given by equation (1). Since the chain is acyclic and the states are already ordered $P^{(PH)}(\tau)$ is upper triangular. Furthermore, the chain repeats its structure every r states. For current states $S_{n,i} \neq S_{K,0}$ the probability only depends on the relative distance to the initial state, i.e.:

$$\pi_{(N,I)(n,i)}(\tau) = \pi_{(N+m,I)(n+m,i)}(\tau) \qquad \text{if N, n, N+m, n+m} < K. \tag{7}$$

The following shows the structure of the matrix $P^{(PH)}(\tau)$.

$$P^{(PH)}(\tau) = \begin{bmatrix} A_1 & A_2 & A_3 & \cdots & A_{K-2} & A_{K-1} & B_1 \\ 0 & A_1 & A_2 & \cdots & A_{K-3} & A_{K-2} & B_2 \\ 0 & 0 & A_1 & \cdots & A_{K-4} & A_{K-3} & B_3 \\ \vdots & & \ddots & \ddots & \vdots & \vdots & \vdots \\ \vdots & & & \ddots & A_1 & A_2 & B_{K-2} \\ 0 & \cdots & \cdots & \cdots & 0 & A_1(1) & B_{K-1}(1) \end{bmatrix} \tag{8}$$

with the submatrices A_n of dimension $r \times r$ $(1 \le n \le K\text{-}1)$:

$$A_n = \begin{bmatrix} \pi_{(1,0)(n,0)}(\tau) & \pi_{(1,0)(n,1)}(\tau) & \cdots & \pi_{(1,0)(n,r-1)}(\tau) \\ \pi_{(1,1)(n,0)}(\tau) & \pi_{(1,1)(n,1)}(\tau) & \cdots & \pi_{(1,1)(n,r-1)}(\tau) \\ \vdots & \vdots & & \vdots \\ \pi_{(1,r-1)(n,0)}(\tau) & \pi_{(1,r-1)(n,1)}(\tau) & \cdots & \pi_{(1,r-1)(n,r-1)}(\tau) \end{bmatrix} \tag{9}$$

and with the vector B_N of dimension r $(1 \le N \le K\text{-}1)$:

$$B_N = \begin{bmatrix} \pi_{(N,0)(K,0)}(\tau) \\ \pi_{(N,1)(K,0)}(\tau) \\ \vdots \\ \pi_{(N,r-1)(K,0)}(\tau) \end{bmatrix} \tag{10}$$

In equation (8) $A_1(1)$ and $B_{K-1}(1)$ denote the first row of the matrix A_1 and the first entry of the vector B_{K-1}, respectively. Since $P^{(PH)}(\tau)$ is upper triangular, also the submatrix A_1 is upper triangular. The transient behavior of the subordinated Markov chain is described by a linear system of differential equations. A row of $P^{(PH)}(\tau)$ is denoted by $\pi_{(N,I)}$, where the indices stand for the corresponding initial state. The entries of the first r rows of $P^{(PH)}(\tau)$ are formally given by the matrix exponential:

$$\pi_{(1,I)}(\tau) = u_{I+1} \cdot e^{Q\tau} \qquad 0 \le I \le r-1 \tag{11}$$

Q denotes the generator matrix of the subordinated Markov chain depicted in Figure 1 and u_{I+1} denotes the I+1-th unity row vector which is employed to filter the I+1-th row of the matrix exponential. Efficient computational formulas for equation (11) have been introduced in [7] using a refined randomization technique. The employment of the randomization technique [6] for calculating the transition probabilities of the EMC has been proposed by Lucantoni and Ramaswami [9]. In the following we show how to employ a refined randomization technique for calculating both the transition probabilities and the conversion factors. Employing this approach for the subordinated Markov chain a scalar q and a matrix A are defined as follows.

$$q = 1.02 \cdot \max_{1 \le j \le (K-1)r+1} |q_{jj}| \tag{12}$$

$$A = \frac{1}{q} Q + I \tag{13}$$

In formula (13) I denotes the identity matrix. Assuming the initial state $S_{1,I}$ the Chapman-Kolmogorov equation of the discrete-time Markov chain with transition probability matrix A denoted by $\Phi(\nu)$ can be efficiently computed by recursive vector-matrix multiplications [6].

$$\Phi(0) = u_{I+1} \tag{14}$$

$$\Phi(\nu+1) = \Phi(\nu) \cdot A \tag{15}$$

The corresponding Poisson probabilities $\beta(\nu)$ have to be calculated for the parameter $q\tau$. Since the probability mass function of the Poisson distribution thins for growing $q\tau$, its computation may get affected by round-off errors in case of large $q\tau$. Thus, the randomization technique introduced in [6] is enhanced by a stable calculation of Poisson probabilities proposed by Fox and Glynn [5]. In case the product $q\tau < 25$ no left-truncation is used. If the product $q\tau > 25$, the summation in (17) starts at some $L > 0$ and the left tail of the Poisson distribution is neglected. The left and right truncation points $L = L(qt, \varepsilon)$ and $R = R(qt, \varepsilon)$ are depending on the pre-defined error tolerance ε. These truncation points are determined according to equation (16).

$$\pi_{(1,I)}(\tau) = \sum_{\nu=L}^{R} \Phi(\nu) \cdot \beta(\nu) \qquad 0 \le I \le r-1 \tag{16}$$

Employing the row vectors $\Phi(\nu)$ and the corresponding Poisson probabilities $\beta(\nu)$ allows the calculation of the entire I+1-th row of the transition probability matrix in one step [7].

$$\pi_{(1,I)}(\tau) = \sum_{\nu=L}^{R} \Phi(\nu)\cdot\beta(\nu) \qquad 0 \le I \le r-1 \tag{17}$$

Since P is a probabilistic matrix, the sum of all entries of a row has value 1. Hence, the remaining entries of the last column of $P^{(PH)}(\tau)$ can be obtained by summing up all entries of a row and subtracting the sum from 1:

$$\pi_{(N,I)(K,0)}(\tau) = 1 - \sum_{n=1}^{K-1}\sum_{i=0}^{r-1} \pi_{(N,I)(n,i)}(\tau) \tag{18}$$

for all pairs (N, I) with $2 \le N \le K-2$, $0 \le I \le r-1$ and with $N = K-1$, $I = 0$.

A sparse implementation of the randomization technique requires an asymptotical computational effort of $O(Kr^2q\tau)$ for calculating all entries of the first r rows of $P^{(PH)}(\tau)$. The number of remaining entries which must be calculated by formula (18) is (K-3)r+1 and the number of entries that can be copied is of asymptotical order $O(K^2r^2)$.

3.2. Calculation of the Submatrix $C^{(PH)}(\tau)$ of the Conversion Matrix C

In order to keep track of state transitions which might occur between two consecutive observation points the steady-state probability vector of the EMC has to be multiplied by a conversion matrix C. Entries of the conversion matrix can be considered as the average sojourn time in a state of the subordinated Markov chain during the deterministic duration of a service period. In general, the submatrix $C^{(PH)}(\tau)$ of the conversion matrix C has the following structure.

$$C^{(PH)}(\tau) = \begin{bmatrix} \tilde{A}_1 & \tilde{A}_2 & \tilde{A}_3 & \cdots & \tilde{A}_{K-1} & \tilde{B}_1 \\ 0 & \tilde{A}_1 & \tilde{A}_2 & \cdots & \tilde{A}_{K-2} & \tilde{B}_2 \\ 0 & 0 & \tilde{A}_1 & \cdots & \tilde{A}_{K-3} & \tilde{B}_3 \\ \vdots & & \ddots & \ddots & \vdots & \vdots \\ \vdots & & & \ddots & \tilde{A}_1 & \tilde{B}_{K-1} \\ 0 & \cdots & \cdots & \cdots & 0 & \tau \end{bmatrix} \tag{19}$$

with the submatrices $\tilde{A}_n$ of dimension $r\times r$ $(1 \le n \le K-1)$:

$$\tilde{A}_n = \begin{bmatrix} \tilde{\pi}_{(1,0)(n,0)}(\tau) & \tilde{\pi}_{(1,0)(n,1)}(\tau) & \cdots & \tilde{\pi}_{(1,0)(n,r-1)}(\tau) \\ \tilde{\pi}_{(1,1)(n,0)}(\tau) & \tilde{\pi}_{(1,1)(n,1)}(\tau) & \cdots & \tilde{\pi}_{(1,1)(n,r-1)}(\tau) \\ \vdots & \vdots & & \vdots \\ \tilde{\pi}_{(1,r-1)(n,0)}(\tau) & \tilde{\pi}_{(1,r-1)(n,1)}(\tau) & \cdots & \tilde{\pi}_{(1,r-1)(n,r-1)}(\tau) \end{bmatrix} \tag{20}$$

and with the vector $\tilde{B}_N$ of dimension r $(1 \le N \le K-1)$:

$$\tilde{B}_N = \begin{bmatrix} \tilde{\pi}_{(N,0)(K,0)}(\tau) \\ \tilde{\pi}_{(N,1)(K,0)}(\tau) \\ \vdots \\ \tilde{\pi}_{(N,r-1)(K,0)}(\tau) \end{bmatrix} \tag{21}$$

The conversion factors $\tilde{\pi}_{(N,I)(n,i)}(\tau)$ can be derived by integration of the transient state transition probabilities. The first r rows of $C^{(PH)}(\tau)$ can formally be expressed as:

$$\tilde{\pi}_{(1,I)}(\tau) = u_{I+1} \cdot \int_0^\tau e^{Qt}\, dt \qquad 0 \le I \le r-1 \tag{22}$$

A computational formula based on the randomization technique has been derived for equation (22) in [7] . An entire row can efficiently be calculated in one step by using the row vectors $\Phi(\nu)$ and the Poisson probabilities $\beta(\nu)$ defined in section 3.1:

$$\tilde{\pi}_{(1,I)}(\tau) = \frac{1}{q} \sum_{\nu=0}^{R} \Phi(\nu) \cdot \left(1 - \sum_{n=L}^{\nu} \beta(n)\right) \qquad 0 \le I \le r-1 \tag{23}$$

Since the stochastic point process is sampled when the service time τ is elapsed, the sum of all entries of a row equals τ. Hence the remaining entries of the last column of $C^{(PH)}(\tau)$ can be obtained by summing up all entries of a row and subtracting the sum from τ:

$$\tilde{\pi}_{(N,I)(K,0)}(\tau) = \tau - \sum_{n=1}^{K-1} \sum_{i=0}^{r-1} \tilde{\pi}_{(N,I)(n,i)}(\tau) \tag{24}$$

for all pairs (N, I) with $2 \le N \le K-1$, $0 \le I \le r-1$ and with $N = K$, $I = 0$.

The calculation of $C^{(PH)}(\tau)$ according to formula (23) requires only little extra computational effort because the $\Phi(\nu)$ and $\beta(\nu)$ have already been computed for calculating $P^{(PH)}(\tau)$. The asymptotical computational effort for calculating the first r rows of $C^{(PH)}(\tau)$ is of order $O(Kr^2qt)$, $(K-2)r+1$ remaining entries of the last column have to be calculated by formula (24) and the number of entries which can be copied is of order $O(K^2r^2)$.

Marie, Reibman, and Trivedi proposed a special-purpose method for the transient analysis of acyclic Markov chains [10] which allows both a symbolic and a numerical calculation of time-dependent quantities. Since the subordinated Markov chain has an acyclic topology, the entries of the submatrix $P^{(PH)}(\tau)$ can be obtained by this algorithm. To calculate the entries of the submatrix $C^{(PH)}(\tau)$ the algorithm is extended by considering zero as a new pole in each state. Experiments have shown that this method yields accurate numerical results in case of few distinct entries in the main diagonal of the generator matrix Q (e.g. for an Erlang interarrival process). In case of many distinct entries in the main diagonal of Q (e.g. for a hypoexponential interarrival process) this algorithm suffers from numerical problems. Moreover, we observed that the absolute computational cost is smaller than randomization only for large values of the product $q\tau$.

4. THE $E_r/D/1/K$ QUEUEING SYSTEM

To illustrate the feasibility of the general solution technique a closed-form solution for the $E_r/D/1/K$ queueing system is derived in this section. Since an Erlang distribution is a special case of the canonical representation of a Coxian distribution with r identical phases, the Markov chain subordinated to the deterministic service process forms simply a sequence of states with state transition rates $r\lambda$. The submatrix $P^{(E_r)}(\tau)$ of the probability matrix P of the EMC has a rather simple structure. Its elements can be specified in an closed-form expression without employing formula (17). The submatrix $P^{(E_r)}(\tau)$ is given by:

$$P^{(E_r)}(\tau) = \begin{bmatrix} a_0 & a_1 & a_2 & \cdots & \cdots & \cdots & a_{(K-1)r-1} & b_{(K-1)r} \\ 0 & a_0 & a_1 & \cdots & \cdots & \cdots & a_{(K-1)r-2} & b_{(K-1)r-1} \\ 0 & 0 & a_0 & \cdots & \cdots & \cdots & a_{(K-1)r-3} & b_{(K-1)r-2} \\ \vdots & & \ddots & \ddots & & & \vdots & \vdots \\ 0 & \cdots & \cdots & 0 & a_0 & \cdots & a_{r-1} & b_r \end{bmatrix} \tag{25}$$

The indices used in formula (25) differ from that used for the submatrices in formula (8) for notational convenience. The entries a_i and b_i of the matrix $P^{(E_r)}(\tau)$ are determined by:

$$a_i = \mathrm{Prob}\{i \text{ phases completed in time } \tau\} = \frac{(r\lambda)^i}{i!} t^i e^{-r\lambda\tau} \tag{26}$$

$$b_i = \mathrm{Prob}\{\text{at least } i \text{ phases completed in time } \tau\} = 1 - \sum_{k=0}^{i-1} a_k \tag{27}$$

The submatrix $C^{(E_r)}(\tau)$ of the weight matrix C is given by:

$$C^{(E_r)}(\tau) = \begin{bmatrix} \tilde{a}_0 & \tilde{a}_1 & \tilde{a}_2 & \cdots & \cdots & \tilde{a}_{(K-1)r-1} & \tilde{b}_{(K-1)r} \\ 0 & \tilde{a}_0 & \tilde{a}_1 & \cdots & \cdots & \tilde{a}_{(K-1)r-2} & \tilde{b}_{(K-1)r-1} \\ 0 & 0 & \tilde{a}_0 & \cdots & \cdots & \tilde{a}_{(K-1)r-3} & \tilde{b}_{(K-1)r-2} \\ \vdots & & \ddots & \ddots & & \vdots & \vdots \\ \vdots & & & \ddots & \ddots & \vdots & \vdots \\ \vdots & & & & \ddots & \tilde{a}_0 & \tilde{b}_1 \\ 0 & \cdots & \cdots & \cdots & \cdots & 0 & 1 \end{bmatrix} \tag{28}$$

The entries of the submatrix $C^{(E_r)}(\tau)$ are determined by:

$$\tilde{a}_i = \int_0^\tau a_i(t)\,dt = \frac{1}{r\lambda} - e^{-r\lambda\tau} \sum_{k=0}^{i} \frac{(r\lambda)^{k-1}}{k!} \tau^k \tag{29}$$

$$\tilde{b}_i = \int_0^t b_i(\tau)\,d\tau = \tau - \frac{i}{r\lambda} + e^{-r\lambda\tau} \sum_{k=0}^{i-1} (i-k) \frac{(r\lambda)^{k-1}}{k!} \tau^k \tag{30}$$

These examples show the benefits of exploiting the symmetries in the probability matrix and the weight matrix. Employing the approach described in section 3 requires to calculate (K-1)r values of a_i and (K-2)r values of b_i. Thus, the number of transient state probabilities which have to be computed for deriving the probability matrix of the EMC and the conversion matrix grows asymptotically linearly with the product rK. A straight-forward derivation of the matrices $P^{(E_r)}(\tau)$ and $C^{(E_r)}(\tau)$ requires the calculation of $O(r^2K^2)$ entries.

5. NUMERICAL RESULTS FOR THE E_r/D/1/K and H_2/D/1/K QUEUEING SYSTEMS

The Erlang distribution of r phases (E_r) and the hyperexponential distribution with two phases (H_2) are important examples of acyclic PH distributions because they can match all first and a wide area of second moments of an arbitrary probability distribution. In this section we derive numerical results for the E_r/D/1/K and H_2/D/1/K queueing systems. In case of an Erlang distribution with r phases the coefficient of variation C_X is determined by the number of phases (i.e. $C_X = 1/\sqrt{r} \leq 1$) and can match values less or equal to one. In case of a hyperexponential distribution with two phases C_X can match values greater than or equal to one. Note, that the interval of possible C_X depends on the ratio q_2/q_1, where q_1 and q_2 denote the switching probabilities corresponding to the two phases of the hyperexponential distribution.

We consider E_{10}/D/1/50, M/D/1/50 and H_2/D/1/50 queueing systems. The ratio q_2/q_1 of the H_2 distribution is set to 9 and its coefficient of variation is set to $\approx \sqrt{10}$. In the experiments the traffic intensity $\rho = \tau \cdot \lambda$ varies from 0.1 to 2.0. Figure 3 shows curves of the average time T a customer spends in the system. Figure 4 shows curves for the expected number of losses L in a time unit versus the utilization factor.

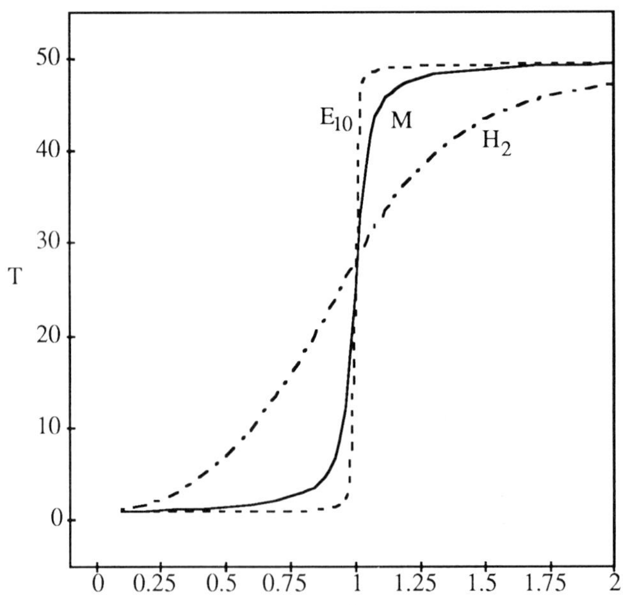

Figure 3: Average delay versus traffic intensity

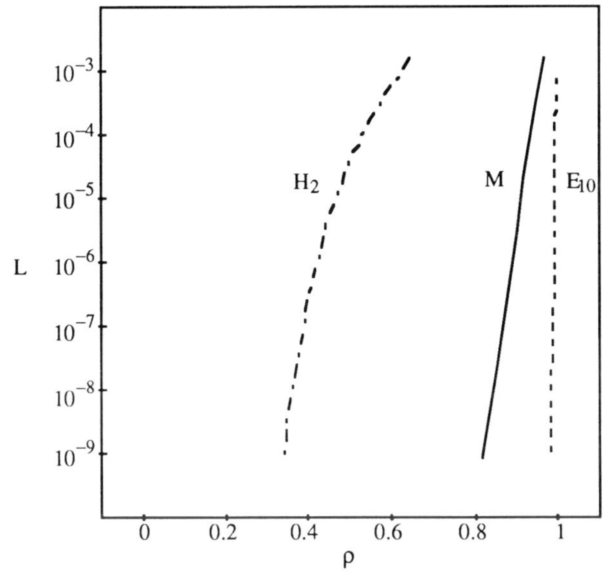

Figure 4: Expected number of losses per time unit versus traffic intensity

These performance measures are derived from the steady-state distribution of customers at an arbitrary instant of time.

$$T = \frac{1}{\rho \cdot (1 - p_K)} \cdot \sum_{n=1}^{K} n \cdot p_n \tag{31}$$

$$L = \rho \cdot p_K \tag{32}$$

As expected in case of the E_{10} distribution the growth of the average system time is more steep whereas in case of the H_2 distribution the growing is less steep than that of the exponential distribution. Similarly, for a given traffic intensity the expected number of losses per time unit is larger for the H_2 arrival distribution than in case of the exponential distribution and smaller than the exponential distribution in case of the E_{10} arrival distribution. These effects are due to the different coefficients of variation associated with the considered distributions.

CONCLUSIONS

In this paper an efficient numerical solution method for single server queueing systems with an acyclic phase-type interarrival process, a deterministic service time, and finite capacity has been presented. The described approach employs a conversion matrix defined by the mean sojourn times in phases of the arrival process and by the mean sojourn times in the subordinated Markov chain during a service period. This conversion matrix is employed for deriving the steady-state distribution of customers in the queue at an arbitrary instant of time from the steady-state solution of the embedded Markov chain. The structure of the probability matrix and the conversion matrix of the embedded Markov chain underlying a PH/D/1/K queueing systems with acyclic phase type distribution has been derived. Since symmetries can be detected in the structure of these matrices, it is sufficient to compute only a part of their entries as the transient state probabilities of the Markov chain subordinated to the deterministic service process. Thus, the numerical approach presented in this paper can be efficiently applied for PH/D/1/K queues with large buffer size (e.g. K = 1000).

A closed-form solution of the distribution of customers of the E_r/D/1/K queueing system has been derived. The presented approach could also be employed to derive closed-form solutions for queueing systems with other acyclic phase-type arrival processes (e.g. a H_2/D/1/K queue). This leads to rather complex terms, thus, in practise it is easier to use the presented procedural approach. Further research will concentrate on extending the presented approach such that general Markovian Arrival Processes as introduced by Lucantoni, Meier-Hellstern, and Neuts [8] can be handled.

REFERENCES

[1] M. Ajmone Marsan and G. Chiola, On Petri Nets with deterministic and exponentially distributed firing times", in: *G. Rozenberg (Ed.) Advances in Petri Nets 1986, Lecture Notes in Computer Science 266*, (Springer 1987) 132-145.

[2] C. Blondia, The finite capacity N/G/1 queue, *Stochastic Models* **5** (1989) 273-294.

[3] D.R. Cox, A use of complex probabilities in the theory of stochastic processes, *Proc. Camb. Phil. Soc. (Math. and Phys. Sciences)* **51** (1955) 313-319.

[4] A. Cumani, On the canonical representation of homogeneous Markov processes modeling, *Microelectronics and Reliability* **22** (1982) 583-602.

[5] B.L. Fox and P.W. Glynn, Computing Poisson probabilities, *Comm. of the ACM* **31** (1988) 440-445 .

[6] D. Gross and D.R. Miller, The randomization technique as a modeling tool and solution procedure for transient Markov processes, *Operations Research* **32** (1984) 345-361.

[7] C. Lindemann, An improved numerical algorithm for calculating steady-state solutions of Deterministic and Stochastic Petri Net models, *Proc. 4th Int. Workshop on Petri Nets and Performance Models,* Melbourne Australia (1991) 176-185.

[8] D.M. Lucantoni, K.S. Meier-Hellstern and M.F. Neuts, A single-server queue with vacations and a class of non-renewal arrival processes, *Advances in Applied Probability* **22** (1990) 676-705.

[9] D.M. Lucantoni and V. Ramaswami, Efficient algorithms for solving the non-linear matrix equation arising in phase-type queues, *Stochastic Models* **1** (1985) 29-51.

[10] R.A. Marie, A.L. Reibman and K.S. Trivedi, Transient analysis of acyclic Markov chains, *Performance Evaluation* **7** (1987) 175-194.

[11] M.F. Neuts, *Structured Stochastic Matrices of M/G/1 Type and Their Applications* (Marcel Dekker 1989).

[12] R.O. Onvural, Survey of closed queueing networks with blocking, *ACM Comput. Surveys* **22** (1990) 83-121.

[13] H.G. Perros, "A bibliography of papers on queueing networks with finite capacity queues", *Performance Evaluation* **10,** (1989) 225-260.

[14] D.D. Yao and J.A. Buzacott, Queueing models for a flexible machining station. Part II: The method of Coxian phases, *Europ. Journal of Operational Research* **19** (1985) 241-252.

Queueing Networks with Finite Capacity
R.O. Onvural and I.F. Akyildiz (Editors)

Analysis of a Tandem Queue with State-Dependent General Blocking: A GSMP Perspective

Dinah W. Cheng

Department of Statistics and Operations Research, New York University, 704 Tisch Hall, 40 West, 4th Street, New York, NY 10003.

Abstract

In this paper, we study a tandem queue where there is a finite number of buffer positions at each stage. The blocking scheme is general in the sense that it includes a number of classical blocking schemes, such as the *communication*, *manufacturing* and *kanban* blockings as special cases. The system considered here differs from the conventional system in two aspects: (1) departure of jobs from the system is determined by an external arrival process of *another* queue lying parallel to the tandem queue; (2) the control parameters of the blocking scheme are *state-dependent* in that they may change value depending on the state of the system.

We model this system as a *generalized semi-Markov process* (GSMP), study the structural properties of its *scheme*, and establish monotonicity and convexity properties of event times with respect to both the clock times and the integer blocking parameters. In particular, we demonstrate that the *state-dependent* scheme is "better", in certain sense, than the corresponding *static* scheme. Our results also recover the structural properties previously established for the classical blockings.

1. INTRODUCTION

This paper is motivated by the study of production/inventory control problems in a multi-stage *make-to-stock* system where production stages are arranged in series; and jobs are processed in anticipation of demands. We model the system as a tandem system with a merge structure at the last stage: there is a tandem queue with a series of m single server stages, indexed $1,...,m$, and a single stage queue, denoted as stage m', lying parallel to it. At each of the stages in the tandem queue, there is a finite number of waiting rooms providing waiting space for both the *raw jobs* (jobs in waiting or in service) and the *finished jobs* (jobs that have completed service). Stage 1 processes jobs from an infinite supply of raw jobs. Each of these jobs goes through operations at stages $1,...,m$ on a FCFS (first come first serve) basis and leave the system via stage m. On the other hand, there is an infinite number of waiting rooms at stage m' that provide waiting space for external arrivals. To distinguish the external arrivals from the jobs that go through operations in the tandem queue, below, we shall refer to the external arrivals as *job orders*.

Job orders interact with the tandem system as follows: Upon the arrival of an order, if there are finished jobs at stage m, the job order removes a finished job from stage m and they both leave the system together. Otherwise, the job order waits at stage m'. By the same token, upon service completion at stage m, a finished job leaves the system if there is an order at stage m'. Otherwise, the finished job waits at stage m. Note that unlike a conventional tandem system where external arrivals are the actual jobs that go through operations in the system, here the external arrivals merely act as synchronizing entities that trigger departure of the processed jobs.

Operations at stages 1,...,m are determined by a control scheme derived from the *general blocking* scheme introduced by Cheng and Yao (1991). To avoid confusion, we shall refer to their scheme *static* scheme, the one here *state-dependent* scheme.

Briefly, for the static scheme, at each stage i, there are three integer parameters (a_i, b_i, k_i) that do not change over time. They are respectively the upper limit on the number of raw jobs, finished jobs and the total number of jobs allowed at stage i. To avoid triviality, it is necessary that

$$1 \leq a_i \leq k_i, \quad 0 \leq b_i \leq k_i, \quad a_i + b_i \geq k_i; \quad i = 1, ..., m. \tag{1.1}$$

Since stage 1 produces from an infinite supply of raw jobs, we assume that

$$a_1 = 1, \quad b_i = k_1. \tag{1.2}$$

The integer parameters $(a, b, k) := (a_i, b_i, k_i)_{i=1}^m$ determine when a finished job moves to the downstream stage and when a server is blocked: Upon service completion at stage i, if the number of raw jobs at stage $i+1$ is less than a_{i+1} and the total number of jobs there is less than k_{i+1}, the finished job will move to the downstream stage, otherwise, it will be blocked and remain at stage i. At that time, if the number of finished jobs at stage i reaches b_i, the server will be blocked and becomes idle. When $b_i = 0$, server i is blocked if either the number of raw jobs at the downstream stage reaches a_{i+1} or the total number of jobs there reaches k_{i+1}. Following Cheng and Yao (1991), we shall refer to this system as a $(\cdot/G/1/(a_i, b_i, k_i)_{i=1}^m)$ system.

Let x_n^{2i-1} and x_n^{2i} denote, respectively, the number of raw jobs and finished jobs at stage i immediately after the nth transition (also referred to as time n below); x_n^{2m+1} denotes the number of orders waiting at stage m' at time n; $x_n := (x_n^i)_{i=1}^{2m+1}$ thus describes the state of the system at time n. Also, let

$$u_{ij} := b_i + \sum_{\ell=i+1}^{j-1} k_\ell + a_j, \qquad i = 1, ..., m, \quad j = i+1, ..., m; \tag{1.3}$$

where $\sum_i^j \equiv 0$, for $j > i$ by convention. u_{ij} is then the upper limit on the total number of jobs that have completed service at stage i but have not yet complete service at stage j. It has been shown that

Lemma 1.1. Server i is blocked if and only if $\exists j$, $j = i+1, ..., m$, such that

$$\sum_{\ell=2i}^{2j-1} x_\ell = u_{ij}, \qquad i = 1, ..., m. \tag{1.4}$$

In other words, server i is *not* blocked if and only if

$$\sum_{\ell=2i}^{2j-1} x_\ell < u_{ij}, \qquad \forall j = i+1, ..., m. \tag{1.5}$$

Based on the above observations, a set of recursive equations is derived for the dynamics of the system (Cheng and Yao (1991)) and structural properties of various performance measures are established with respect to the arrival process, the service process, and the integer parameters (a, b, k) (see, e.g., Cheng (1991a,b,c), Cheng and Yao (1991), Cheng and Zhu (1991), Glasserman and Yao (1991).) It has also been shown that by varying the parameters (a, b) (for given k), one can recover the following blocking mechanisms previously studied in the literature:

- communication blocking: set $a_i = k_i$ and $b_i = 0$;
- manufacturing blocking: set $a_i = k_i$ and $b_i = 1$;
- kanban blocking: set $a_i = b_i = k_i$;
- the variation of kanban blocking considered by Buzacott (1989) and Zipkin (1990): set $b_i = k_i$.

(Note that the communication and manufacturing blockings above are also commonly referred to as "block-before-service" (BBS-SO) and "block-after-service" (BAS) respectively (see, e.g., Perros (1989), Akyildiz and Perros, and Onvural (1990).) Also, when service times are exponentially distributed, communication blocking is equivalent to "blocking-before-service" with repetitive service (RS) discussed in Onvural (1990).)

2. STATE-DEPENDENT GENERAL BLOCKING

The state-dependent scheme considered here is an extension of the above scheme. It is motivated by the concern of reducing waiting time of the external arrivals. We consider the following performance measures: let

$$IP_n^i := \sum_{\ell=2i}^{2m} x_n^\ell - x_n^{2m+1}, \qquad i = 1, ..., m; \tag{2.1}$$

$$SS_n^i := (\sum_{\ell=2i}^{2m} x_n^\ell - x_n^{2m+1})^+, \qquad i - 1, ..., m; \tag{2.2}$$

and

$$OO_n^j := (x_n^{2m+1} - \sum_{\ell=2j}^{2m} x_n^\ell)^+, i = 1, ..., m; \tag{2.3}$$

where $y^+ = \max(y, 0)$. In inventory theory, IP^i, SS^i, and OO^i can be respectively referred to as *echelon inventory position*, *echelon safety stocks* and *echelon outstanding orders* at stage i. Note that from the above, we have

$$SS^i = (IP^i)^+, \qquad OO^i = (-IP^i)^+, \qquad \text{and} \qquad IP^i = SS^i - OO^i. \tag{2.4}$$

From order waiting time point of view, high level of OO^i is undesirable since it means high order response time. In order to improve order response time (and reduce the number of outstanding orders), it is necessary that a job be processed and released to the downstream stage as soon as possible if there is an order waiting for it. However, in a static scheme, since the control parameters do not respond to the number of waiting orders in the system, conceivably a server at some upstream stage j may be blocked and become idle, due to the blocking parameters, at the time when $OO^j_n > 0$ and jobs are needed to fill outstanding orders. This will obviously create unnecessary delay. Although one may increase (a, b, k) to improve order response time in a static scheme, that will also increase inventory after the number of outstanding orders in the system is reduced.

The idea behind the state-dependent scheme is to take into account the number of waiting orders in the system and to adjust the control parameters accordingly so that the level of OO^i will be reduced, but at the same time without creating excessive inventory when it is not needed. To be specific, let

$$v_i = b_i + \sum_{\ell=i+1}^{m} k_\ell, \qquad i = 1, ..., m, \tag{2.5}$$

be the upper limit on the total number of jobs that have completed service at stage i but have not yet left stage m. From (2.1) and the dynamics of the system, $IP^i \in (-\infty, v_i]$. Since $SS^i = (IP^i)^+$, it follows that

$$SS^i \leq v_i. \tag{2.6}$$

Hence v_i can be viewed as the upper limit on the level of *echelon safety stock* allowed at stage i in the static scheme. In the state-dependent scheme, one of our objectives is to adjust the parameters in such a way that (2.6) is not violated.

To allow more generality, we shall devise the control mechanism as follows. At each stage i, there is an integer parameters r^i. r^i satisfies

$$-\infty < r^i < r^{i+1}; \qquad r^i \leq v_i; \qquad i = 1, ...m. \tag{2.7}$$

Immediately after each state transition, if IP^i falls below r^i, then both the b_i and k_i will be increased by $(r^i - IP^i)$, otherwise, the control parameters remain intact. In other words, if we let $\Delta^i_n := (r^i_n - IP^i_n)^+$, the value of the control parameters a^i_{n+}, b^i_{n+}, and k^i_{n+} immediately after time n are

$$a^i_{n+} = a_i; \qquad b^i_{n+} = b_i + \Delta^i_n; \qquad \text{and} \qquad k^i_{n+} = k_i + \Delta^i_n; \qquad i = 1, ..., m. \tag{2.8}$$

This way, a server who is originally blocked at time n may now serve jobs at time n^+. (In a more general setting where the transit time between two consecutive stages is significant, we may also set the value of $a^i_{n+} = a_i + \Delta^i_n$. In that case, a finished job that is originally blocked at time n may then transit to the downstream stage at time n^+. However here, since we (implicitly) assume that there is no transit time between two stages and there is no distinction between the finished jobs at stage i and the raw job at stage $i+1$, the adjustment of a_i would be superfluous.) Note that since

$$IP^i_{n+} \leq IP^i_n + \Delta^i_n \leq r^i, \tag{2.9}$$

which in turn is less than v_i (cf. 2.7). From (2.4), it follows that $SS^i_n \leq v_i$. That is, under this control scheme, the safety stock limit is the same as the static scheme, thus satisfying our goal alluded earlier.

Two special cases of the state-dependent scheme are noteworthy:

(i) $r^i = -\infty, \quad i = 1, ..., m$:
In this case, $\Delta^i_n = 0, i = 1, ..., m, \forall n$. Hence, this is equivalent to the corresponding static scheme.

(ii) $r^i = 0, \quad i = 1, ..., m$:
In this case, $\Delta^i_n = OO^i$. This corresponds to an "ideal" situation where the state-dependent control is triggered as soon as there are echelon outstanding orders, i.e., when $OO^j > 0$ for some j.

3. A GENERALIZED SEMI-MARKOV PROCESS FORMULATION

We now analyze the above system as a *generalize semi-Markov process* (GSMP). For general exposition of GSMP, the readers are referred to Schassberger (1978), Whitt (1980) and Glasserman and Yao (1990a,b). Below, we shall follow closely the formalism in Glasserman and Yao (1990a,b).

The *deterministic* structure of a generalized semi-Markov process is specified by its *scheme*, denoted $\mathcal{G} := (\mathbf{S}, \mathbf{A}, \mathcal{E}, \phi, \gamma)$, where $\mathbf{S}$ is the state space, $\mathbf{A}$ is a set of all possible events, $\mathcal{E}$ is a mapping that gives a set of *active events* at state $s \in \mathbf{S}$, $\phi(s, \alpha)$ is a transition mapping that determines the (new) state to which the system moves from state s immediately after the occurrence of event α, and γ is the clock speed. For simplicity, we shall assume that clocks run at unit speed throughout.

Note that for a given system, specification of its scheme may not be unique. However, its choice can be crucial in facilitating analysis. As an example, for our system, $x := (x^\ell)_{\ell=1}^{2m+1}$, which is a natural state description defined in §1, would be a legitimate state variable. However, it turns out that a better choice would be $y_n := (y^\ell_n)_{\ell=1}^{m+1}$, where $y^i_n = x^{2i-2}_n + x^{2i-1}_n$, $i = 1,...m$, is the number of jobs that have completed service at stage $i-1$ but have not yet completed service at stage i at time n, and $y^{m+1}_n = x^{2m}_n - x^{m+1}_n$. $\sum_{\ell=i+1}^{m+1} y^\ell_n$ is then the echelon inventory position at stage i (cf. (2.1)). Taking into account (2.8), the state space of y_n is then given by

$$\mathbf{S} := \{Y \in Z^m_+ \times Z : \sum_{\ell=i}^{j} y^\ell \leq u_{ij} + \sum_{\ell=i}^{j-1} (r^\ell - \sum_{\ell=i+1}^{m+1} y^\ell)^+\}; \tag{3.1}$$

with Z^m_+ denotes the set of m-dimensional vectors of non-negative integer components; and Z is the set of integers.

Let α_i be the service completion event at stage i and α_{m+1} be the order arrival event.

$$\mathbf{A} := \{\alpha_i, i = 1, ..., m+1\}. \tag{3.2}$$

is thus the set of possible events.

Since it is assumed that stage m' has infinite number of waiting rooms, α_{m+1} is always active; on the other hand, α_i is active if and only if server i is not blocked and there are jobs to be served in stage i. That is, the set of active events at time n when the system is at stage s is

$$\mathcal{E}(s) = \{\alpha_{m+1}\} \cup \{\alpha_i \in \mathbf{A} : y_n^{i-1} > 0 \text{ and } \sum_{\ell=i+1}^{j} y_n^{\ell} < u_{ij}, \forall j = i+1, ..., m\}$$

$$\cup \{\alpha_i \in \mathbf{A} : y_n^{i-1} > 0 \text{ and } \sum_{\ell=i+1}^{m+1} y_n^{\ell} < r^i\} \tag{3.3}$$

where the last two subsets are due to Lemma 1.1. and the state-dependent scheme.

The active event α^* that has the smallest clock time triggers the next state transition. The new state to which the system moves from state s is given by

$$\phi(s, \alpha^*) = \begin{cases} s - e^i + e^{i+1}; & \text{for } \alpha^* = \alpha_i = 1, ..., m; \\ s - e^{m+1} & \text{for } \alpha^* = \alpha_{m+1}; \end{cases} \tag{3.4}$$

where e^i denote the ith unit vector. In the first case, for $i = 1, ..., m-1$, it is because a job completion at stage i transforms a raw job into a finished job. For $i = m$, the finished job will either fill an order or increase the number of finished job at stage m. In any case, y^{i-1} is reduced by 1 and y^i is increased by 1. For the case when $i = m+1$, the arrival enters stage m' if there is no finished job at stage m (i.e., $y^{m+1} \leq 0$), otherwise, it removes one finished job from stage m. For both cases, y^{m+1} is reduced by 1.

The (deterministic) scheme $\mathcal{G}$ described above, together with the (stochastic) job order arrival and the service processes thus completely characterize the dynamics of the GSMP.

4. CONDITIONS (M), (C), (CX), AND SUBSCHEME

Several important structural properties of $\mathcal{G}$ can be established using the notions of *string* σ, *language* $\mathcal{L}$, and *score space* $\mathcal{N}$. These are structural elements that relate the system to a monotone structure known as *antimatroid* (see Glasserman and Yao (1990a,b) for details). Briefly, a string $\sigma = \beta_1, \beta_2, ...$ is an order sequence of events. It is "feasible" if it describes a feasible evolution of event sequences. A language $\mathcal{L}$ is a set of all feasible strings generated by $\mathcal{G}$. The score $N_\alpha(\sigma)$ is the number of occurrence of event α in the string $\sigma \in \mathcal{L}$. Denote $N(\sigma) := (N_\alpha(\sigma))_{\alpha\in\mathbf{A}}$, the score space $\mathcal{N}$ is the set of all feasible scores

$$\mathcal{N} := \{N(\sigma) \in Z_+^{|\mathbf{A}|} : \sigma \in \mathcal{L}\}; \tag{4.1}$$

where $|\mathbf{A}|$ denote the cardinality of $\mathbf{A}$. The following conditions are crucial in establishing various stochastic properties of a GSMP (Glasserman (1988), and Glasserman and Yao (1990a,b)):

Definition 4.1.
- *non-interruption*: $\forall s \in \mathbf{S}$, α, $\beta \in \mathcal{E}(s)$, $\alpha \neq \beta \Rightarrow \beta \in \mathcal{E}(\phi(s,\alpha))$;
- *permutability*: $\forall\sigma_1$, $\sigma_2 \in \mathcal{L}$, $N(\sigma_1) = N(\sigma_2) \Rightarrow \mathcal{E}(\phi(s_0,\sigma_1)) = \mathcal{E}(\phi(s_0,\sigma_2))$;
- *strong permutability*: $\forall\sigma_1$, $\sigma_2 \in \mathcal{L}$, $N(\sigma_1) = N(\sigma_2) \Rightarrow \phi(s_0,\sigma_1) = \phi(s_0,\sigma_2)$;
- *min closure*: $\forall\sigma_1,\sigma_2 \in \mathcal{L}$, there exists $\sigma \in \mathcal{L}$ such that $N(\sigma) = N(\sigma_1) \wedge N(\sigma_2)$;

where $\wedge$ denotes the *min* operation in the componentwise sense between two vectors. $\phi(s_0,\sigma)$ denotes the state reached by the string from the initial state s_0.

In words, non-interruption requires that no event be interrupted before it is completed; permutability refers to the fact that two strings with the same score lead to the same event list (regardless of the sequence of occurrences of events). Strengthening the requirement of the "same event list" to the same *state* leads to strong permutability. A combination of the above conditions in turn give the following properties:

Definition 4.2.
- Condition (C): non-interruption and strong permutability;
- Condition (M): non-interruption and permutability;
- Condition (CX): Condition (M) and min-closure.

Note that while both (C) and (CX) imply (M), they do not imply each other.

Lemma 4.1. The scheme $\mathcal{G}$ in §3 (cf. (3.1)-(3.4)) satisfies the non-interruption and strong permutability conditions, the min-closure property is not satisfied in general except for the case where $r^i = -\infty$, $\forall i$. Therefore, generally, the scheme satisfies (M) and (C), but not (CX).

Proof. From the description of the system, non-interruption is obvious. For strong permutability, the following relationship between scores and state is immediate: for any $\sigma \in \mathcal{L}$, if $y^i = \phi(s_0,\sigma)$, then

$$y^i = N_{\alpha_{i-1}}(\sigma) - N_{\alpha_i}(\sigma) + s_{0,i}, \qquad i = 2,...,m+1; \tag{4.2}$$

where $s_{0,i}$ is the ith component of s_0 (i.e., y^i at time zero); and $y^1 = +\infty$ by assumption. From the above relation, it is clear that if two strings have the same score, they must lead to the same state. Hence strong permutability is satisfied. This, together with the non-interruption property, implies that condition (C), and hence condition (M), is satisfied.

For the second part in the lemma, from the state space **S** (cf. (3.1)), the score space can be written as:

$$\mathcal{N} := \{N \in Z_+^m \times Z : N_{\alpha_i} \geq N_{\alpha_{i+1}}; N_{\alpha_i} - N_{\alpha_j} \leq u_{ij}$$

$$+\sum_{\ell=i}^{j}[r^\ell - (N_{\alpha_i} - N_{\alpha_m} - N_{\alpha_{m+1}})]^+, \quad i = 1,...,m;\ j = i+1,...,m\}. \tag{4.3}$$

(Note the quantity $(\cdot)$ above denotes the echelon inventory position.) When $r^i = -\infty$, $\forall i$, $[\cdot]^+ \equiv 0$, i.e., we have the static scheme. In that case, it is straightforward to verify that $\mathcal{N}$ is min-closed, i.e., if N^1, $N^2 \in \mathcal{N}$, then $N^1 \wedge N^2 \in \mathcal{N}$. However, for the state-dependent scheme, this is *not* true in general as can be seen by the following counter-example: Suppose

$$N^1 = (n_1, ..., n_{m-2}, 3, 2, 4), \qquad \text{and} \qquad N^2 = (n_1, ..., n_{m-2}, 4, 0, 7), \tag{4.4}$$

i.e., N^1 and N^2 only differ in $N_{\alpha_{m-1}}$, N_{α_m} and $N_{\alpha_{m+1}}$. Suppose $n_1 = \cdots = n_{m-2} = 4$. Consider the system where $a_i = k_i = 1$, $b_i = 0$ and $r^{m-1} = 0$, $i = 1, ..., m$. Then clearly both N^1 and N^2 are feasible since $n_i - n_j$, $i = 1, ..., m-2$, $j = i+1, ..., m$ satisfy (4.3). Also

$$N_{m-1}^1 - N_m^1 = 1 \leq k_m + b_{m-1} + 1 = 2; \tag{4.5}$$

and

$$N_{m-1}^2 - N_m^2 = 4 \leq k_m + b_{m-1} + 3 = 4. \tag{4.6}$$

However,

$$N' := N^1 \wedge N^2 = (n_1, ..., n_{m-2}, 3, 0, 4) \notin \mathcal{N}; \tag{4.7}$$

since

$$N'_{\alpha_{m-1}} - N'_{\alpha_m} = 3 > k_m + b_{m-1} + 1 = 2 \tag{4.8}$$

which violates the second constraint in (4.3). *Q.E.D.*

Having consider the scheme of the GSMP, we now relate the above properties to the effects of changing the (stochastic) clock times on system performance. We first recall the following definitions

Definition 4.3. (see, e.g., Ross (1983), Stoyan(1983) and more generally, Kamae, Krengel, and O'Brien (1977)) A stochastic process $\{X_t\}$ is dominated by another s-tochastic process $\{Y_t\}$ in the sense of stochastic ordering, denoted $\{X_t\} \leq_{st} \{Y_t\}$, if there exist on a common probability space two processes $\{\hat{X}_t\}$ and $\{\hat{Y}_t\}$ that are equal in probability law to $\{X_t\}$ and $\{Y_t\}$ respectively, and for all t, $\hat{X}_t \leq \hat{Y}_t$ almost surely (a.s.).

Definition 4.4. (Shanthikumar and Yao (1989, 1991)) A collection $\{X(\theta), \theta \in \Theta\}$ of random variables with a convex parameter set $\Theta \subseteq \Re^m$ is said to be strongly s-tochastically (increasing and) convex in θ, denoted $\{X(\theta)\} \in$ SSCX (SSICX), if $X(\theta)$ is (increasing and) convex in θ almost surely (a.s.). When θ is a vector of integer components, convexity is similarly defined.

Let $\omega := \{\omega_\alpha(n) : \alpha \in \mathbf{A}\}$ be an input sequence to the scheme where $\omega_\alpha(n)$ is the nth clock time for event α, let $T(n) := (T_{\alpha_1}(n), ..., T_{\alpha_m}(n))$, where $T_\alpha(n)$ is the event time of the nth occurrence of event α. From Lemma 4.1, and Theorem 3.3 of Glasserman and Yao (1990a), and Theorem 4.1 of Glasserman and Yao (1990b), we have

Proposition 4.2. For the system with the scheme defined in §3, if $\omega^1 \leq_{st} \omega^2$, then for any n, $T(n, \omega^1) \leq_{st} T(n, \omega^2)$. Also, $T(n, \omega) \in$ SSCX in ω if $r^i = -\infty$ for all $i = 1, ..., m$. Consequently, E $fT(n)$ is increasing (and convex for $r^i = -\infty$) in ω for any increasing (and convex) function f where E$(\cdot)$ denotes expectation.

To show that the state-dependent scheme is "better" than the static scheme, we need the notion of subscheme in Glasserman and Yao (1990a).

Definition 4.5. The scheme $\mathcal{G}^1$ is a sub-scheme of $\mathcal{G}^2$, denoted $\mathcal{G}^1 \subseteq \mathcal{G}^2$, if
(i) $\mathbf{S}^1 \subseteq \mathbf{S}^2$,
(ii) $\mathcal{E}^1(s) \subseteq \mathcal{E}^2(s), \forall s \in \mathbf{S}^1$, and
(iii) $\phi^1(s, \alpha) = \phi^2(s, \alpha), \forall s \in \mathbf{S}^1$, $\forall \alpha \in E^1(s)$.

Now consider two systems with parameters (a^1, b^1, k^1, r^1) and (a^2, b^2, k^2, r^2) and denote the corresponding generalized semi-Markov schemes (GSMS) with the super-scripts 1 and 2 respectively. From (3.1) and (3.2), it is obvious that increasing any components of the control parameters (a, b, k, r) will enlarge the state space as well as the active event list, but this will not affect the transition function. Hence, the three conditions in Definition 4.5 are trivially satisfied. In other words, $\mathcal{G}^1 \subseteq \mathcal{G}^2$ if $(a^1, b^1, k^1, r^1) \leq (a^2, b^2, k^2, r^2)$. It now follows from Lemma 4.1 and Theorem 4.3 of Glasserman and Yao (1990a) that

Proposition 4.3. For the system in §3, $T_{\alpha_i}(n)$, $i = 1, ..., m$, $n = 1, ...$, are (stochastically) decreasing in the control parameters (a, b, k, r). In particular, this implies that for any n, $T_{\alpha_i}(n)$, $i = 1, ..., m$, for the state-dependent scheme are smaller than those of the static scheme almost surely. In this sense, the state-dependent control is better than the static scheme.

In summary, we have analyzed a tandem system controlled using a fairly complex mechanism. We have demonstrated that by recasting the system as a GSMP (with an appropriately chosen state space), and by studying its deterministic elements, important structural properties can be established which would otherwise be difficult to obtain.

REFERENCES

Akyildiz, I. F., and Perros, H. G. (Eds.). Special Issue of Performance Evaluation Journal on Queueing Networks with Finite Buffers.

Buzacott, J. A. 1989. Generalized Kanban/MRP System. Working paper, Department of Management Sciences, University of Waterloo, Ontario, Canada.

Cheng, D. W. 1991a. Tandem Queues with General Blocking: Stochastic Comparisons and Structural Properties. Ph.D. Thesis, Graduate School of Business, Columbia University, New York.

Cheng, D. W. 1991b. Second Order Properties in a Tandem Queue with General Blocking. (To appear in *OR Letters.*)

Cheng, D. W. 1991c. Line Reversibility of Tandem Queues with General Blocking. (Submitted for publication.)

Cheng, D. W. and Yao, D. D. 1991. Tandem Queues with General Blocking: A Unified Model and Comparison Results. (Submitted for publication.)

Cheng, D. W. and Zhu Y. Optimal Order of Servers in a Tandem Queue With General Blocking. (Submitted for publication.)

Glasserman, P., and Yao, D. D. 1990a. Monotonicity in Generalized Semi-Markov Process. *Math. O.R.*, **17** 1-21.

Glasserman, P., and Yao, D. D. 1990b. Generalized Semi-Markov Process: Antimatroid Structure and Second-Order Properties. *Math. O.R.*, **17** 444-469.

Glasserman, P., and Yao, D.D. 1991. Structured Buffer-Allocation Problems in Production Lines. (Submitted for publication.)

Kamae T., Krengel U. and O'Brien, G. 1977. Stochastic Inequalities on Partially Ordered Spaces. *Ann. Prob.*, **5**, 899-912.

Onvural, R. O., 1990. Survey of Closed Queueing Networks with Blocking. *ACM Computing Surveys*, **22** 83-121.

Perros, H. G. 1986. A Survey of Queueing Networks with Blocking. Technical Report, Department of Computer Science, North Carolina State University, Raleigh, NC27695.

Ross, S. 1983. *Stochastic Processes.* New York, NY: Wiley.

Schassberger, R 1978. Insensitivity of Steady State Distribution of Generalized Semimarkov Processes with Speeds. *Advances in Applied Probability*, **10**, 836-851.

Shanthikumar J. G. and Yao D. D. 1989. Second-Order Stochastic Properties in Queueing Systems. *Proceedings of the IEEE*, Special issue on Dynamics of Discrete Event Systems, **77**, 162-170.

Shanthikumar J. G. and Yao D. D. 1991. Strong Stochastic Convexity: Closure Properties and Applications. *Journal of Applied Probability*, **28**, 131-145.

Stoyan, D. 1983. *Comparison Methods for Queues and other Stochastic Models.* Wiley, New York.

Whitt, W. 1980. Continuity of Generalized Semi-Markov Process. *Mathematics of Operations Research*, **5**, 494-501.

Zipkin, P. 1989. A Kanban-like Production Control System: Analysis of Simple Models. Research working paper 89-1, Graduate School of Business, Columbia University.

Queueing Networks with Finite Capacity
R.O. Onvural and I.F. Akyildiz (Editors)

A FINITE CAPACITY QUEUING NETWORK WITH SINGLE AND MULTIPLE PROCESSING NODES

S. Chakravarthy

Department of Science and Mathematics, GMI Engineering & Management Institute, Flint, MI 48504-4898.

Abstract

We consider a queuing network consisting of two nodes connected in series. Jobs (or customers) arrive at node 1 according to a Poisson process of rate λ. There is a finite buffer in front of node 1 and any arrival that finds the buffer full is considered lost. Jobs are processed one at a time in node 1 by a single server and proceed to node 2, where they are served in groups of varying size by another server. A dynamic service rule in terms of a pre-specified number is used at node 2, which also has a finite buffer. The service times of the jobs in both nodes are assumed to be independent and exponentially distributed with (possibly) different parameters. We study this queuing network using Markov process by obtaining expressions for the steady-state queue length densities at arbitrary time points and at departures, which are the main ingredients for computing various system performance measures such as the throughput, fraction of time the server in node 1 is blocked, fraction of time the servers are idle and the fraction of lost jobs. Efficient algorithmic procedures for computing the steady-state queue length densities and other system performance measures are discussed. Some numerical examples are presented.

Key Words: Queuing network, finite capacity, Markov process, and algorithmic probability.

1. INTRODUCTION AND MODEL DESCRIPTION

We consider a finite capacity queuing network consisting of two nodes. Node i has its own finite capacity of size K_i, for $i = 1,2$. We assume that arrivals at node 1 occur according to a homogeneous Poisson process of rate λ. Any arrival that finds the buffer full is considered to be lost. Upon completion of a service at node 1, the job proceeds to node 2 if there is a space available. Otherwise the server at node 1 is blocked. The server at node

2 processes the jobs in groups of size at least L, where the pre-specified value $L \geq 1$, called threshold operates as follows. The server at node 2 will begin service to all jobs present in node 2 if the number of jobs is at least L; otherwise the server waits until the queue length reaches L for the first time. The services at node i are assumed to be exponentially distributed with parameter μ_i, $i = 1,2$. The services at both nodes are independent of each other.

The above queuing network has potential applications in areas such as flexible manufacturing systems, where the jobs may go through two stations. In station 1 these are processed one at a time. For example, the jobs at station 1 need individual processing such as boring, grinding, pinning, and cutting. In this case the jobs are considered to be heterogeneous. However, in the second station these jobs may all require some standard processing such as sand blasting or heat treatment or flushing with the same coolant. In this situation the jobs may now be considered to be homogeneous, so that they can all be processed efficiently by grouping them together. Similar applications can be obtained in industrial lasers, automated production process including machine vision, package sorting, and in chemical plants that treat hazardous petrochemical and petroleum wastes.

This class of models in the context of finite capacity single server queuing systems, such as GI/PH/1/N and MAP/G/1/N, was first studied by Chakravarthy [2-3]. The potential applications of these models were outlined in those papers. However, here we concentrate on the usefulness of such models in the area of finite capacity queuing networks. Although we assume Poisson arrivals and exponential services, the methodology presented here can be extended to include, for example, Markovian arrival process and with phase type services. In this case the state space of the Markov process governing the queuing network will be large. In order to minimize the notation and present the main ideas of the model, we have chosen to study the simplest case. The model is pictorially described in Figure 1 below.

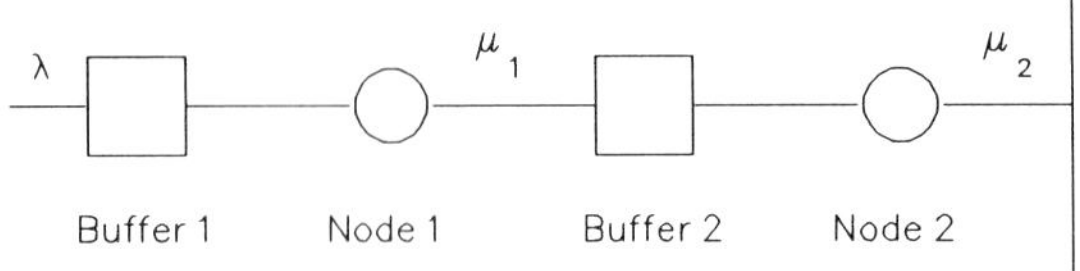

Figure 1

The paper is organized as follows. The Markov process governing the model under study and the joint steady-state queue length densities at arbitrary time points and at departure epochs are given in Section 2. Algorithmic procedures for the computation of the steady-state vector are also presented in Section 2. Several system performance measures useful in qualitative interpretation of the model are given in Section 3. Selected numerical examples are discussed in Section 4 and concluding remarks, including some variants of the model are given in Section 5.

2. THE JOINT QUEUE LENGTH DENSITIES

2.1 The Markov Process: The model discussed in Section 1 can be studied as a Markov process with $\{(K_1 + 2)[L + (K_2 + 1)(K_2 - L + 1)] + (K_1 + 1)(K_2 - L + 1)\}$ states. These states are partitioned into three groups as follows.

$$\mathbf{i'} = \{(i,j): 0 \le j \le L-1\},\ 0 \le i \le K_1+1,$$

$$\mathbf{i''} = \{(i,j,k): 0 \le j \le K_2,\ L \le k \le K_2\},\ 0 \le i \le K_1+1,$$

$$\mathbf{i^*} = \{(i^*, K_2, k): L \le k \le K_2\},\ 1 \le i \le K_1+1.$$

Note that the level **i'** corresponds to the set of states in which there are i jobs (including one, if any, in service) in node 1 and j jobs are waiting for service in node 2, while the server in that node is idle; the level **i''** corresponds to the set of states in which i jobs are present in node 1, j jobs are waiting in queue in front of node 2 and that the current service in node 2 has k jobs. The set of states $\mathbf{i^*}$ corresponds to the case where server in node 1 is blocked. Our definition of blocking is as follows. The server in node 1 will be blocked when upon completion of the current job there is no space available in buffer in front of node 2. That is, the server in node 1 will be blocked when upon completion of the current job there are already K_2 jobs waiting in front of node 2.

In the following the notation **e** will stand for a column vector of 1's; $\mathbf{e}_i$, a unit column vector with 1 in the i-th position and 0 elsewhere; and I, an identity matrix of appropriate dimensions. When there is a need to identify the order of these quantities, they will be denoted within parentheses. For example $\mathbf{e}_i(r)$ will denote a unit vector of order r.

The infinitesimal generator Q of the Markov process governing the system is given by

$$Q = \begin{bmatrix}
-\lambda I & \lambda I & 0 & \cdots & 0 & 0 & 0 & 0 & 0 & \cdots & 0 & 0 & 0 & 0 & 0 & \cdots & 0 & 0 \\
C_2 & C_0 & \lambda I & \cdots & 0 & 0 & C_1 & 0 & 0 & \cdots & 0 & 0 & 0 & 0 & 0 & \cdots & 0 & 0 \\
\vdots & \vdots & \vdots & \cdots & \vdots & \vdots & \vdots & \vdots & \vdots & \cdots & \vdots & \vdots & \vdots & \vdots & \vdots & \cdots & \vdots & \vdots \\
0 & 0 & 0 & \cdots & \lambda I & 0 & 0 & 0 & 0 & \cdots & 0 & 0 & 0 & 0 & 0 & \cdots & 0 & 0 \\
0 & 0 & 0 & \cdots & C_0 & \lambda I & 0 & 0 & 0 & \cdots & 0 & 0 & 0 & 0 & 0 & \cdots & 0 & 0 \\
0 & 0 & 0 & \cdots & C_2 & -\mu_1 I & 0 & 0 & 0 & \cdots & C_1 & 0 & 0 & 0 & 0 & \cdots & 0 & 0 \\
D_1 & 0 & 0 & \cdots & 0 & 0 & D_0 & \lambda I & 0 & \cdots & 0 & 0 & 0 & 0 & 0 & \cdots & 0 & 0 \\
0 & D_1 & 0 & \cdots & 0 & 0 & A_1 & A_0 & \lambda I & \cdots & 0 & 0 & A_2 & 0 & 0 & \cdots & 0 & 0 \\
\vdots & \vdots & \vdots & \cdots & \vdots & \vdots & \vdots & \vdots & \vdots & \cdots & \vdots & \vdots & \vdots & \vdots & \vdots & \cdots & \vdots & \vdots \\
0 & 0 & 0 & \cdots & D_1 & 0 & 0 & 0 & 0 & \cdots & A_0 & \lambda I & 0 & 0 & 0 & \cdots & A_2 & 0 \\
0 & 0 & 0 & \cdots & 0 & D_1 & 0 & 0 & 0 & \cdots & A_1 & A_0+\lambda I & 0 & 0 & 0 & \cdots & 0 & A_2 \\
0 & 0 & 0 & \cdots & 0 & 0 & B_1 & 0 & 0 & \cdots & 0 & 0 & B_0 & \lambda I & 0 & \cdots & 0 & 0 \\
0 & 0 & 0 & \cdots & 0 & 0 & 0 & B_1 & 0 & \cdots & 0 & 0 & 0 & B_0 & \lambda I & \cdots & 0 & 0 \\
\vdots & \vdots & \vdots & \cdots & \vdots & \vdots & \vdots & \vdots & \vdots & \cdots & \vdots & \vdots & \vdots & \vdots & \vdots & \cdots & \vdots & \vdots \\
0 & 0 & 0 & \cdots & 0 & 0 & 0 & 0 & 0 & \cdots & 0 & 0 & 0 & 0 & 0 & \cdots & B_0 & \lambda I \\
0 & 0 & 0 & \cdots & 0 & 0 & 0 & 0 & 0 & \cdots & B_1 & 0 & 0 & 0 & 0 & \cdots & 0 & -\mu_1 I
\end{bmatrix},$$

where the entries of the generator are given in the Appendix.

Before we proceed further we note the following observation. We assume that at the moment when server 1 is unblocked, the number of jobs that enter into service in node 2 is K_2. It is very trivial to modify this assumption to include the case when $K_2 + 1$ jobs (including the processed one in node 1) can enter into service in node 2. In this case the state space of the Markov process will have $\{(K_1 + 2)[L + (K_2 + 1)(K_2 - L + 2)] + (K_1 + 1)(K_2 - L + 2)\}$ states. Only few entries of the generator Q need to be modified and the details are omitted.

2.2. The joint queue length density at an arbitrary time: Let $\mathbf{x} = (\mathbf{u}, \mathbf{v}, \mathbf{w})$ denote the stationary probability vector of Q. That is, $\mathbf{x}$ is the unique solution to

$$\mathbf{x}Q = \mathbf{0} \quad \text{and} \quad \mathbf{xe} = 1. \tag{2}$$

Partitioning the vectors $\mathbf{u}$, $\mathbf{v}$, and $\mathbf{w}$ into $\mathbf{u} = (\mathbf{u}(0), \ldots, \mathbf{u}(K_1+1))$, $\mathbf{v} = (\mathbf{v}(0), \ldots, \mathbf{v}(K_1+1))$ and $\mathbf{w} = (\mathbf{w}(1), \ldots, \mathbf{w}(K_1+1))$, the equations in (2) can be expanded as

$$-\lambda\, \boldsymbol{u}(0) + \boldsymbol{u}(1)\, C_2 + \boldsymbol{v}(0)\, D_1 = \mathbf{0}, \tag{3}$$

$$\lambda\, \boldsymbol{u}(i-1) + \boldsymbol{u}(i)\, C_0 + \boldsymbol{u}(i+1)\, C_2 + \boldsymbol{v}(i)\, D_1 = \mathbf{0}, \; 1 \le i \le K_1, \tag{4}$$

$$\lambda\, u(K_1) - \mu_1 u(K_1 + 1) + v(K_1 + 1) D_1 = \mathbf{0}\,, \tag{5}$$

$$u(1) C_1 + v(0) D_0 + v(1) A_1 + w(1) B_1 = \mathbf{0}\,, \tag{6}$$

$$u(i+1) C_1 + \lambda\, v(i-1) + v(i) A_0 + v(i+1) A_1 + w(i+1) B_1 = \mathbf{0}\,,\ 1 \le i \le K_1\,, \tag{7}$$

$$\lambda\, v(K_1) + v(K_1 + 1)\,[A_0 + \lambda I] = \mathbf{0}\,, \tag{8}$$

$$v(1) A_2 + w(1) B_0 = \mathbf{0}\,, \tag{9}$$

$$v(i) A_2 + \lambda\, w(i-1) + w(i) B_0 = \mathbf{0}\,,\ 2 \le i \le K_1\,, \tag{10}$$

$$v(K_1 + 1) A_2 + \lambda\, w(K_1) - \mu_2 w(K_1 + 1) = \mathbf{0}\,. \tag{11}$$

Before we discuss algorithmic procedures for computing the vector **x**, which will be the main ingredient for the qualitative interpretation of the model under study, we will prove a number of results relating the vectors **u**, **v** and **w**.

THEOREM 1: The vector **w** is explicitly given in terms of **v** as

$$w(1) = \frac{\mu_1}{\lambda + \mu_2}\, v_{K_2}(1)\,,$$

$$w(i) = \left[\frac{\lambda}{\lambda + \mu_2}\right]^{i-1} \frac{\mu_1}{\lambda + \mu_2}\, v_{K_2}(1) + \sum_{\nu=2}^{i} \left[\frac{\lambda}{\lambda + \mu_2}\right]^{i-\nu} \frac{\mu_1}{\lambda + \mu_2}\, v_{K_2}(\nu)\,,\ for\ 2 \le i \le K_1\,, \tag{12}$$

$$w(K_1 + 1) = \frac{\lambda}{\mu_2}\left[\frac{\lambda}{\lambda + \mu_2}\right]^{K_1 - 1} \frac{\mu_1}{\lambda + \mu_2}\, v_{K_2}(1) + \sum_{\nu=2}^{K_1} \left[\frac{\lambda}{\lambda + \mu_2}\right]^{K_1-\nu} \frac{\mu_1}{\lambda + \mu_2}\, v_{K_2}(\nu) + \frac{\mu_1}{\mu_2}\, v_{K_2}(K_1 + 1)\,.$$

PROOF: By exploiting the special structure of the coefficient matrix A_2 appearing in Equations (9-11), we get the following recursive relations on **w**(i).

$$w(1) = \frac{\mu_1}{\lambda + \mu_2} v_{K_2}(1),$$

$$w(i) = \frac{1}{\lambda + \mu_2} [\mu_1 v_{K_2}(i) + \lambda w(i-1)], \; 2 \le i \le K_1, \tag{13}$$

$$w(K_1 + 1) = \frac{1}{\mu_2} [\mu_1 v_{K_2}(K_1 + 1) + \lambda w(K_1)],$$

from which, after some simplifications, the stated Equations (12) follow.

REMARK: Although Equations (12) give explicit expressions for **w** in terms of **v**, computationally it is efficient to use the recursive relations given in (13) for the evaluation of **w**.

THEOREM 2: We have the following relations between **u** and **v**.

$$u_0(0) = \frac{\mu_2}{\lambda} v_0(0) e,$$

$$u_0(i) = \frac{1}{\lambda + \mu_1} [\lambda u_0(i-1) + \mu_2 v_0(i) e], \; 1 \le i \le K_1, \tag{14}$$

$$u_0(K_1 + 1) = \frac{1}{\mu_1} [\lambda u_0(K_1) + \mu_2 v_0(K_1 + 1) e],$$

and for $1 \le j \le L - 1$,

$$u_j(0) = \frac{\mu_1}{\lambda} u_{j-1}(1) + \frac{\mu_2}{\lambda} v_j(0) e,$$

$$u_j(i) = \frac{1}{\lambda + \mu_1} [\lambda u_j(i-1) + \mu_1 u_{j-1}(i+1) + \mu_2 v_j(i) e], \; 1 \le i \le K_1 \tag{15}$$

$$u_j(K_1 + 1) = \frac{1}{\mu_1} [\lambda u_j(K_1) + \mu_2 v_j(K_1 + 1) e].$$

PROOF: Once again by exploiting the special structure of the coefficient matrices C_2 and D_1 appearing in Equations (3-5), we get

$$\lambda u_0(0) = \mu_2 v_0(0) e, \quad \lambda u_j(0) = \mu_1 u_{j-1}(1) + \mu_2 v_j(0) e, \; 1 \le j \le L-1,$$

$$(\lambda + \mu_1) u_0(i) = \lambda u_0(i-1) + \mu_2 v_0(i) e, \; 1 \le i \le K_1, \tag{16}$$

$$(\lambda + \mu_1) u_j(i) = \lambda u_j(i-1) + \mu_1 u_{j-1}(i+1) + \mu_2 v_j(i) e, \; 1 \le j \le L-1, \; 1 \le i \le K_1, \tag{17}$$

$$\mu_1 u_j(K_1 + 1) = \lambda u_j(K_1) + \mu_2 v_j(K_1 + 1) e, \; 1 \le j \le L-1. \tag{18}$$

The stated equations in (14) and (15) follow from (16-18).

REMARK: From expressions in (14) it is readily apparent that **u** can also be written explicitly in terms of **v** and the arduous details are omitted. However, for efficient computation of **u** one has to use the recursive schemes given in Equations (16-18).

THEOREM 3: We have

$$\mu_2 \sum_{i=1}^{K_1+1} w(i)e = \mu_1 \sum_{i=1}^{K_1+1} v_{K_2}(i)\, e\,, \tag{19}$$

$$\mu_1 \sum_{i=1}^{K_1+1} u_j(i) = \mu_2 \sum_{\nu=0}^{j} \sum_{i=0}^{K_1+1} v_\nu(i)\, e\,,\ 0 \le j \le L-1\,. \tag{20}$$

PROOF: Equation (19) follows by adding Equations in (13) over $1 \le i \le K_1 + 1$. To get (20) we first add Equations (3-5) over i to get

$$-\mu_1 \sum_{i=1}^{K_1+1} u(i) + \sum_{i=1}^{K_1+1} u(i)\, C_2 + \sum_{i=0}^{K_1+1} v(i)\, D_1 = \mathbf{0}\,. \tag{21}$$

Post-multiplying Equation (21) by **e** and using the special structure of C_2 and D_1, we get the stated equation.

REMARKS: The equations (19) and (20) are intuitively obvious. For example the LHS of (19) gives the rate at which the server in node 1 will be unblocked. The RHS of the same equation gives the rate at which the server in node 1 will be blocked.

2.3. Computation of v vector: In view of Theorems 1 and 2, we need to evaluate only the vector **v** to obtain the steady-state probability vector **x** of Q. Once again by exploiting the special structure of the coefficient matrices appearing in Equations (6-8), these equations can be rewritten in such a way that they are well suited for solution by Gauss-Seidel iterative procedure. The necessary equations are given in the Appendix.

2.4. The joint queue length density at a departure epoch: Let $\mathbf{y} = (\mathbf{y}(0), \ldots, \mathbf{y}(K_1+1))$ denote the joint queue length density of the number of jobs in nodes 1 and 2 at a departure epoch. The j-th component of $\mathbf{y}(i)$, $y_j(i)$, $0 \le i \le K_1 + 1$, $0 \le j \le K_2$, gives the steady-state probability that a departure from the system (namely from node 2) will see exactly i jobs in node 1 and j jobs in node 2. The following theorem gives an expression for **y**.

THEOREM 4: We have

$$\begin{aligned} y_j(0) &= c\, v_j(0)\, e\,,\ 0 \le j \le K_2\,, \\ y_j(i) &= c\, v_j(i)\, e\,,\ 1 \le i \le K_1+1\,,\ 0 \le j \le K_2 - 1\,, \\ y_{K_2}(i) &= c\, [\, v_{K_2}(i)\, e + w(i)\, e\,]\,,\quad \le i \le K_1+1\,, \end{aligned} \tag{22}$$

where the normalizing constant c is given by $c = (1 - \boldsymbol{u}\,\boldsymbol{e})^{-1}$.

PROOF: follows immediately from the definition of **y**.

2.5. The density of the number of jobs processed in Node 2: Suppose the N denotes the number of jobs processed during a service in node 2. Then the following theorem gives an expression for the conditional probability density of N.

THEOREM 5: The conditional probability density $f_B(.)$, of the random variable N given B = {the server in node 2 is busy} is given by

$$f_B(L) = \sum_{i=0}^{K_1+1} \sum_{j=0}^{L} y_j(i) ,$$
$$f_B(n) = \sum_{i=0}^{K_1+1} y_n(i) , \; L+1 \leq n \leq K_2 . \tag{23}$$

PROOF: follows immediately from Equations in (22) and noting that P(B) = 1 - **ue**.

2.6. The Throughput: The *throughput*, γ, defined as the rate at which the jobs depart from the system is given by

$$\gamma = \mu_2 \left[\sum_{i=0}^{K_1+1} \sum_{j=0}^{K_2} \sum_{k=L}^{K_2} k \, v_{j,k}(i) + \sum_{i=1}^{K_1+1} \sum_{k=L}^{K_2} k \, w_k(i) \right] . \tag{24}$$

The following theorem gives an alternative expression for γ.

THEOREM 6: We have

$$\gamma = \mu_2 \, \mu_N (1 - \mathbf{u} \, \mathbf{e}) , \tag{25}$$

where μ_N denotes the mean number of jobs processed during a service.
PROOF: From Equations (A1-A10) given in the appendix, we see that

$$(\lambda + \mu_2) \sum_{r=L}^{K_2} r \, \mathbf{v}_{0,r}(0) = L \, \mu_1 \, \mathbf{u}_{L-1}(1) + \mu_2 \sum_{r=L}^{K_2} r \, \mathbf{v}_r(0) \, \mathbf{e} , \tag{26}$$

$$(\lambda + \mu_1 + \mu_2) \sum_{r=L}^{K_2} r \, \mathbf{v}_{0,r}(i) = L \, \mu_1 \, \mathbf{u}_{L-1}(i+1) + \lambda \sum_{r=L}^{K_2} r \, \mathbf{v}_{0,r}(i-1)$$
$$+ \mu_2 \sum_{r=L}^{K_2} r \, \mathbf{v}_r(i) \, \mathbf{e} , \; 1 \leq i \leq K_1 , \tag{27}$$

$$(\mu_1 + \mu_2) \sum_{r=L}^{K_2} r \, \mathbf{v}_{0,r}(K_1+1) = \lambda \sum_{r=L}^{K_2} r \, \mathbf{v}_{0,r}(K_1) + \mu_2 \sum_{r=L}^{K_2} r \, \mathbf{v}_r(K_1+1) \, \mathbf{e} , \tag{28}$$

$$(\lambda + \mu_2) \sum_{r=L}^{K_2} r \, \mathbf{v}_{1,r}(0) = \mu_1 \sum_{r=L}^{K_2} r \, \mathbf{v}_{0,r}(1) + K_2 \, \mu_2 \, \mathbf{w}(1) \, \mathbf{e} , \tag{29}$$

$$(\lambda + \mu_2) \sum_{r=L}^{K_2} r\, v_{j,r}(0) = \mu_1 \sum_{r=L}^{K_2} r\, v_{j-1,r}(1) \,,\ 2 \le j \le K_2 \,, \tag{30}$$

and for $1 \le i \le K_1$

$$\begin{aligned}(\lambda + \mu_1 + \mu_2) \sum_{r=L}^{K_2} r\, v_{1,r}(i) &= \lambda \sum_{r=L}^{K_2} r\, v_{1,r}(i-1) + K_2 \mu_2 w(i+1)\, e \\ &\quad + \mu_1 \sum_{r=L}^{K_2} r\, v_{0,r}(i+1) \,,\end{aligned} \tag{31}$$

$$\begin{aligned}(\lambda + \mu_1 + \mu_2) \sum_{r=L}^{K_2} r\, v_{j,r}(i) &= \lambda \sum_{r=L}^{K_2} r\, v_{j,r}(i-1) \\ &\quad + \mu_1 \sum_{r=L}^{K_2} r\, v_{j-1,r}(i+1) \,,\ 2 \le j \le K_2 \,,\end{aligned} \tag{32}$$

and

$$(\mu_1 + \mu_2) \sum_{r=L}^{K_2} r\, v_{j,r}(K_1 + 1) = \lambda \sum_{r=L}^{K_2} r\, v_{j,r}(K_1) \,,\ 1 \le j \le K_2 \,. \tag{33}$$

Adding Equations (26-33), we get

$$\begin{aligned}&\mu_2 \sum_{i=0}^{K_1+1} \sum_{j=0}^{K_2} \sum_{r=L}^{K_2} r\, v_{j,r}(i) + \mu_1 \sum_{i=1}^{K_1+1} \sum_{r=L}^{K_2} r\, v_{K_2,r}(i) \\ &= L \mu_1 \sum_{i=1}^{K_1+1} u_{L-1}(i) + \mu_2 \sum_{i=0}^{K_1+1} \sum_{r=L}^{K_2} r\, v_r(i)\, e + K_2 \mu_2 \sum_{i=1}^{K_1+1} w(i)\, e \,.\end{aligned} \tag{34}$$

From Equations in (13) it can be verified that

$$\mu_2 \sum_{i=1}^{K_1+1} w(i) = \mu_1 \sum_{i=1}^{K_1+1} v_{K_2}(i) \,,$$

from which we get

$$\mu_2 \sum_{i=1}^{K_1+1} \sum_{r=L}^{K_2} r\, w_r(i) = \mu_1 \sum_{i=1}^{K_1+1} \sum_{r=L}^{K_2} r\, v_{K_2,r}(i) \,. \tag{35}$$

Now using Equation (20) with $j = L-1$ and (35) in (34), we get

$$\gamma = \mu_2 \left[L \sum_{i=0}^{K_1+1} \sum_{j=0}^{L} v_j(i) + \sum_{i=0}^{K_1+1} \sum_{r=L+1}^{K_2} r\, v_r(i) + K_2 \sum_{i=1}^{K_1+1} w(i)\, e \right] . \tag{36}$$

It is very easy to verify from Equations (22) and (23) that the expression given within square brackets in (36) is $\mu_N(1 - \mathbf{ue})$. This completes the proof of the theorem.

REMARK: Note that the result (25) can be obtained through a probabilistic interpretation. The right-hand side is the product of three quantities:the service rate of server in node 2, the fraction of time that server is busy and the average number of jobs processed by the server during a service. This product clearly gives the throughput of the system.

3. SYSTEM PERFORMANCE MEASURES

In this section some selected but useful system performance measures apart from the ones discussed earlier, are listed along with the corresponding formulas. These measures, obtained from the knowledge of the steady-state probability vector $\mathbf{y}_1$ and $\mathbf{y}_2$, are useful in the interpretation of the physical behavior of the model.

1. The fraction of time the jobs are lost: The fraction η of time the jobs are lost is given by $\eta = \mathbf{u}(K_1+1)\mathbf{e} + \mathbf{v}(K_1+1)\mathbf{e} + \mathbf{w}(K_1+1)\mathbf{e}$.

2. The fraction of time node 1 is blocked: The proportion, P_{block} of time the server in node 1 is blocked is given by $P_{block} = \mathbf{we}$.

3. The fraction of time the server i is idle: The fraction of time the server i, i=1,2, is idle is given by $P_{idle}^{(1)} = u(0)\, e + v(0)\, e$ *and* $P_{idle}^{(2)} = u\, e$.

4. The queue length densities of the number of jobs in queue at nodes 1 and 2 at arbitrary time points: The marginal stationary probability density of the number of jobs in the queue at node i, i=1,2, at arbitrary time points is given by

$$\varphi_0^{(1)} = u(0)\, e + v(0)\, e\,, \quad \varphi_i^{(1)} = u(i)\, e + v(i)\, e + w(i)\, e\,, \quad 1 \le i \le K_1 + 1.$$

$$\varphi_j^{(2)} = \sum_{\nu=0}^{K_1+1} u_j(\nu) + \sum_{\nu=0}^{K_1+1} \sum_{r=L}^{K_2} v_{j,r}(\nu)\,, \quad 0 \le j \le L-1\,,$$

$$\varphi_j^{(2)} = \sum_{\nu=0}^{K_1+1} \sum_{r=L}^{K_2} v_{j,r}(\nu)\,, \quad L \le j \le K_2 - 1\,, \quad \varphi_{K_2}^{(2)} = \sum_{\nu=0}^{K_1+1} \sum_{r=L}^{K_2} v_{K_2,r}(\nu) + \sum_{\nu=1}^{K_1+1} \sum_{r=L}^{K_2} w_r(i)\,,$$

Other routine measures such as the means and standard deviations of the number of jobs in Nodes 1 and 2, and the number of jobs processed during a service in Node 2, can be computed. The corresponding formulas can be obtained from the densities given above. The details are omitted.

4. NUMERICAL EXAMPLES

In order to test the feasibility of the algorithmic procedures proposed in this paper, a FORTRAN code was written and a number of examples were run. In this section we will discuss two examples.

EXAMPLE 1: Here we examine the effect of varying the threshold parameter L (by fixing all other parameters) on the system performance measures. We have chosen $\lambda = 1$, $\mu_1 = 1$, $\mu_2 = 0.1$, $K_1 = 20$ and $K_2 = 30$. The throughput, the fractions of times the servers 1 and 2 are idle, the fraction of lost customers, the fraction of time the server 1 is blocked, the means and standard deviations of the number of jobs present in node 1 and buffer 2 and the mean and the standard deviation of the number of jobs processed during a service in node 2 are plotted as functions of L, in Figures 2-7. An examination of these figures reveal the following observations, which are somewhat intuitively clear.
(a) The throughput and the fractions of times the servers are idle appear to increase with increasing values of L.
(b) As is to be expected the fraction of time the server 1 is blocked and the fraction of lost jobs appear to decrease with increasing L.
(c) It is interesting to note that the mean as well as the standard deviation of the number of jobs in Node 1 appear to decrease with increasing values of L.
(d) The mean number of jobs in buffer 2 appears to decrease initially and then increase with increasing values of L. However, if one looks at the mean number of jobs in Node 2, which is the sum of the mean number in buffer 2 and the mean number in service, then this measure appears to increase with increasing L.
(e) While the mean number of jobs served during a service in Node 2 appears to increase with L, the standard deviation of the number of jobs served decreases with L.

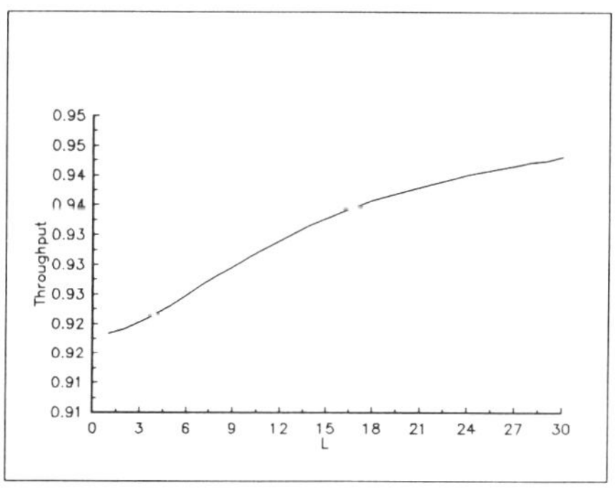

FIGURE 2

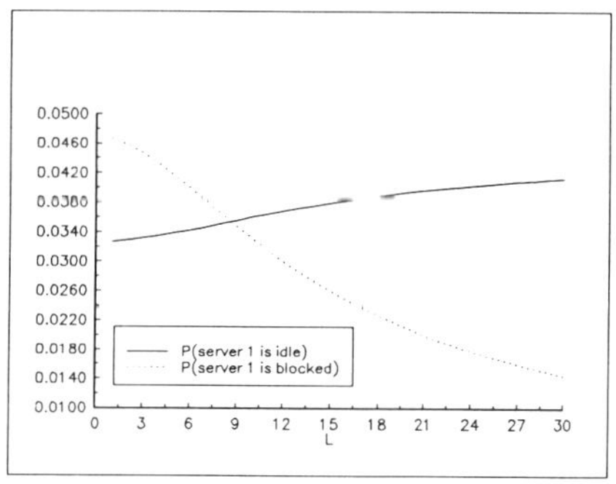

FIGURE 3

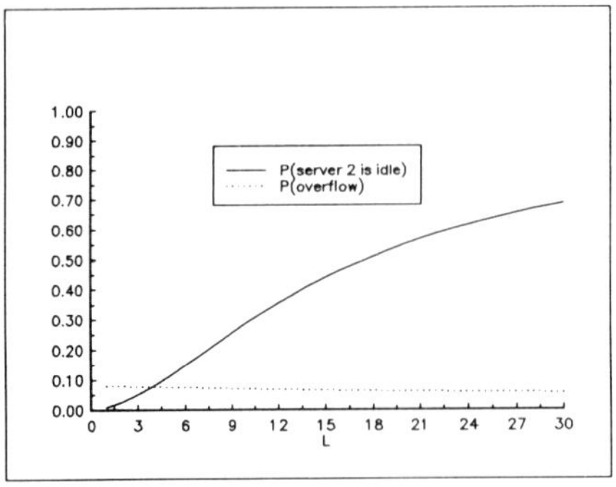

FIGURE 4

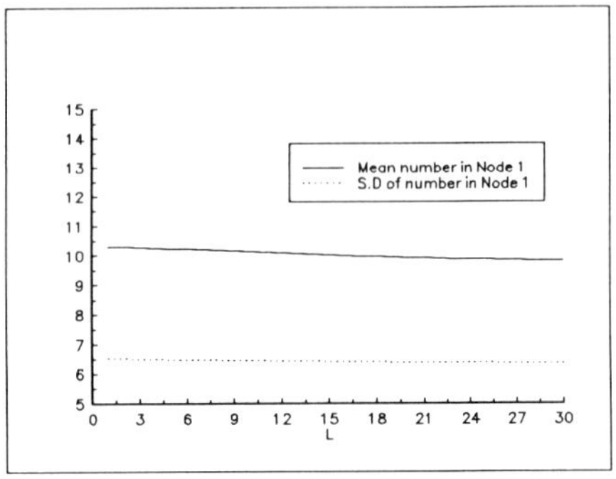

FIGURE 5

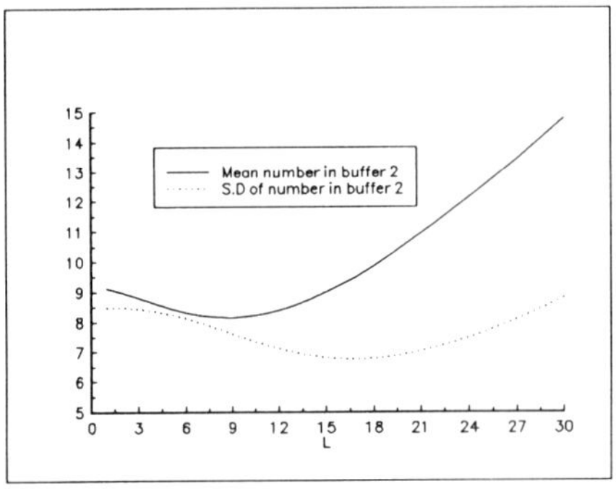

FIGURE 6

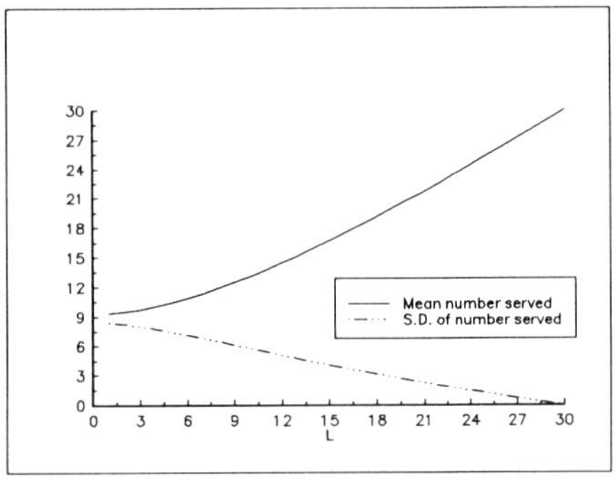

FIGURE 7

The purpose of the next example is to see how the performance measures are affected by varying K_2 for selected values of L and by fixing all other parameters.

EXAMPLE 2: Here we take $\lambda = 1$, $\mu_1 = 1$, $\mu_2 = 0.1$ and $K_1 = 20$. We varied K_2 from L to 30 for L = 1, 4, 7 and 10. The throughput, the blocking probability, the mean number of jobs in Node 1, the expected number of jobs in buffer 2, the mean number of jobs served and the mean number of jobs in Node 2 are plotted in Figures 8-13. Again the following observations are intuitively clear.

(a) For all values of L, the throughput appears to increase with increasing values of K_2.

(b) The blocking probability approaches zero as K_2 increases, for all values of L.

(c) It is interesting to observe that the mean number of jobs in Node 1 appears to increase and then decrease as K_2 increases. This measure appears to approach a common value for all values of L.

(d) While the mean number of jobs waiting in buffer 2 appears to increase with decreasing values of L, the mean number of jobs in Node 2 increases with increasing values of L, for a fixed value of K_2. This is because the mean number of jobs served increases with L and the amount of increase is such that the mean number of jobs present in Node 2 increases with L for any fixed value of K_2. Also the mean number of jobs present in Node 2 approaches a steady-state value as K_2 increases.

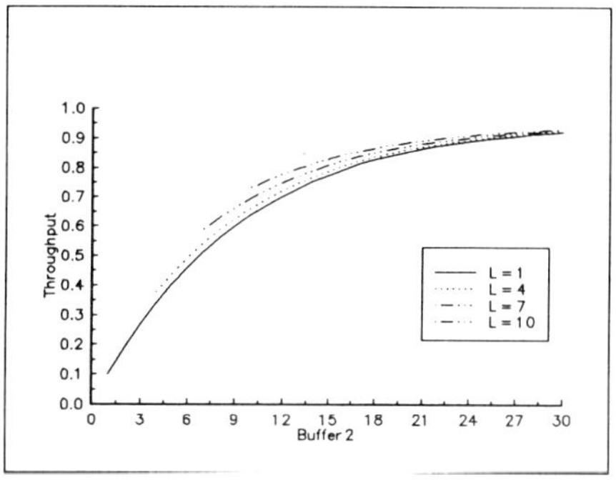

FIGURE 8

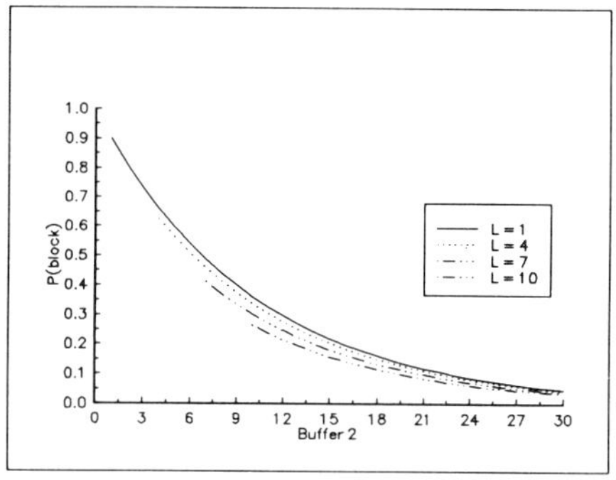

FIGURE 9

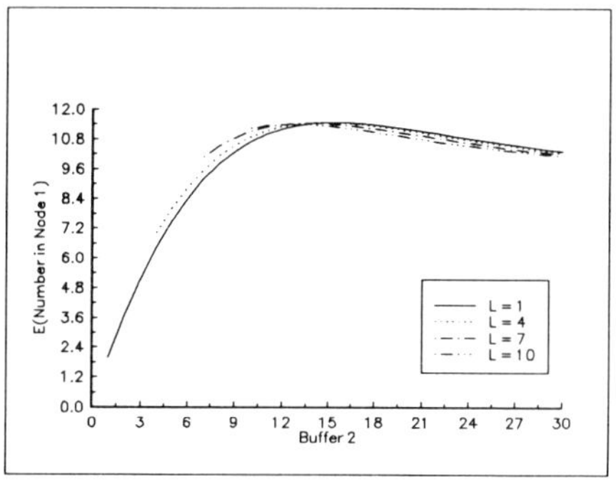

FIGURE 10

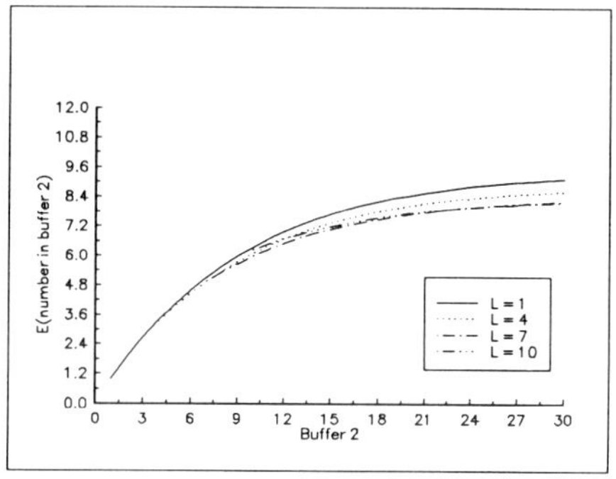

FIGURE 11

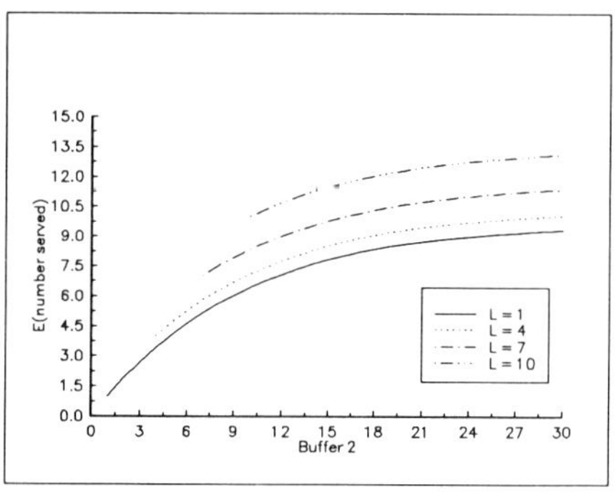

FIGURE 12

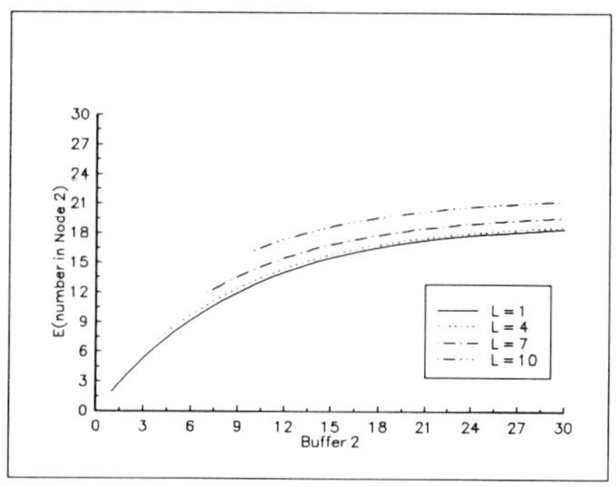

FIGURE 13

5. CONCLUDING REMARKS

In this paper we considered a finite capacity queuing network with two nodes. The services in node 1 are performed singly and in node 2 services are rendered to a group of size at least L. With Poisson arrivals and exponential services, steady-state analysis of the model was done using Markov chain theory. Our algorithmic procedure can be adopted to study the following variants of the present model.

(a) Suppose that the service rate of the server in Node 2 depends on the number in service. This assumption will change the entries corresponding to μ_2 in Q. The generator of the Markov process corresponding to this variant is obtained from the generator of the current model by replacing μ_2 by appropriate multiples of μ_2. The computational procedures for the evaluation of the steady-state vectors will remain the same as the special structure of the coefficient matrices are still preserved. However, these matrices may not conveniently be written in a compact form.

(b) Suppose that arrivals are governed by a Markovian arrival process and that the service times in both the nodes follow (possibly) different phase type distributions[5]. This variant considerably increases the dimension of the problem and the entries of the generator Q will be matrices of higher order. However, by exploiting the sparsity of the matrices we can analyze this variant along the similar lines. The details are omitted.

In some follow-up papers we will discuss some optimization problems useful in the practical design of such finite capacity queuing networks.

6. REFERENCES

[1] Bellman, R. (1960). **Introduction to Matrix Analysis**, McGraw Hill, New York.

[2] Chakravarthy, S. (1992). A Finite Capacity GI/PH/1 queue with group services, **Naval Research Logistics, 39, 345-357.**

[3] Chakravarthy, S. (1991). Analysis of a Finite MAP/G/1 Queue with Group Services, submitted for publication.

[4] Marcus, M and Minc, H. (1964). **A Survey of Matrix Theory and Matrix Inequalities**, Allyn and Bacon, Boston, MA.

[5] Neuts, M.F. (1981). **Matrix-Geometric Solutions in Stochastic Models - An Algorithmic Approach**, The Johns Hopkins University Press, Baltimore, MD.

7. APPENDIX

Here the elements of the generator Q given in (2) are displayed and also the equations that are needed in the Gauss-Seidel iterative procedure are given.In the sequel $\otimes$ denotes Kronecker product of two matrices. For details on the Kronecker products, we refer the reader to Bellman [1] or Marcus and Minc [4].

$$C_0 = -(\lambda + \mu_1)I, \quad D_0 = -(\lambda + \mu_1)I + E_0, \; A_0 = -(\lambda + \mu_1 + \mu_2)I + E_0,$$

$$B_0 = -(\lambda + \mu_2)I, \quad B_1 = \mu_2[\, e_2'(K_2 + 1) \otimes e_{K_2-L+1}'(K_2 - L + 1) \otimes e(K_2 - L + 1)\,],$$

$$C_1 = \mu_1\{\, e_L(L) \otimes e_1'[\,(K_2 + 1)(K_2 - L + 1)\,]\}, \quad A_2 = \mu_1[\, e_{K_2+1}(K_2 + 1) \otimes I\,],$$

$$A_1 = \begin{bmatrix} 0 & \mu_1 I & 0 & \cdots & 0 & 0 \\ 0 & 0 & \mu_1 I & \cdots & 0 & 0 \\ \vdots & \vdots & \vdots & \cdots & \vdots & \vdots \\ 0 & 0 & 0 & \cdots & 0 & \mu_1 I \\ 0 & 0 & 0 & \cdots & 0 & 0 \end{bmatrix}, \quad C_2 = \begin{bmatrix} 0 & \mu_1 & 0 & \cdots & 0 & 0 \\ 0 & 0 & \mu_1 & \cdots & 0 & 0 \\ \vdots & \vdots & \vdots & \cdots & \vdots & \vdots \\ 0 & 0 & 0 & \cdots & 0 & \mu_1 \\ 0 & 0 & 0 & \cdots & 0 & 0 \end{bmatrix},$$

$$D_1 = \mu_2 \begin{bmatrix} I(L) \otimes e(K_2 - L + 1) \\ 0 \end{bmatrix}, \quad E_0 = \begin{bmatrix} 0 & 0 \\ \mu_2 (I \otimes e) & 0 \end{bmatrix}.$$

By exploiting the sparsity of the coefficient matrices, Equations (6-8) are now recast in a form that is well suited for solution by Gauss-Seidel iterative procedure.

$$v_{0,L}(0) = \frac{1}{\lambda + \mu_2} [\mu_1 u_{L-1}(1) + \mu_2 \boldsymbol{v}_L(0) \boldsymbol{e}], \tag{A1}$$

$$v_{0,r}(0) = \frac{\mu_2}{\lambda + \mu_2} \boldsymbol{v}_r(0) \boldsymbol{e}, \; L + 1 \le r \le K_2, \tag{A2}$$

$$\boldsymbol{v}_1(0) = \frac{1}{\lambda + \mu_2} [\mu_1 \boldsymbol{v}_0(1) + \mu_2 (\boldsymbol{w}(1) \boldsymbol{e}) \boldsymbol{e}'_{K_2 - L + 1}(K_2 - L + 1)], \tag{A3}$$

$$v_j(0) = \frac{\mu_1}{\lambda + \mu_2} \boldsymbol{v}_{j-1}(1), \; 2 \le j \le K_2, \tag{A4}$$

and for $1 \le i \le K_1$,

$$v_{0,L}(i) = \frac{1}{\lambda + \mu_1 + \mu_2} [\mu_1 u_{L-1}(i+1) + \lambda v_{0,L}(i-1) + \mu_2 \boldsymbol{v}_L(i) \boldsymbol{e}], \tag{A5}$$

$$v_{0,r}(0) = \frac{1}{\lambda + \mu_1 + \mu_2} [\lambda v_{0,r}(i-1) + \mu_2 \boldsymbol{v}_r(i) \boldsymbol{e}, \; L + 1 \le r \le K_2, \tag{A6}$$

$$v_1(i) = \frac{1}{\lambda + \mu_1 + \mu_2} [\lambda \boldsymbol{v}_1(i-1) + \mu_1 \boldsymbol{v}_0(i+1) + \mu_2 (\boldsymbol{w}(i+1) \boldsymbol{e}) \boldsymbol{e}'_{K_2 - L + 1}(K_2 - L + 1)], \tag{A7}$$

$$v_j(i) = \frac{1}{\lambda + \mu_1 + \mu_2} [\lambda \boldsymbol{v}_j(i-1) + \mu_1 \boldsymbol{v}_{j-1}(i+1), \; 2 \le j \le K_2, \tag{A8}$$

and

$$v_{0,r}(K_1 + 1) = \frac{\lambda}{\mu_1 + \mu_2} \left[v_{0,r}(K_1) + \frac{\mu_2}{\mu_1 + \mu_2} \boldsymbol{v}_r(K_1) \boldsymbol{e} \right], \; L \le r \le K_2, \tag{A9}$$

$$\boldsymbol{v}_j(K_1 + 1) = \frac{\lambda}{\mu_1 + \mu_2} \boldsymbol{v}_j(K_1), \; 1 \le j \le K_2. \tag{A10}$$

Queueing Networks with Finite Capacity
R.O. Onvural and I.F. Akyildiz (Editors)

Two-stage, Multi-product Production/Inventory Systems *

Goang An Shiue and Tayfur Altiok
Department of Industrial Engineering
Rutgers University, Piscataway, N.J. 08855

Abstract

A production/inventory system consisting of two single-stage facilities with a finite capacity buffer in between and a warehouse at the downstream end, producing multiple product types is considered. The demand process for each product is assumed to follow a Poisson distribution and the excess demand is backordered. Products have a priority structure and the processor is shared according to a switching rule. (R,r) continuous-review inventory control policy is used to start and stop production of each product at the second facility and a slightly modified version of it is used at the first facility. Production of a product continues until its target level is reached. Then, a switch-over takes place to the highest priority product amongst the ones that need the processor's attention. A set-up takes place every time a switch-over occurs. Both processing and set-up times are assumed to have exponential distributions. Our objective is to obtain the steady-state average inventory and backorder levels for each product. We have developed an approximation procedure to characterize the delay each product type experiences in receiving the processor's attention.

1 Introduction

We consider a production/inventory system consisting of two stages with a finite capacity buffer in between and a finished product warehouse as shown in Figure 1, producing different types of products. For each type of product, demand arrives at the warehouse and the demand arrival process is assumed to be Poisson and independent of the demand arrival processes of

*This research is fully supported by the National Science Foundation Grant No. NSF-ECS88-08772 and No. NSF-DDM-9014868 and by the North Atlantic Treaty Organization Grant No. NATO-CRG-900580.

the other products. Unit processing times are independent and exponentially distributed. The unsatisfied demand for any product is assumed to be backordered. Also, there exists a priority structure according to which the second processor's attention is allocated to the products in the warehouse. Production of a particular product requires the stages to go through a set-up process. Every product type has its own distinct set-up process. The set-up times are assumed to be independent of the processing times of the products, independent of the sequence in which the products are processed and exponentially distributed.

In general, finite target inventory levels are assumed for each product and low levels of inventories are desired. The stages have to switch back and forth among the products due to the multi-product nature of the system.

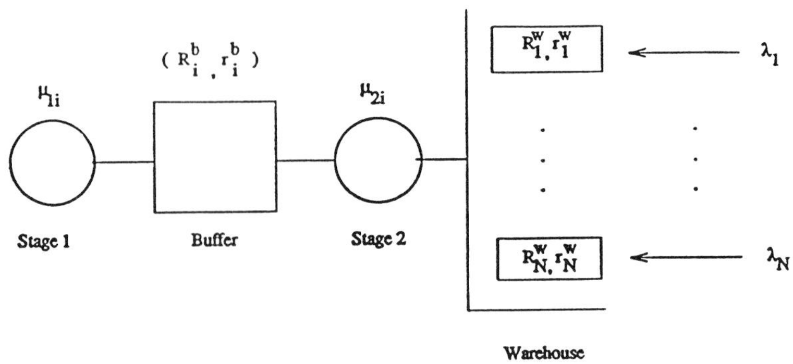

Figure 1: A graphical representation of the production/inventory system

The production of a particular product comes to an end when its inventory level reaches its target value R_i for product i. Restarting its production is deferred until a sufficient number to produce accumulates. In both storage locations, a request to initiate the production of a product is made as soon as its inventory level drops on or below a threshold called the reorder point denoted by r_i for product i. Hence, (R,r) continuous-review policy is used to control the production as well as the inventory level of each product in both the buffer and the warehouse. Stage 2 responds to the requests for production according to the priority structure of the products. Once the processor is ready to start processing a new product, the highest-priority product that needs the processor's attention is selected and the respective set-up process takes place. Production starts after the set-up is completed and it continues until the inventory level of the selected product reaches its target level. While processing, the processor is assumed to be totally dedicated to the current product and it ignores the status of the other products. When processing of a product is completed, in case none of the other products need to be produced, the processor will remain inactive until a product requests its attention. A similar discipline has been referred to as the "exhaustive service" discipline in the context of queueing theory. Clearly, there are other polices such as the nonexhaustive set of polices one can apply to the above production/inventory system. We choose to start with the exhaustive policy. Hence, in the above scenario, the (R,r) control policy and the priority structure form the mechanism to decide "when to switch to a new product" and "which product to switch to", the two most important elements of the dynamic nature of the above scheduling problem in the warehouse. The inventory level in the intermediate buffer is also controlled by using an (R,r) policy for each product. However, the concept of priority does not exist in the buffer due to our assumption of $Q_i^w = R_i^w - r_i^w > R_i^b$ for every product. That is, at any point in time, there can be only one product requesting the first stage's attention. The optimality of the (R,r) policy for single-product systems with set-ups were shown by Heyman (1968), Sobel (1969) and Bell (1971), among others. It is well understood in the literature that when a set-up is involved every time production is to be initiated, it is done so only after a sufficient number ($R-r$ in our case) of units to be produced accumulates, which suggests a two-level control policy.

Very little has been done in the literature on production/inventory systems with multiple stages and multiple products. Single-product (R, r) continuous-review inventory control policy has been investigated by many authors with varying assumptions. Avi-Itzhak, Maxwell and Miller (1965) studied a queueing system with alternating priorities, which is equivalent to a two-product production/inventory system with (0,-1) continuous-review inventory control policy applied to both products allowing backorders. Takács (1968), through a different approach, studied the same queueing system and came up with the distribution of the waiting time of a customer. Syke (1970) studied an alternating priority queueing model with set-up times. Eisenberg (1972,1979) reported on queues with periodic service and changeover times as well as two queues with alternating service. Hodgson (1972) considered a two-product problem with set-ups and lost sales. Graves (1980) studied the multi-product production cycling problem. Gavish and Graves (1980,1981) studied (R,r) production/inventory systems and found cost optimal values of R and r. Hofri and Ross (1987) recently studied a two-queue system with set-ups and a threshold-switching policy, which can be viewed as a two-product production/inventory system with (R, r) continuous-review inventory control policy with backorders. Zheng and Zipkin (1989) considered a two-product problem with backorders and with Markovian assumptions. Altiok (1989) studied a single-product (R,r) production/inventory system with both backorder and lost sale cases. Altiok and Shiue (1991) studied a single-stage, multi-product system that is equivalent to the second stage with the warehouse presented in this paper. Altiok and Shiue decomposed the system into a set of single-product systems by incorporating into the analysis the delay each product experiences in getting the processor's attention. In this study, we are interested in obtaining the stead-steady average inventory and backorder levels of each product in the warehouse and the buffer. For this purpose, we have developed an approximation method that decomposes the system into a set of subsystems each representing the behavior of the inventory of a product in the buffer or the warehouse. A scheme of nested fixed-point iterations relates these subsystems to each other.

2 Analysis of The System

Let us consider the above production/inventory system with products, $1, 2, ..., N$ with product 1 being the highest-priority product. Let μ_{1i}, μ_{2i} be the processing rate of product i at stages 1 and 2, respectively. The system operates in the following manner. The demand for product i arrives at the warehouse with rate λ_i and eventually depletes the inventory level down to r_i initiating a request at stage 2 to start producing product i. Stage 2 switches among the products according to the priority structure in warehouse with the following switching rule. The second stage can not be set up for the same high priority product while a low priority product is waiting for the processor's attention. Otherwise, in the absence of this assumption, a low priority product may wait for long periods before being produced. We have also incorporated into the system a simplifying assumption of $R_i^w - r_i^w > R_i^b$ for all i. The superscript w stands for the warehouse and b for the buffer. This assumption states that the total number of units of product i in the buffer is always less than the number necessary to fill the warehouse once a production request is placed. It implies that while stage 1 is producing a product, there may only be one other product asking for its attention and not more. This is because stage 2 may finish producing i and switch to j, while stage 1 may still be producing product i. In case the inventory level of product j in the buffer drops to zero, stage 2 will simply wait until more of product j appears in the buffer. During this period, stage 2 does not demand other products in the buffer until it finishes producing product j, and therefore product j remains to be the only product in the buffer requesting attention from stage 1. This in turn eliminates the priority concept in the buffer. In the absence of this simplifying assumption, the problem becomes highly complicated to study analytically. In this study, we use a decomposition method demonstrated in Figure 2. The system is decomposed into two major sets of subsystems, one representing the buffer and the other representing the warehouse. Each subsystem is further decomposed into several processes each representing a single-product system denoted by $\Omega(j, i)$ with j=1,2 and i=1,...,N. Each subsystem is constructed in a manner to incorporate the interactions among the products and that of the stages into the analysis. We decompose the two-satge system into two single-stage subsystems and analyze each of them in isolation incoorperating information from the other subsystem. Subsystem 1 consists of the first processor and the buffer and subsystem 2 consists of the second processor and the warehouse. Below, we briefly describe the construction and the analysis of the subsystems starting from the downstream end.

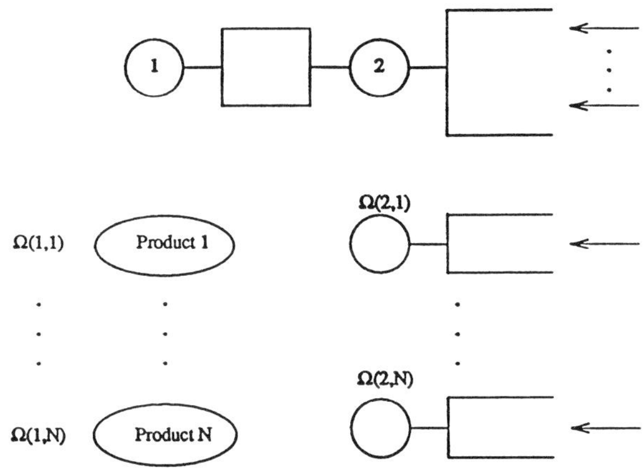

Figure 2: The decomposition

Characterization of $\Omega(2,i)$

We analyze the second subsystem following the exact same method we have used to study the single-stage, multi-product systems [see Altiok and Shiue (1991)]. $\Omega(2,i)$ is constructed in such a way that the impact of the other products in the warehouse as well as the impact of the idleness from the first stage are taken into account in the analysis. The impact of the other products in the warehouse appear in the form of a delay that a product experiences from the time of the request for the processor's attention to the time it actually gets it. On the other hand, the impact of stage 1 appears in the modified unit processing times of the products at stage 2. The analysis of $\Omega(2,i)$ is carried out utilizing independence assumptions and regeneration arguments. The imaginary processor at $\Omega(2,i)$ goes through idle, delay, set-up and busy periods. We have developed an approximation for the expected delay that product i experiences due to sharing the processor at stage 2. An iterative procedure relates $\Omega(2,.)s$ to each other.

Characterization of $\Omega(1,i)$

$\Omega(1,i)$ represent the inventory process of product i and takes into account the delays in the buffer due to sharing the processor at stage 1 and the demand process from stage 2. Let us select a particular product, say product 1, and explain how $\Omega(1,1)$ is modeled and analyzed in detail. First, let us examine the demand arrival process of product 1 in the buffer. The demand will come only if the second stage is producing product 1 and during that time period the time between successive arrivals is the processing time of product 1 at stage 2, which is exponentially distributed. During other times while stage 2 is not producing product 1, there are no arrivals for product 1 in the buffer. This time period is composed of the idle time, delay and set-up time of product 1 at stage 2, which have phase-type representations. Consequently, the period during which no arrivals occur for product 1 has a phase-type distribution with appropriate parameters.

The production of product 1 at stage 1 is controlled by the parameters R_1^b and r_1^b for product 1 at the buffer. When the inventory level of product 1 in the buffer reaches its target level R_1^b, if the second processor is still producing product 1, that is the arrival process for product 1 in the buffer is still on, we simply let the first processor keep producing product 1 whenever necessary to bring its inventory level up to its target value. This is because, during this period, there are no demand arrivals for other products to the buffer anyway. So, the first stage can quit producing product 1 only after the second stage finishes producing product 1.

Let us focus on the stochastic behavior of the inventory process of product 1. Suppressing the index for product 1, consider the stochastic process $\{N_t, M_t, t \geq 0\}$ where $N_t = 0, 1, ..., R_1^b$ representing the inventory level of product 1 in buffer 1 at time t and M_t representing the status of the first stage. The particular realizations of M_t may be explained as follows.

The states of this stochastic process can be divided into 6 groups. The first group of states represents that the arrival process is on and the inventory level is dropping. M_t takes the value of 1 over these states. As soon as the inventory level drops to r_1^b, product 1 initiates a request for the processor's attention. However, it may experience a delay before the processor starts setting up for product 1 due to the multi-product environment. The second group of states corresponds to this delay period and M_t takes the value of 2. We have assumed that this delay is exponentially distributed with an appropriate mean. Once the delay period is over, the processor starts to set up for product 1 and the third group of states is introduced where M_t is 3. After the set-up, the production starts and the arrival process is still on, represented by the fourth group where M_t is 4. The fifth group of states captures the behavior of the inventory process while the arrival process is stopped and the production still continues until the inventory level reaches its target value in the buffer. In this group of states, M_t =5. The last group of states is dedicated to the fact that an arrival may occur when the inventory level is zero and therefore block the production of product 1 at station 2, where M_t =6. $\{N_t, M_t, t \geq 0\}$ is a finite-state Markov chain which possesses a unique steady-state distribution. The probabilities of this Markov chain are obtained in a recursive manner.

So the solution of the Markov chain not only provides us with an approximation for the average inventory level of product 1 in the buffer but also provides us with an approximation for the idleness probabilities, based on which we can approximate the effective process completion time for product 1 at stage 2.

Let us next focus on how to approximate the expected delay for each product at stage 1. Once the steady-state distribution of the Markov chain for each product is obtained, the approximate probability that the first stage is idle can be computed as

$$P(idle) = 1 - \sum_{i=1}^{N} P(processor\ \ is\ \ set-up\ \ for\ \ i\ \ or\ \ producing\ \ i)$$

When product 1 initiates a request for the first processor's attention, the delay period that it may experience depends on the status of the first processor at the time of the request. The processor may be idle or it may be producing one of the N products, including product 1, in which case the delay is zero. Note that if the processor is producing product i at the time product 1 places a request for production, then we know that the Markov chain of the inventory process for product i must be in a state from group 5, i.e. the production is on but the arrival process is off. From the steady-state distribution of the Markov chain for each product, it is possible to obtain the probability that the first processor is producing product i while there are no arrivals for product i to the buffer, call it P_i, and the expected time till the inventory level of product i reaches its target level under the same situation, call it H_i. Then, an approximation for the expected delay of product 1 may be computed as follows:

$$E(D_1) = \sum_{i \neq 1} P(processor\ is\ producing\ i\ at\ the\ time\ of\ product 1's\ request) \cdot H_i$$

where for $i \neq 1$,

$P(processor\ \ is\ \ producing\ \ i\ \ at\ \ the\ \ time\ \ of\ \ product1's\ \ request)$ can be approximated as

$$\frac{P_i}{P(idle) + \sum_{i=1}^{N} P_i}$$

In a similar manner, we can obtain the approximate expected delay for all the products. An iterative procedure that starts with initial guesses of the expected delays and obtains the steady-state probabilities of the Markov chains in the order of $\Omega(1,1)$,...,$\Omega(1,N)$. We execute a few iterations (3-5 for instance) among $\Omega(1,.)$s without waiting until it converges to a desired level. Then, with the new values of the idleness probabilities, a fixed-point iteration

Input	Product 1		Product 2		Product 3		Product 4		Product 5		Product 6		Product 7		Product 8	
(R,r)	(20,10)		(8,5)		(15,10)		(4,2)		(20,10)		(8,5)		(15,10)		(4,2)	
$u^{\rceil}$	2		1		2		3		2		1		2		3	
$x^{\rceil}$	2		3		1		1		2		3		1		1	
Output	Appr.	Sim.	Appr.	Sim.	Appr.	Sim.	Appr.	Sim.	Appr.	Sim.	Appr.	Sim.	Appr.	Sim.	Appr.	Sim.
I	19.58	19.62	7.87	7.84	14.89	14.89	3.92	3.91	19.58	19.62	7.87	7.85	14.89	14.88	3.92	3.92
I	19.01	19.11	7.86	7.84	14.89	14.88	3.91	3.91	19.08	19.12	7.87	7.84	14.89	14.89	3.92	3.91
I	18.52	18.54	7.87	7.84	14.89	14.88	3.92	3.91	18.61	18.59	7.88	7.85	14.90	14.85	3.92	3.91
I	19.58	19.62	7.66	7.64	14.89	14.88	3.92	3.92	19.59	19.62	7.67	7.66	14.89	14.89	3.92	3.92
I	19.59	19.62	7.87	7.84	14.71	14.71	3.92	3.91	19.59	19.62	7.87	7.85	14.71	14.72	3.92	3.92
I	19.58	19.62	7.87	7.84	14.89	14.88	3.84	3.84	19.58	19.62	7.87	7.84	14.89	14.88	3.84	3.85

Table 1: The approximation results for the buffer

Input	Product 1		Product 2		Product 3		Product 4		Product 5		Product 6		Product 7		Product 8	
(R,r)	(40,15)		(20,10)		(50,30)		(10,5)		(40,15)		(20,10)		(50,30)		(10,5)	
$u^{\rceil}$	3		2		4		5		3		2		4		5	
$x^{\rceil}$	4		5		2		1		4		5		2		1	
Output	Appr.	Sim.	Appr.	Sim.	Appr.	Sim.	Appr.	Sim.	Appr.	Sim.	Appr.	Sim.	Appr.	Sim.	Appr.	Sim.
ρ	0.06		0.05		0.05		0.05		0.06		0.05		0.05		0.05	
I	27.77	27.80	15.18	15.17	40.35	40.34	7.85	7.83	27.74	27.72	15.14	15.03	40.33	40.29	7.83	7.75
ρ	0.15		0.05		0.05		0.05		0.15		0.05		0.05		0.05	
I	27.19	27.32	14.97	14.95	40.24	40.25	7.74	7.71	26.95	27.01	14.79	14.54	40.15	40.07	7.64	7.47
ρ	0.3		0.05		0.05		0.05		0.3		0.05		0.05		0.05	
I	24.18	23.73	13.97	13.90	39.74	39.92	7.18	7.22	21.31	22.58	12.32	12.32	38.87	39.17	6.30	6.26
ρ	0.06		0.15		0.05		0.05		0.06		0.15		0.05		0.05	
I	27.71	27.74	14.20	13.98	40.28	40.28	7.78	7.75	27.62	27.58	13.76	13.19	40.19	40.12	7.68	7.50
ρ	0.06		0.05		0.15		0.05		0.06		0.05		0.15		0.05	
I	27.63	27.68	14.96	14.95	39.78	39.85	7.72	7.69	27.51	27.47	14.77	14.57	39.50	39.48	7.60	7.39
ρ	0.06		0.05		0.05		0.1		0.06		0.05		0.05		0.1	
I	27.75	27.76	15.14	15.13	40.33	40.32	7.65	7.59	27.69	27.65	15.06	14.91	40.29	40.23	7.57	7.33

Table 2: The approximation results for the warehouse

is carried out among $\Omega(2,.)$s until the convergence in the expected delay is obtained at stage 2. Next, we repeat the iteration among $\Omega(1,.)$s, and this way we iterate back and forth between stages 1 and 2 until a desired level of convergence is obtained. The measures of performance of our interest, the average inventory and the backordering levels can easily be obtained from the above subsystems. The iterative procedure always converges and it does so in a few steps in all the cases we studied. We present numerical examples in the next section.

3 Numerical Results

The above approximation procedure was coded using the Fortran 77 compiler on AT&T 3B2/310 machine and results were compared against the simulation results that were obtained using the GPSS simulation package.

The numerical examples are designed to report on the impact of the congestion level on the accuracy of the approximation. All of the examples involve eight products. The input in each problem consists of the values of $R_{1i}, r_{1i})$, R_{2i} and r_{2i}, the first moment of the processing times $\bar{u}^1$ and the set-up times $\bar{x}^1$ for each product. Products 1 and 5 have the same parameters, products 2 and 6 have the same parameters and so on. We altered ρ for each product at the warehouse by changing its arrival rate. We report the results in two tables, the first is for the buffer, the second is for the warehouse, and the orders in the two tables are compatible. The output in each example consists of the average inventory level $\bar{I}$ for each product. As shown in Table 1 and Table 2, the error level in the approximation of $\bar{I}$ is highly acceptable.

Our experience shows that the magnitudes of R and r have negligable impacts on the accuracy of the approximation.

References

[1] T. Altiok. (R, r) production/inventory systems. *Operations Research*, 37:266–276, 1989.

[2] T. Altiok and G. A. Shiue. Single-stage, multi-product production/inventory systems with lost sales. Technical report, Department of Industrial Engineering, Rutgers University, 1990.

[3] T. Altiok and G. A. Shiue. Control policies for production/inventory systems. American Control Conference, June 1991, Boston, MA.

[4] B. Avi-Itzhak, W. L. Maxwell and L. W. Miller. Queueing with alternating priorities. *Operations Research*, 13:306–318, 1965.

[5] C. Bell. Characterization and computation of optimal policies for operating an M/G/1 queueing system with removable server. *Operations Research*, 19:208–218, 1971.

[6] M. Eisenberg. Queues with periodic service and changover times. *Operations Research*, 20:440–451, 1972.

[7] M. Eisenberg. Two queues with alternating service. *SIAM J. Appl. Math*, 36:287–303, 1979.

[8] S.W. Fuhrmann and R. Cooper. Stochastic decompsitions in the M/G/1 queue with generalized vacations. *Operations Research*, 33:1117–1129, 1985.

[9] B. Gavish and S. C. Graves. A one-product production/inventory problem under continuous review policy. *Operations Research*, 28:1228–1236, 1980.

[10] B. Gavish and S. C. Graves. Production/inventory systems with a stochastic production rate under a continuous review policy. *Computers and Operations Research*, 8:169–183, 1981.

[11] S. C. Graves. The multi-product production cycling problem. *AIEE Transctions*, 12:233–240, 1980.

[12] D.P. Heyman. Optimal operating policies for M/G/1 queueing systems. *Operations Research*, 16:362–382, 1968.

[13] T. Hodgson. An analytical model of a two-product, one-machine, production-inventory system. *Management Science*, 19:391–405, 1972.

[14] M. Hofri and K. W. Ross. On the optimal control of two queues with server set up times and its analysis. *SIAM J. Computation*, 16:399–420, 1987.

[15] D. R. Manfield. Analysis of a priority polling system for two-way traffic. *IEEE Transactions on Communications*, Com-33:1001–1006, 1985.

[16] M. Sobel. Optimal average cost policy for a queue with start-up and shut-down costs. *Operations Research*, 17:145–162, 1969.

[17] J. S. Syke. Simplified analysis of an alternating priority queueing model with set up times. *Operations Research*, 18:1182–1192, 1970.

[18] L. Takács. Two queues attended by a single sever. *Operations Research*, 16:639–650, 1968.

[19] H. Takagi. Queueing analysis of polling models : An update. In H. Takagi (Editor) *Stochastic Analysis of Computer and Communication Systems.* Elsevier Science Publishers, Amsterdam, 1990.

[20] P. Tran-Gia and T. Raith. Performance analysis of finite capacity polling systems with nonexhaustive service. *Performance Evaluation*, 9:1–16, 1988.

[21] M. Yadin and P. Naor. Queueing systems with a removable service station. *Operations Research Quarterly*, 14:393–405, 1963.

[22] Y. Zheng and P. Zipkin. A queueing model to analyze the value of centralized inventory information. Research Working Paper No. 86–6 (Revised), Department of Industrial Engineering and Operations Research, Columbia University, 1989.

Queueing Networks with Finite Capacity
R.O. Onvural and I.F. Akyildiz (Editors)

AN APPROXIMATION MODEL FOR A MANUFACTURING CELL ATTENDED BY A MATERIAL HANDLING SYSTEM

Samuel Y. Ruan and Mohsen A. Jafari

Department of Industrial and Systems Engineering, Rutgers, The State University of New Jersey, New Brunswick, New Jersey

Abstract

Cellular manufacturing systems have been recognized to be advantageous in a number of ways. In this article, we develop a performance model for a manufacturing cell. Two basic approximation techniques for the analysis of general closed queueing networks (aggregation method and Marie's method) are utilized in this research work. A set of numerical results are obtained and compared to the simulation results. Also, a possible extension to a multi-cell manufacturing system is proposed.

1. Introduction

Conventional manufacturing systems can no longer keep up with ever increasing demand for higher productivity and efficiency. Consequently, these systems have been gradually replaced by more advanced flexible manufacturing systems (FMS). In many circumstances, however, the conversion process needs to take place in several steps due to economical and/or technological reasons (Tompkins and White [1984]). This has promoted the concept of starting manufacturing automation with flexible manufacturing cell (FMC) rather than FMS (Hyer [1985], Kusiak [1987], Ham [1985]). Here, automated work-cells are built one at a time by converting a group of workstations into a FMC. Such an evolution would eventually result in a cellular layout for the whole system.

In a FMS with cellular layout, the entire system is divided into several work-cells. Each work-cell could consist of a number of workstations (machine tools, assembly or inspection stations) connected by a local material transporter (e.g., a robot). Within the work-cell, there may be a storage buffer to avoid unnecessary machine starvation or blockage. Also, there could be another loading/unloading buffer in each work-cell used as a connecting point between the work-cells. The work-cells are connected by some kind of material handling system (MHS), e.g., an automated guided vehicle (AGV). If necessary, the parts can be transferred from one work-cell to another via the MHS.

A proper analytical model of such a system can be helpful in its design and planning stage. Some typical system design parameters are: the number of workstations and the buffer storage capacity in each cell, and the number of material transporters in the MHS. The way the MHS is serving different cells could also be an issue to discuss in the planning stage. Having a performance model would permit us to study the system performance as a function of these parameters or policies. Though, in this article we present a performance model for part of this system, we believe that it could be extended for the whole system without much difficulty.

Manufacturing systems have been modeled as open networks of queues or closed networks of queues by many researchers. When using a closed queueing network (CQN), it is assumed that a fixed number of parts are circulating in the system according to specified routing probabilities. Finished parts are immediately replaced with new parts at a loading/unloading station. The earlier efforts directly related to the modeling and analysis of manufacturing systems are: Solberg [1977], Buzacott and Shanthikumar [1980], Yao and Buzacott [1985a,b,c,d, 1986], Buzacott and Yao [1985], Dallery [1985], and Dallery and David [1983], among others. In these studies, the MHS is either ignored or modeled as a central station. When modeling the MHS as a central station, two problems may arise. First, in a real system the MHS cannot start serving a station unless there is a material transfer request generated by that workstation. This is in contrast to the central server model of the MHS. Second, modeling the MHS as a central station could result in circumstances where a part completed on a workstation next goes to the same workstation.

2. System Description

The system considered here consists of a work-cell and a material handling system (MHS) which is a batch-processing transporter (i.e., transports a batch of parts at one time). In the work-cell, there are K workstations with one sever in each station, a material transporter (e.g., a robot), a common buffer with finite capacity N, and an output buffer with infinite capacity. The common buffer is used to store the WIP inventory and the output buffer is to store the finished parts leaving the cell. Though, K and N are independent of each other, it seems that a system of practical value would be designed with $K < N$. There is a fixed number $\Omega = K + N$ of parts within the system. Note that the common buffer will eliminate the circumstance that a workstation is blocked by another workstation. Consider the situation where the common buffer is full and workstation i has a completed part. There are two possibilities to consider: either there is a part somewhere in the work-cell waiting for workstation i, or there is not such a part. In the first case, it is reasonable to assume that the completed part at workstation i is swapped with the part waiting for this workstation. In the second case, since there is not any part waiting for workstation i, we can think of workstation i being starved. Within the work-cell, the material transfer between workstations is done by a robot. We assume that the robot transfer times are small compared to the processing times at stations. The MHS is assumed to be a batch-server which can transfer at most R parts at a time.

The routing scheme within the cell follows a Markov chain. A newly arrived part will go to station i ($i = 1,...,K$) with probability p_{0i}. After a part being processed at workstation i, it will go to workstation j ($j \neq i$) with probability p_{ij}, or go to the output buffer with probability p_{i0}. The processing time for a part at workstation i is S_i which follows two-stage Coxian distribution with mean μ_i^{-1} and squared coefficient of variation (SCV) C_i^2 ($C_i^2 \geq 0.5$).

When the number of finished parts in the output buffer reaches a limit r_0 ($r_0 \leq R$), the cell sends a signal to the MHS requesting a material transfer. In the case that the MHS is not busy, it removes a random number B of parts from the output buffer and transfers them to an storage area which is not explicitly modeled here. We note that the possible values for B are within the range $[r_0, R]$. The same number of parts are then brought back to the cell, thus keeping the number of parts in the work-cell including the MHS fixed at Ω. We assume that new parts are always available when needed to be delivered to the work-cell. The service time T_0 for a round-trip material transfer is assumed to follow a two-stage Coxian distribution with mean μ_0^{-1} and SCV C_0^2. The time it takes to store finished parts in the storage area and retrieve new parts from that area is assumed to be included in T_0.

Our ultimate goal is to model a system with several cells which share the MHS. To this end, we could use the model presented here as a base model (this will be discussed more in section 5). Such an extension would require us to consider the material transfer requests from other cells. Therefore, we assume that there are C more cells (not including the cell in our model) which can also generate requests for the material transfer. Since the nature of requests from the other cells is the same as far as the transfer times are concerned, we would consider only one type of request generated by all other cells. It will be assumed that the time between two consecutive arrival of requests form other cells follows an exponential distribution with rate λ. Note that there will be at most one request from each cell at any time, which makes the total number of awaiting requests not to exceed C. We assume that there is a queue for these awaiting requests. The service discipline in this queue is assumed to be probabilistic. If there is an outstanding request from this cell (modeled here) and one or more requests from the other work-cells, with probability P the MHS will serve this cell next. Following the completion of unloading the output buffer of the work-cell (being modeled here), it is possible that the remaining number of parts in the output buffer still exceeds r_0. In such a case, a service request will be generated immediately. This request joins the queue awaiting for the service from the MHS.

The model described above has a number of features and shortcomings. An important feature in the model is an explicit representation of a MHS to move material out of or into the work-cell. More importantly, the MHS operation is synchronous as it works based on requests from the workcell. Furthermore, it is associated with two general parameters, r_0 and R. Another feature is that the processing times are assumed to follow Coxian distribution. An attractive aspect of this distribution is that it closely approximates a mixture-of-constants distribution (Yao and Buzacott [1985b]). Consider a system where there is a large number of part types to be processed. Each type is associated with a fixed processing time on each workstation and a part mix ratio which indicates the ratio of the number of parts of this type relative to the whole part population. Suppose that we mix all these part types together and then randomly select a part from this population. The processing time of the selected part on a workstation is a random variable having a mixture-of-constants distribution, with probabilities defined by the part mix ratios. Another attractive feature of this distribution is that it models service times of stations that are subject to breakdowns (Altiok and Stidham [1983]). Thus, our model could implicitly consider workstation breakdowns.

The failure to explicitly consider workstation breakdowns could be considered as a shortcoming to our model. The assumption of unlimited part availability could also be considered a shortcoming depending on whether or not one likes to study the impact of, say different ordering policies on system performance. This adds a dynamic dimension to the model but at the same time complicates the analysis. Our objective is not to study such a dynamic behavior. We rather eliminate this external factor and concentrate on the internal system factors or parameters and study how these influence the system performance. The assumption of a single part type is also a shortcoming mainly because the interaction between different part types could influence the system performance. Another shortcoming which requires more attention is the queueing policy associated with the MHS. Presently, we have random queueing policy. A more realistic scenario would perhaps require selecting the work-cell which has the highest number of parts in its output buffer. Currently, we are working on extending our model to eliminate some of these shortcomings.

3. Solution Procedure

There are two basic approximation techniques for the analysis of general closed queueing networks, namely the aggregation method and Marie's method (Baynat and Dallery [1991]). Both methods are based on the idea that a subsystem (a subnetwork) is replaced by a flow-equivalent single server with load-dependent service rates. The parameters of the equivalent server is determined in such a way that the behavior of the equivalent network becomes close to the behavior of the original network. Since the subsystem is replaced by a single server, the network is simplified and is easier to analyze. In the aggregation method the parameters of the equivalent server are obtained by analyzing the subsystem in isolation as a closed network with different populations, while in Marie's method the parameters are obtained by analyzing the subsystem in isolation under a load-dependent arrival process. Jafari and Ruan [1991] had developed an approximate performance model for the same system based on Marie's method. The system is decomposed into two networks: a part network and an information network. The part network contains the workstations and the MHS with parts being the only dynamic entities in the system. The information network possesses a special structure with load/unload requests being the dynamic entities. Though, their decomposition is quite natural, there are problems regarding the part network. To use the Marie's method for this network requires that the stations are all single processors. Therefore, the MHS in their model is substituted by a single processor which works b (the average batch size) times faster than the MHS but moves parts one at a time. The calculation of b happens to be quite involved. There are problems with this approximation which are discussed by Jafari and Ruan [1991].

Here, we investigate another approach to incorporate the batch processing in the performance model. We decompose the system into two subnetworks: A and A^c. A contains all the workstations in the work-cell. A^c represents the rest of the system. We use Marie's method for the subnetwork A. Then this network is replaced by a single aggregate station or server in another network which also includes A^c. That is, in a sense, we combine two basic approximation methodologies that exist for queueing networks. One problem with this approach is that the transitions from A^c to A involve more than one customers. Despite this problem, we find that the approximation works reasonably accurate as discussed in section 4.

To replace the network A by a single aggregate server, we need to compute the load dependent service rates for the aggregate server. To do this, subsystem A is analyzed in isolation under a constant load Ω', for $\Omega' = 1,...,\Omega$. For each Ω', A is modeled as a CQN. Let A^{Ω} denote this CQN which is obtained by setting the service rate of the MHS (in the original network) to ∞, in other words, the rest of the system is short circuited. To completely define this CQN, we have to calculate its routing probabilities to be called aggregate routing probabilities, hereafter. Let $P = [p_{ij}]$ denote the routing matrix in the original system and $P^A = [p^A_{ij}]$ the aggregate routing matrix. The probability p^A_{ij} can be divided into two parts. One part is the probability that the part leaving station i goes to station j directly, and the other part is the probability that the part goes to station j through A^c (Baynat and Dallery [1991]). That is:

$$p^A_{ij} = p_{ij} + q_{ij}\,, \quad \text{for } i,j = 1,...,K$$

where

p_{ij} : the probability that the part leaving station i goes to station j directly,

q_{ij} : the probability that the part goes to station j through A^c.

Let us express the routing matrix P in the following form

$$P = \left[\begin{array}{cc} 0 & D \\ \hline E & F \end{array}\right]$$

where submatrices D and E have the routing probabilities from A^c to A and vice versa, and submatrix F has thc routing probabilities within A. Then

$$q_{ij} = p_{i0}\, p_{0j}$$

where 0 is the index used for A^c. The above equation can be expressed in a matrix form:

$$Q = E D$$

Given Q the routing matrix for A^Q, we analyze A^Q using Marie's method (Marie [1979]) conditioning on that there are Ω' parts in A^Q, for $\Omega' = 1,...,\Omega$. For each value of Ω', the following steps should be performed:

1). Formulate the network as a CQN with closed product form solution - Given load independent service rates μ_k, $k = 1,...,K$, compute the state dependent arrival rates $\lambda_k(n)$ to workstation k. Also compute $Ca^2{}_k$ the SCV of the arrival process to workstation k.

2). For each workstation k in isolation, $k = 1,...,K$, model it as a $PH_n/PH/1/\Omega'$ queue where PH_n stands for the state dependent arrival process with phase type interarrival time distributions. The parameters of this distribution are obtained using $\lambda_k(n)$ and $Ca^2{}_k$. PH is the phase type service process. Compute the steady state probabilities $p_k(n)$ of having n, $n = 0,...,\Omega'$ parts in station k, $k = 1,...,K$.

3). Compute the load dependent service rates

$$\mu_k(n)=\lambda_k(n-1)\frac{p_k(n-1)}{p_k(n)}\ , \ k=0,...K\ , \ n=1,...,\Omega$$

Note that the above formulation relates to the conservation of flow from/to a state for each station.

4). Go back to step 1 using load dependent service rates.

The termination criteria for the above algorithm is similar to the one described by Marie [1979]. The iteration ends when the total number of parts in all stations is close enough to Ω' and the number of parts in each station is close to the one obtained in the previous iteration.

From the above formulation we obtain the load-dependent throughput rates for the equivalent server:

$$TH(\Omega')=\sum_{k=1}^{K}\bar{\lambda}_k\, p_{k0}^{A}\ , \quad \Omega'=0,...,\Omega$$

where

$$\overline{\lambda}_k = \sum_{n=1}^{\Omega'} \mu_k(n)\, p_k(n), \quad k = 1,\ldots,K$$

Also, we obtain the average number of the parts in the stations, $\overline{n_i}(\Omega')$, the average number of parts in the queues, $q_i(\Omega')$, and the utilization, $u_i(\Omega')$, for $i = 1,\ldots,K$ and $\Omega' = 0,\ldots,\Omega$, through the following equations.

$$\overline{n_k}(\Omega') = \sum_{i=1}^{\Omega'} i\, p_k(i), \; k = 1, \ldots, K \; ; \; \Omega' = 1, \ldots, \Omega$$

$$\overline{q_k}(\Omega') = \sum_{i=2}^{\Omega'} (i-1)\, p_k(i), \; k = 1, \ldots, K \; ; \; \Omega' = 1, \ldots, \Omega$$

$$u_k(\Omega') = \sum_{i=1}^{\Omega'} p_k(i), \; k = 1, \ldots, K \; ; \; \Omega' = 1, \ldots, \Omega$$

The next step is to consider the queueing network N^* which contains A^c, an aggregate server substituting A^Q, and a MH-request station which models the process by which requests to MHS are generated from the other C cells.

The reader notes that the distribution of the service times of the aggregate server is unknown. We can just compute the first two moments of these service times. It is well known that the phase type distributions are quite general in their application scope and that they have been commonly used by many to approximate unknown distributions given the first two moments. Following the same reasoning, we shall use a phase type distribution (more specifically, two-phase Coxian distribution) to approximate the service times of the aggregate server.

Let $[(n_1, ph_1), n_2, n_v, ph_2]$ be the state of the network, where

n_1: the number of parts in the aggregate server (including both the server and the queue)
ph_1: the service phase of the aggregate server
n_2: the number of parts in the MHS queue
n_v: the number of outstanding requests from the other cells
ph_2: the service phase of the MHS

Note that when $n_1 + n_2 = \Omega$, the MHS is serving other cells or is idle, while $n_1 + n_2 < \Omega$ implies that the MHS is serving this cell with batch size $\Omega - n_1 - n_2$.

The above queueing network behaves according to a Markov process. The balance equations are not listed here. Let $p_s[(n_1, ph_1), n_2, n_v, ph_2]$ be the steady state probabilities.

We can obtain the following marginal probabilities:

$$p\{n_1 = i\} = \sum_{ph_1} \sum_{n_2} \sum_{n_v} \sum_{ph_2} p_s[(i,ph_1),n_2,n_v,ph_2]$$

$$p\{n_2 = i\} = \sum_{n_1} \sum_{ph_1} \sum_{n_v} \sum_{ph_2} p_s[(n_1,ph_1),i,n_v,ph_2]$$

$$p\{ph_2 = i\} = \sum_{n_1} \sum_{ph_1} \sum_{n_2} \sum_{n_v} p_s[(n_1,ph_1),n_2,n_v,i]$$

Using these probabilities, we can compute different performance measures, e.g.:

the average number of parts in workstation $i = \sum_{j=0}^{\Omega} \bar{n}_i(j)\ p\{n_1=j\},\ \ i = 1,\dots,K$

the average number of parts in queue $i = \sum_{j=0}^{\Omega} \bar{q}_i(j)\ p\{n_1=j\},\ \ i = 1,\dots,K$

the utilization of workstation $i = \sum_{j=0}^{\Omega} u_i(j)\ p\{n_1=j\},\ \ i = 1,\dots,K$

To calculate the performance measures of the MHS, we recall that in state $[(n_1, ph_1), n_2, n_v, ph_2]$, n_1 is the number of parts in the aggregate station and n_2 is the number of parts in the MHS queue. Thus $\Omega - n_1$ is the number of parts in the MHS (including both transporter and queue), and $\Omega - n_1 - n_2$ is the number of parts being served by the MHS transporter. With this in mind, the performance measures of the MHS can be obtained as follows:

the average number of parts in the MHS = $\sum_{j=0}^{\Omega} (\Omega - j)\ p\{n_1=j\}$

the average number of parts in the MHS queue = $\sum_{j=0}^{\Omega} j\ p\{n_2=j\}$

the utilization of the MHS = $p\{ph_2 = 1\} + p\{ph_2 = 2\}$

Another measurement of interest is $\bar{b}$ the average batch size for the MHS. To obtain $\bar{b}$, we first recognize the probability $p_b(i) = p\{\bar{b} = i\}$ is the same as $p\{\Omega - n_1 - n_2 = i\}$, for $i = 0, r_0, \dots, R$. Note that $p_b(0)$ is the proportion of times that the MHS is either idle or serving the other cells. Thus, for this cell, we have to normalize $p_b(i)$, $i = r_0, \dots, R$, by the normalization constant G, where

$$G = \sum_{i=r_0}^{R} p_b(i)$$

and

$$\bar{b} = \frac{\sum_{i=r_0}^{R} i \, p_b(i)}{G}$$

The solution procedure can be summarized by the following algorithm.

Algorithm

Step 1. Construct the queueing network model A^Q from the original system by short circuiting the MHS.

Step 2. Calculate the aggregate routing probabilities within A^Q.

Step 3. Analyze the subsystem A^Q conditioning on that there are total of Ω' parts in A^Q, for $\Omega' = 1,...,\Omega$. Using Marie's method, find the load-dependent throughput rate TH(Ω'), the average number of parts in each station $n_i(\Omega')$, the average number of parts in each queue $q_i(\Omega')$, and the utilization for each station $u_i(\Omega')$, for $\Omega' = 1,...,\Omega$.

Step 4. Solve the network N^* and find the steady state probabilities.

Step 5. Calculate the performance measures.

4. Numerical Results

The analytical model has been implemented in a C program and runs on a unix-based computer. A set of case studies have been tested and the results are presented in Tables 1 to 3. To check the accuracy of the analytical model, a simulation program is also developed in MODSIM II, an object oriented programming language. Note that in the tables, the simulation output is placed in the parentheses right below the results of the analytical model. In Table 1, all the stations have the same service rate and equal routing probabilities. This make the performance measures of all stations the same so that they are presented with only one number. In Tables 2 to 3, the measures for the stations are listed in the same column starting from station 1 to K.

The test cases can be divided into different groups. Within each group, only one parameter is changed while all the others remain the same. Our observation is that the error for the average number of parts in the MHS is larger compared to other measures (in some cases, the error could be more than 30%). For the workstations, the error for each performance measure is less than 5% in most cases. For the MHS, the error of utilization is also less than 5% and the error of the expected batch size is less that 10%, in most cases.

Table 1

C	K	N	r_0	R	P	(1-K) sys #	MHS sys #	(1-K) q #	(1-K) uti	MHS uti	$\bar{b}$
2	4	6	1	5	0.9	2.22	1.09	1.48	0.74	0.86	1.41
						(2.13)	(1.48)	(1.40)	(0.73)	(0.91	(1.63)
2	4	6	2	5	0.9	2.14	1.44	1.41	0.74	0.81	2.35
						(2.07)	(1.72)	(1.35)	(0.72	(0.84)	(2.51)
2	4	6	3	5	0.9	2.03	1.88	1.31	0.73	0.77	3.30
						(1.96)	(2.16)	(1.25)	(0.71)	(0.80)	(3.41)
2	4	6	4	5	0.9	1.91	2.36	1.20	0.71	0.75	4.21
						(1.83)	(2.68)	(1.14)	(0.69)	(0.77)	(4.28)
2	4	6	2	3	0.9	2.13	1.47	1.40	0.74	0.81	2.25
						(2.04)	(1.84)	(1.32)	(0.72)	(0.84)	(2.34)
2	4	6	2	4	0.9	2.14	1.44	1.40	0.74	0.81	2.32
						(2.05)	(1.80)	(1.33)	(0.72)	(0.84)	(2.46)
2	4	6	2	5	0.9	2.14	1.44	1.41	0.74	0.81	2.35
						(2.07)	(1.72)	(1.35)	(0.72)	(0.84)	(2.51)
2	4	4	2	5	0.9	1.66	1.36	0.98	0.68	0.80	2.31
						(1.59)	(1.64)	(0.92)	(0.67)	(0.83)	(2.45)
2	4	6	2	5	0.9	2.14	1.44	1.41	0.74	0.81	2.35
						(2.07)	(1.72)	(1.35)	(0.72	(0.84)	(2.51)
2	4	8	2	5	0.9	2.63	1.49	1.85	0.77	0.81	2.37
						(2.53)	(1.88)	(1.77)	(0.76)	(0.84)	(2.54)
1	4	6	2	5	0.9	2.17	1.34	1.43	0.74	0.68	2.25
						(2.09)	(1.64)	(1.36)	(0.73)	(0.72)	(2.38)
2	4	6	2	5	0.9	2.14	1.44	1.41	0.74	0.81	2.35
						(2.07)	(1.72)	(1.35)	(0.72	(0.84)	(2.51)
3	4	6	2	5	0.9	2.13	1.48	1.40	0.74	0.87	2.39
						(2.04)	(1.84)	(1.32)	(0.72)	(0.89)	(2.57)
2	2	6	2	5	0.9	3.39	1.23	2.52	0.87	0.79	2.21
						(2.99)	(2.02)	(2.16)	(0.83)	(0.86)	(2.59)
2	3	6	2	5	0.9	2.48	1.38	1.70	0.78	0.80	2.31
						(2.39)	(1.83)	(1.62)	(0.77)	(0.84)	(2.53)
2	4	6	2	5	0.9	2.14	1.44	1.41	0.74	0.81	2.35
						(2.07)	(1.72)	(1.35)	(0.72	(0.84)	(2.51)
2	4	6	2	5	0.2	2.05	1.82	1.32	0.72	0.81	2.62
						(1.92)	(2.32)	(1.22)	(0.70)	(0.84)	(2.83)
2	4	6	2	5	0.5	2.10	1.61	1.37	0.73	0.81	2.48
						(2.00)	(2.00)	(1.29)	(0.71)	(0.84)	(2.67)
2	4	6	2	5	0.9	2.14	1.44	1.41	0.74	0.81	2.35
						(2.07)	(1.72)	(1.35)	(0.72	(0.84)	(2.51)

Note: $p_{ij} = 1/K$, for $i \neq j; i,j = 0,\ldots,K$
$p_{ii} = 0$, $i = 0,\ldots,4$
$\mu_i = 1.0$, $i = 0,\ldots,4$
$C_i^2 = 1.0$, $i = 0,\ldots,4$
$\lambda = 1.0$
sys #: average number of parts in both the server and the queue
q #: average number of parts in the queue
uti: utilization
$\bar{b}$: expected batch size

Table 2

C	K	N	r_0	R	P	(1-K) sys #	MHS sys #	(1-K) q #	(1-K) uti	MHS uti	$\bar{b}$
2	4	6	1	5	0.9	3.77	0.56	2.90	0.86	0.83	1.20
						(3.73)	(0.82)	(2.87)	(0.86)	(0.86)	(1.34)
						0.29		0.07	0.22		
						(0.29)		(0.08)	(0.22)		
						1.63		0.98	0.65		
						(1.57)		(0.93)	(0.64)		
						3.73		2.87	0.87		
						(3.59)		(2.73)	(0.86)		
2	4	6	2	5	0.9	3.58	1.01	2.73	0.85	0.77	2.17
						(3.58)	(1.19)	(2.72)	(0.85)	(0.78)	(2.28)
						0.28		0.07	0.21		
						(0.30)		(0.08)	(0.22)		
						1.58		0.94	0.64		
						(1.53)		(0.90)	(0.63)		
						3.55		2.69	0.86		
						(3.40)		(2.55)	(0.85)		
2	4	6	3	5	0.9	3.37	1.49	2.53	0.85	0.74	3.16
						(3.30)	(1.68)	(2.45)	(0.84)	(0.76)	(3.24)
						0.28		0.07	0.21		
						(0.30)		(0.08)	(0.21)		
						1.53		0.89	0.64		
						(1.52)		(0.89)	(0.63)		
						3.34		2.49	0.85		
						(3.20)		(2.35)	(0.84)		
2	4	6	4	5	0.9	3.15	1.99	2.32	0.83	0.72	4.13
						(3.08)	(2.23)	(2.24)	(0.83)	(0.74)	(4.19)
						0.28		0.07	0.21		
						(0.29)		(0.08)	(0.21)		
						1.48		0.84	0.63		
						(1.48)		(0.86)	(0.62)		
						3.13		2.28	0.84		
						(2.92)		(2.09)	(0.83)		

Note: $p_{ij} = 1/K$, for $i \neq j; i, j = 0, \ldots, K$
$p_{ii} = 0$, $i = 0, \ldots, 4$
$\mu_1 = 1.0$,
$\mu_1 = 0.5, \mu_2 = 2.0, \mu_3 = 0.667, \mu_4 = 0.5$
$C_0^2 = 0.8$,
$C_1^2 = 1.2, C_2^2 = 1.5, C_3^2 = 0.9, C_4^2 = 1.0$
$\lambda = 1.0$
sys #: average number of parts in both the server and the queue
q #: average number of parts in the queue
uti: utilization
$\bar{b}$: expected batch size

Table 3

C	K	N	r_0	R	P	(1-K) sys #	MHS sys #	(1-K) q #	(1-K) uti	MHS uti	$\bar{b}$
2	4	6	1	5	0.9	7.82	0.37	6.83	1.00	0.79	1.12
						(7.63)	(0.60)	(6.63)	(1.00)	(0.83)	(1.26)
						0.21		0.04	0.17		
						(0.22)		(0.05)	(0.17)		
						0.98		0.48	0.51		
						(1.00)		(0.49)	(0.50)		
						0.54		0.19	0.35		
						(0.55)		(0.20)	(0.35)		
2	4	6	2	5	0.9	7.37	0.83	6.38	1.00	0.74	2.11
						(7.17)	(1.05)	(6.17)	(1.00)	(0.77)	(2.23)
						0.21		0.04	0.17		
						(0.22)		(0.05)	(0.17)		
						0.98		0.48	0.51		
						(1.01)		(0.51)	(0.50)		
						0.54		0.19	0.35		
						(0.55)		(0.20)	(0.35)		
2	4	6	3	5	0.9	6.90	1.32	5.91	0.99	0.72	3.11
						(6.64)	(1.58)	(5.65)	(0.99)	(0.74)	(3.21)
						0.21		0.04	0.17		
						(0.22)		(0.06)	(0.17)		
						0.98		0.47	0.50		
						(1.01)		(0.52)	(0.49)		
						0.53		0.18	0.35		
						(0.55)		(0.20)	(0.35)		
2	4	6	4	5	0.9	6.42	1.82	5.43	0.99	0.71	4.09
						(6.09)	(2.11)	(5.10)	(0.99)	(0.72)	(4.16)
						0.21		0.04	0.17		
						(0.23)		(0.06)	(0.16)		
						0.97		0.46	0.50		
						(1.03)		(0.53)	(0.50)		
						0.53		0.18	0.35		
						(0.54)		(0.19)	(0.35)		

Note: $p_{ii} = 0, i = 0, \ldots, 4$
$p_{i0} = 0.25$, for $i = 1, \ldots, 4$
$p_{i1} = 0.45$, for $i = 0,2,3$
$p_{i2} = 0.25$, for $i = 0,1,3,4$
$p_{i3} = 0.25$, for $i = 0,1,2,4$
$p_{i4} = 0.05$, for $i = 0,2,3$
$p_{14} = p_{41} = 0.25$
$\mu_1 = 1.0$,
$\mu_1 = 0.5, \mu_2 = 2.0, \mu_3 = 0.667, \mu_4 = 0.5$
$C_0^2 = 0.8$,
$C_1^2 = 1.2, C_2^2 = 1.5, C_3^2 = 0.9, C_4^2 = 1.0$
$\lambda = 1.0$
sys #: average number of parts in both the server and the queue
q #: average number of parts in the queue
uti: utilization
$\bar{b}$: expected batch size

We also noticed that the model is sensitive to the change in the input parameters.

One major shortcoming of this analytical model is that the number of states in N^* grows rapidly as the values of the system parameters (e.g. number of stations, common buffer capacity, number of other cells) increase. Presently, we are investigating a way of approximating the N^* network by a simpler network.

5. Extension

Our ultimate objective is to develop a performance model for a system with an arbitrary number of work-cells, a MHS with a number of transporters, and a storage/retrieval system. The model developed in this article is general enough to be extended to a multi-cell system. To model a multi-cell system, the idea is to analyze each cell separately (like the model in this article) with an aggregate request rate from all other cells. Two issues require attention when we extend this model to a multi-cell system. First one is the procedure to compute the aggregate request rate which requires that request rate for individual cells be computed. We note that every time the number of parts in the output buffer reaches to r_0 a request is generated by the corresponding cell. We also note that the time between two consecutive requests for any given cell could be zero. Thus, The process of generating requests for a cell is complex and needs attention. The second issue is the service policy for the request queue, that is, the policy by which the MHS would serve different cells. Presently, we are working on these issues.

6. Summary

In this article, we developed an approximate analytical model for a FMS with a work-cell and a material transporter. It utilized a combination of the aggregation and Marie's methods commonly used in the analysis of queueing networks. The numerical results show that the accuracy of the analytical model is satisfactory. The model can be extended to a more complex cellular system.

REFERENCES

1. Altiok, T. and Stidham, S., "The Allocation of Interstage Buffer Storage Capacities in Production Lines", *IIE Transactions*, 15, pp. 292-299, 1983.
2. Baynat, B. and Dallery, Y., "A Unified View Of Product-Form Approximation Techniques For General Closed Queueing Networks", Working paper, Laboratoire MASI, UA 818 CNRS, 4 place Jussieu, 75252 Paris Cedex 05, France.
3. Buzacott, J.A., and Shanthikumar, J. G., "Models for Understanding Flexible Manufacturing Systems," *IIE Transactions*, 12, pp. 339-350, 1980.
4. Buzacott, J.A and Yao, D.D., "On Queueing Network Models of Flexible Manufacturing Systems", Technical Report, Department of Management Science, Univ. of Waterloo, Ontario, Canada, 1985.
5. Dallery, Y. and David, R., "A New Approach Based On Operational Analysis For Flexible Manufacturing System Performance Evaluation," *IEEE Conf. on Dec. and control*, San Antonio, TX, Dec. 1983"
6. Dallery, Y., "A Queueing Network Model of Flexible Manufacturing Systems Consisting of Cells", Technical Report, Harvard University, Cambridge, 1985.
7. Ham, I., Hitomi, K., and Yoshida, T., Group Technology, Kluwer-Nijhoff Publication Co., Boston, MA, 1985.
8. Hyer, N. L., Group Technology at Work, SME, Dec. 1985.
9. Jackson, J. R., "Jobshop-Like Queueing Syatems," Management Science, Vol. 10, pp. 131-142, 1963.
10. Jafari, M.A. and Ruan, S., "A Performance Model For Cellular Manufacturing Syatem: Part I," Submitted to IIE Transaction, Working Paper, Dept. of IE, Rutgers University, P.O. Box 909, Piscataway, NJ 08855, 1991.
11. Kusiak, A., "AI and OR in FMS," *INFOR*, 25, 1, 1987.
12. Marie, R. A., "An Approximate Analytical Method for General Queueing Network," *IEEE Transactions On Software Engineering*, Vol. SE-5, Sep. 1979, pp. 530-538.
13. Solberg, James J., "A Mathemetical Model Of Computerized Manufacturing System" *Proc. 4th Int. Conf. Prod. Res., Tokyo, Japan, 1977*
14. Tompkins, J.A. and White, J.A., Facilities Planning, John Wiley, 1984.
15. Yao, D.D. and Buzacott, J.A., "Modelling The Performance Of Flexible Manufacturing System," *Int'l. J. PRO. RES.*, 1985, Vol. 23, No. 5, 29, pp. 945-959.
16. Yao, D.D. and Buzacott, J.A., "Queueing Model For a Flexible Machining Station: The Method Of Coxian Phases," *Europ. J. Oper. Res.*, 19, 1985, pp. 241-252.
17. Yao, D.D. and Buzacott, J.A., "The Exponentialization Approach To Flexible Manufacturing System Models With General Processing Times," *Europ. J. Oper. Res.*, 22, 1985.
18. Yao, D.D. and Buzacott, J.A., "Modeling A Class Of State-Dependent Routing In Flexible Manufacturing Syatem," *Annals of Oper. Res.*, 1985.
19. Yao, D.D. and Buzacott, J.A., "Model Of Flexible Manufacturing Systems With Limited Local Buffers," *Int'l J. Prod. Res.*, 25, 1986 pp. 107-118.

Queueing Networks with Finite Capacity
R.O. Onvural and I.F. Akyildiz (Editors)

An Error Bound for Infinite Approximations of Queueing Networks with Large Finite Stations

L. Decreusefond[a], H. Korezlioglu[a] and N. Van Dijk[b]

[a]Ecole Nationale Supérieure des Télécommunications, 46 rue Barrault,75634 Paris, France

[b]University of Amsterdam, Departement of Econometrics, Roeterstraat 11, Amsterdam, The Netherlands

Abstract Queueing networks with stations having finite but relatively large buffers are studied. A simple infinite product approximation is suggested to evaluate the system. An analytic error bound is established for the accuracy of this approximation. This error bound is shown to be of the order of the steady state probability to exceed the buffer limits computed by product form expressions for the infinite system. This probability can often be thought as being quite small and thus provides a practical error bound. Some numerical support is provided. To establish the error bound, a new method consisting of introducing an intermediate model is used. This approach seems of interest for wider application to obtain error bounds.

1 Motivation

Queueing network modelling has become a general tool to evaluate the performances of manufacturing, computer and communication networks. Unfortunately, simple closed form expressions such as Jackson's celebrated product form are usually destroyed by practical phenomena like resource contentions, breakdowns, dynamic routing, class priorities and, especially, blocking due to finite capacity constraints. Given the practical importance to evaluate such systems for design, optimization or on-line control purposes, and given the fact that numerical computations very rapidly become prohibitively expensive due to the astronomical number of states that is usually involved, approximation techniques have been widely investigated. Unfortunately, these approximation techniques are also computationally expensive. Moreover, error bounds on the accuracy of such approximations are usually not available. See [1], [3], [5] and references in [4] for a detailed presentation of the subject.

1.1 Simple estimates

In many practical situations, in contrast, one may be interested in somewhat less accurate but simple and secure estimates to obtain quick insights into qualitative behaviour or orders of magnitude. Analytic error bounds on the accuracy of such estimates are thereby most welcome, so as to guarantee service levels, to employ worst case analysis or to determine stopping criteria for more expensive approximations. We note that such error bound results would provide formal justification for practical engineering approaches.

1.2 Estimates by infinite capacities

Finite capacity constraints in queueing networks are quite natural and the capacities chosen are generally large enough to avoid high blocking probabilities. Typically, in the application areas mentioned, such probabilities can be thought of as being within the order of 1% or less. For example, telecommunication systems are designed according to standards for blocking probabilities of .1% at most, the so-called service level, while in manufacturing, congestions are also avoided since they have a cumulative negative effect on upstream stations. By disregarding finite constraints and assuming infinite capacities, simple product form approximations become appealing for quick engineering applications. Error bounds on the accuracy of such approximations are thus of interest.

1.3 Error bound results

In [9], a method was proposed to establish error bound results for perturbations of Markov chains based on the Markov reward theory, provided bounds can be established for the so-called bias-terms. (A more instructive presentation of this method, which also includes an extension to truncations, can be found in [6]). The crucial step of this method, however, is the bounding of these bias-terms. Based on the special Markov structure of queueing networks, in which only one job can move at a time, the method could be successfully exploited for several queueing network applications, for example in order to propose a simple throughput bound for closed networks with a large number of jobs.

1.4 Directly related reference

Along the same lines, in [7], error bounds were recently established for infinite approximation of networks with finite stations under a loss protocol according to which blocked jobs clear the system. Practically, however, the loss protocol is limited to special situations while protocols under which jobs are kept in the system are much more common. In manufacturing, for example, a congestion may block upstream stations; in packet switched computer networks, data transmissions are set up in intermediate stages with delays due to saturated common limited resources and, in telecommunication networks, the so-called *holding blocked jobs* protocols are known ever since classical teletraffic engineering.

1.5 Objective

This paper intends to extend the results in [7] to finite networks with protocol consisting of *routing jobs to non saturated nodes* and becoming thus equivalent to the em repetitive service random destination protocol (see e.g. [3]). Technically, this extension tends to special complications that are far from trivial. In fact, inductive proofs no longer apply to one-step Markov reward equations under the protocol considered here (see relations (8),(15) and Remark 3.2). Only in the special case of a two-station tandem line could this step still be applied, but based on its special structure and with some technical complications (cf. [8]).

1.6 Approach

In this paper a new approach is suggested and applied to circumvent such technical complications. To be exact, the required bounds are obtained by using an intermediate model, considered in [7], for which these complications do not arise. By establishing a comparison of the intermediate model with the models of interest and combining these comparisons in a single relation, the desired result is obtained. Though only exploited here for the system under consideration, this approach, which has not been used before, seems to be promising for further extension to establish analytic error bound results.

1.7 Outline

First, in Section 2, the model is presented. In Section 3, we establish the two-step comparison result. Section 4, finally, presents some numerical results to illustrate the practical usefulness.

2 Presentation of the model

As announced above, the study of the network in which the jobs are routed only to non-saturated nodes, will be done by considering two auxiliary networks. These three networks have the following common structure : they all consist of N stations, the service at station i has an exponential distribution with parameter $\mu_i(n_i)$ where n_i is the number of customers at station i and the arrival streams are independent Poisson processes with intensity λ_i at station i. It is furthermore assumed that μ_i is a nondecreasing function of n_i, i.e.,

$$\mu_i(n_i + 1) \geq \mu_i(n_i)$$

The exit from the network is represented by 0. For each $i = 1, \ldots, N$, p_{ij} defines a probability on $\{0, 1, \ldots, N\}$.

Network 1 (finite buffers, ignoring downstream saturated nodes)
Buffers are of size B_i at station i which may be finite or infinite. When a job completes service at station i, it routes to station j , $(j \neq i)$ with probability $p_{ij}(\overline{n})$ which is taken to be 0 if the buffer at j is saturated (i.e., $n_j = B_j$) and with a fixed probability p_{ij}

otherwise. The probability that a customer, leaving station i, goes back to station i is then

$$p'_{ii}(\overline{n}) = 1 - p_{i0} - \sum_{j:j\neq i} p_{ij}\mathbf{1}_{\{n_j<B_j\}}(\overline{n})$$

that is to say, the more downstream stations are blocked the more customers keep busy the same station. In fact, the routing matrix here coincides with the one in the case of the repetitive service random destination protocol (cf. [2], [3], for the equivalence of this mechanism with other blocking protocols).

Remark 2.1 :
If a station is an input station, its buffer may be finite or infinite. In the former case, once the buffer is full no customers are admitted into the station; therefore, they can be considered as being lost. In the case where some input buffers are infinite, we always suppose that the network is stable.

For purely technical convenience, we introduce here Network 2. We refer to Remark 3.2 for the justification of its use.

Network 2 (finite buffers, blocked jobs lost)
This network will appear as an intermediate one, its difference with the previous scheme lies in the fact that after completing a service, a customer routed to a station with a saturated buffer leaves the network and is considered as being lost.

Network 3 (infinite buffers, i.e., product form network)
In this case, the buffers are supposed to be infinite at all the stations and the routing matrix consists of the probabilities p_{ij} mentioned above. We suppose that the stability conditions of this network are satisfied. Consequently, its invariant distribution exists and is the well known product form distribution.

We are interested in comparing the throughputs of Networks 1 and 3 and we use Network 2 as an intermediate stage for this comparison.

3 Result

3.1 Statement

Let π^s and θ^s be the equilibrium probability and the corresponding throughput of Network s, $s = 1, 2, 3$. S^s is the state space of the Markov process describing Network s. As in [7], we assume that

$$Q := \sum_{i=1}^{N} \lambda_i + \sup_{\overline{n}} \sum_{i=1}^{N} \mu_i(n_i) < \infty \tag{1}$$

For notational convenience, we put $h = Q^{-1}$. We wish to derive a bound for $|\theta^3 - \theta^1|$ when the probability that the state of Network 3 goes beyond Network 1 buffer sizes is small. The main result of this work is expressed in

Theorem 3.1

$$|\theta^3 - \theta^1| \le \mathbf{E}^3 \left(\sum_{i=1}^{N} \lambda_i \mathbf{1}_{\{n_i \ge B_i\}} + \sum_{i,j} \mu_i(n_i) p_{ij} \mathbf{1}_{\{n_j \ge B_j\}} \right) \le Q.\beta \tag{2}$$

where $\mathbf{E}^3$ *is the expectation under* π^3 *and* β *is the probability of* $B = \cup_i \{\overline{n} | n_i \ge B_i\}$ *under* π^3.

Remark 3.1 :

In case all the entrance buffers are infinite and the network is stable, under the stationary probability the throughput of Network 1 is equal to its input flow and hence coincides with that of Network 3. Therefore formula (2) is useful only when the network has some finite input buffers.

3.2 Proof

The throughput is given by

$$\theta^s = \sum_{\overline{n}} \pi^s(\overline{n}) \sum_j \mu_j(n_j) p_{j0} \quad s = 1, 2, 3$$

For a state $\overline{n}$, let $\overline{n} + e_i$ (resp. $\overline{n} - e_i$) denote the same state with one more job (resp. one less job) at station i. We put $e_0 = \overline{0}$. According to the Markov reward theory, a transition matrix p^s corresponding to Network s is defined as follows : for $i \ne 0$, $j \ne 0$ and $j \ne i$, put

$$\begin{cases} p^1(\overline{n}, \overline{n} + e_i) & = h\lambda_i \mathbf{1}_{\{n_i < B_i\}}(\overline{n}) \\ p^1(\overline{n}, \overline{n} - e_i) & = h\mu_i(n_i) p_{i0} \\ p^1(\overline{n}, \overline{n} - e_i + e_j) & = h\mu_i(n_i) p_{ij} \mathbf{1}_{\{n_j < B_j\}}(\overline{n}) \quad \text{for } j \ne i \end{cases}$$

$$\begin{cases} p^2(\overline{n}, \overline{n} + e_i) & = h\lambda_i \mathbf{1}_{\{n_i < B_i\}}(\overline{n}) \\ p^2(\overline{n}, \overline{n} - e_i) & = h\mu_i(n_i) \left(1 - \sum_{j=1}^{N} p_{ij} \mathbf{1}_{\{n_j < B_j\}}(\overline{n})\right) \\ p^2(\overline{n}, \overline{n} - e_i + e_j) & = h\mu_i(n_i) p_{ij} \mathbf{1}_{\{n_j < B_j\}}(\overline{n}) \quad \text{for } j \ne i \end{cases}$$

$$\begin{cases} p^3(\overline{n}, \overline{n} + e_i) & = h\lambda_i \\ p^3(\overline{n}, \overline{n} - e_i) & = h\mu_i(n_i) p_{i0} \\ p^3(\overline{n}, \overline{n} - e_i + e_j) & = h\mu_i(n_i) p_{ij} \end{cases}$$

and

$$p^s(\overline{n}, \overline{n}) = 1 - \sum_{i=0}^{N} \sum_{\substack{j=0 \\ j \ne i}}^{N} p^s(\overline{n}, \overline{n} - e_i + e_j) \quad \forall s = 1, 2, 3$$

Note that the probability for a customer to leave Network 2 at station i is

$$p'_{i0}(\overline{n}) = 1 - \sum_{j=1}^{N} p_{ij} \mathbf{1}_{\{n_j < B_j\}}(\overline{n})$$

Define the positive operators T^s and T^s_t for all $t \in \mathbb{N}$ on real valued functions $f : S^s \to \mathbb{R}$ by

$$\begin{cases} T^s f(\overline{n}) &= p^s(\overline{n},\overline{n}) f(\overline{n}) + \sum_{i=0}^{N} \sum_{\substack{j=0 \\ j\neq i}}^{N} p^s(\overline{n}, \overline{n} - e_i + e_j) f(\overline{n} - e_i + e_j) \\ T^s_0 f &= f \\ T^s_{t+1} f &= T^s(T^s_t f) \end{cases}$$

We put

$$r(\overline{n}) = h \sum_{i=1}^{N} \mu_i(n_i) p_{i0}$$

and

$$\begin{cases} V^s_t(\overline{n}) &= \sum_{k=0}^{t-1} T^s_k r(\overline{n}) \\ V^s_0(\overline{n}) &= 0 \end{cases} \qquad \overline{n} \in S^s, \;\; s = 1,2,3$$

We then have

$$V^s_{t+1} = r + T^s V^s_t \tag{3}$$

and

$$\theta^s \;=\; \lim_{t\to\infty} \frac{Q}{t} V^s_t(\overline{n}) \tag{4}$$

Let V and V' be defined as in (3) starting from two transition operators T and T' with state spaces S and S' such that $S' \subset S$. We extend T' to S as follows : for a function f on S, we put

$$\overline{T}' f(\overline{n}) = \begin{cases} \sum_{\overline{n} \in S'} P'(\overline{n}, \overline{n}') f(\overline{n}') & if \quad n \in S' \\ 0 & if \quad n \neq S' \end{cases}$$

$\overline{T}'_k$ is defined similarly. We put $\overline{T}'_0 = \mathrm{Id}_S$. Then the following formula holds :

$$V_t - V'_t = (T - \overline{T}')V_{t-1} + \sum_{k=1}^{t-1} \overline{T}'_k (T - \overline{T}') V_{t-k-1} \tag{5}$$

In the case of $S' = S$, we can write

$$V_t - V'_t \;=\; \sum_{k=0}^{t-1} T_k (T - T') V'_{t-k-1} \tag{6}$$

$$V_t - V'_t \;=\; \sum_{k=0}^{t-1} T'_k (T - T') V_{t-k-1} \tag{7}$$

We need the following differences, for $\overline{n} \in S^1 = S^2$:

$$T^3 f(\overline{n}) - T^1 f(\overline{n}) = h\sum_{i=1}^{N} \lambda_i \mathbf{1}_{\{n_i=B_i\}}(\overline{n})[f(\overline{n}+e_i) - f(\overline{n})] + h\sum_{i=1}^{N}\sum_{j=1}^{N} \mu_i(n_i)p_{ij}\mathbf{1}_{\{n_j=B_j\}}(\overline{n})[f(\overline{n}-e_i+e_j) - f(\overline{n})] \tag{8}$$

$$T^3 f(\overline{n}) - T^2 f(\overline{n}) = h\sum_{i=1}^{N} \lambda_i \mathbf{1}_{\{n_i=B_i\}}(\overline{n})[f(\overline{n}+e_i) - f(\overline{n})] + h\sum_{i=1}^{N}\sum_{j=1}^{N} \mu_i(n_i)p_{ij}\mathbf{1}_{\{n_j=B_j\}}(\overline{n})[f(\overline{n}-e_i+e_j) - f(\overline{n}-e_i)] \tag{9}$$

$$T^1 f(\overline{n}) - T^2 f(\overline{n}) = h\sum_{i=1}^{N}\sum_{j=1}^{N} \mu_i(n_i)p_{ij}\mathbf{1}_{\{n_j=B_j\}}(\overline{n})[f(\overline{n}) - f(\overline{n}-e_i)] \tag{10}$$

Remark 3.2 :

Note here that the whole problem stands on the rhs of (8). In fact rewriting the difference $f(\overline{n}-e_i+e_j) - f(\overline{n})$ as

$$f(\overline{n}-e_i+e_j) - f(\overline{n}) = [f(\overline{n}-e_i+e_j) - f(\overline{n}-e_i)] - [f(\overline{n}) - f(\overline{n}-e_i)] \tag{11}$$

we could, by Lemma 3.1, prove monotonicity results, as in [7], for each term of the rhs of (11) with f substituted by V^3, but it would not lead us to a concluding result since the two terms would operate in 'opposite directions'. Hence to break this difficulty, we are led to use Network 2 as an intermediate one.

Using the equations (5),(6),(7), we can write

$$(V_t^3 - V_t^2)(\overline{n}) = \sum_{k=0}^{t-1} \overline{T}_k^2 (T^3 - \overline{T}^2) V_{t-k-1}^3(\overline{n}) \quad , \quad \overline{n} \in S^2 = S^1 \tag{12}$$

$$(V_t^1 - V_t^2)(\overline{n}) = \sum_{k=0}^{t-1} T_k^2 (T^1 - T^2) V_{t-k-1}^1(\overline{n}) \quad , \quad \overline{n} \in S^2 = S^1 \tag{13}$$

From (8)-(13),

$$V_t^3 - V_t^1 = (V_t^3 - V_t^2) - (V_t^1 - V_t^2) + \sum_{k=0}^{t-1} \overline{T}_k^2 [(T^3 - \overline{T}^2)V_{t-k-1}^3 - (T^1 - T^2)V_{t-k-1}^1] \tag{14}$$

and, for $k \geq 0$ and $\overline{n} \in S^1 = S^2$,

$$((T^3 - \overline{T}^2)V_k^3 - (T^1 - T^2)V_k^1)(\overline{n}) =$$

$$h\sum_{i=1}^{N} \lambda_i \mathbf{1}_{\{n_i=B_i\}}(\overline{n})[V_k^3(\overline{n}+e_i) - V_k^3(\overline{n})] + h\sum_{i=1}^{N}\sum_{j=1}^{N} \mu_i(n_i)p_{ij}\mathbf{1}_{\{n_j=B_j\}}(\overline{n}) \Big\{[V_k^3(\overline{n}-e_i+e_j) - V_k^3(\overline{n}-e_i)] - [V_k^1(\overline{n}) - V_k^1(\overline{n}-e_i)]\Big\} \tag{15}$$

Remark 3.3 :

Observe that the role of Network 2 has been of short duration since it is now canceled, we can therefore use Lemma 3.1.

According to this lemma, we have

$$|[V_k^3(\overline{n} - e_i + e_j) - V_k^3(\overline{n} - e_i)] - [V_k^1(\overline{n}) - V_k^1(\overline{n} - e_i)]| \le 1 \tag{16}$$

Therefore, (14),(15) and (16) give

$$|V_t^3 - V_t^1| \le h \sum_{k=0}^{t-1} \bar{T}_k^2 \varphi$$

where

$$\varphi(\overline{n}) = \sum_{i=1}^{N} \lambda_i \mathbf{1}_{\{n_i \ge B_i\}}(\overline{n}) + \sum_{i=1}^{N}\sum_{j=1}^{N} \mu_i(n_i) p_{ij} \mathbf{1}_{\{n_j \ge B_j\}}(\overline{n}) \tag{17}$$

Now using Lemma 3.2 below, we can write for all $\overline{n} \in S^2 = S^1$

$$\begin{aligned}
\lim_{t \to +\infty} \frac{Q}{t} |V_t^3(\overline{n}) - V_t^1(\overline{n})| &\le \lim_{t \to +\infty} \frac{1}{t} \sum_{k=0}^{t-1} T_k^3 \varphi(\overline{n}) \\
&= \sum_{\overline{m} \in S^3} \pi^3(\overline{m}) \varphi(\overline{m})
\end{aligned}$$

This achieves the proof of the theorem provided that the following two lemmas are proved .

Lemma 3.1

$$0 \le V_t^3(\overline{n} + e_j) - V_t^3(\overline{n}) \le 1 \quad \overline{n} \in S^3 \tag{18}$$

$$0 \le V_t^1(\overline{n}) - V_t^1(\overline{n} - e_i) \le 1 \quad \overline{n} \in S^2 = S^1 \tag{19}$$

Lemma 3.2

$$\forall k \ge 0, \quad \forall \overline{n} \in S^2, \quad \bar{T}^2 \varphi(\overline{n}) \le T^3 \varphi(\overline{n})$$

Proof of Lemma 3.1. (18) was proved in [7], we prove (19) by induction on t. It is clearly true for $t = 0$ and we assume it is true for some $t > 0$. Due to the increasingness of μ_i's, the positivity of the difference is deduced from

$$V_{t+1}^1(\overline{n}) - V_{t+1}^1(\overline{n} - e_i) = h[\mu_i(n_i) - \mu_i(n_i - 1)]p_{i0} + T^1[V_t^1(\overline{n}) - V_t^1(\overline{n} - e_i)] \tag{20}$$

To prove that the difference is not greater than 1, we write, isolating the term with index i, and denoting $\overline{n}-e_i$ by m,

$$
\begin{aligned}
T^1 f(\overline{n}) = & \\
& h\sum_{k=1}^{N}\lambda_k \mathbf{1}_{\{n_k<B_k\}}(\overline{n}) f(\overline{n}+e_k) \\
+ & \; h\sum_{k=1}^{N}\sum_{l=1,l\neq k}^{N}\mu_k(m_k)p_{kl}\mathbf{1}_{\{n_l<B_l\}}(\overline{n}) f(\overline{n}-e_k+e_l)\mathbf{1}_{\{n_k>0\}}(\overline{n}) \\
+ & \; h\sum_{k=1}^{N}\mu_k(m_k)p_{k0} f(\overline{n}-e_k)\mathbf{1}_{\{n_k>0\}}(\overline{n}) \\
+ & \; h[\mu_i(n_i)-\mu_i(n_i-1)]\sum_{l=1,l\neq i}^{N} p_{il}\mathbf{1}_{\{n_l<B_l\}}(\overline{n}) f(\overline{n}-e_i+e_l)\mathbf{1}_{\{n_i>0\}}(\overline{n}) \\
+ & \; h[\mu_i(n_i)-\mu_i(n_i-1)]p_{i0} f(\overline{n}-e_i)\mathbf{1}_{\{n_i>0\}}(\overline{n}) \\
+ & \; f(\overline{n})\Bigg\{1-h\sum_{k=1}^{N}\lambda_k\mathbf{1}_{\{n_k<B_k\}}(\overline{n}) - h\sum_{k=1}^{N}\sum_{l=1,l\neq k}^{N}\mu_k(m_k)p_{kl}\mathbf{1}_{\{n_l<B_l\}}(\overline{n})\mathbf{1}_{\{n_k>0\}}(\overline{n}) \\
- & \; h\sum_{k=1}^{N}\mu_k(m_k)p_{k0}\mathbf{1}_{\{n_k>0\}}(\overline{n}) - h[\mu_i(n_i)-\mu_i(n_i-1)]\left(\sum_{l=1,l\neq i}^{N} p_{il}\mathbf{1}_{\{n_l<B_l\}}(\overline{n})+p_{i0}\right)\Bigg\}
\end{aligned}
$$

From this we deduce,

$$
\begin{aligned}
V_{t+1}^1(\overline{n}) - V_{t+1}^1(\overline{n}-e_i) = & \\
& h[\mu_i(n_i)-\mu_i(n_i-1)]p_{i0} \\
+ & \; h\sum_{k=1}^{N}\lambda_k\mathbf{1}_{\{n_k<B_k\}}(\overline{n})[V_t^1(\overline{n}+e_k)-V_t^1(\overline{n}-e_i+e_k)] \\
+ & \; h\sum_{k=1}^{N}\sum_{l=1,l\neq k}^{N}\mu_k(m_k)p_{kl}\mathbf{1}_{\{n_l<B_l\}}(\overline{n})\mathbf{1}_{\{n_k>0\}}(\overline{n})[V_t^1(\overline{n}-e_k+e_l)-V_t^1(\overline{n}-e_i-e_k+e_l)] \\
+ & \; h\sum_{k=1}^{N}\mu_k(m_k)p_{k0}\mathbf{1}_{\{n_k>0\}}(\overline{n})[V_t^1(\overline{n}-e_k)-V_t^1(\overline{n}-e_i-e_k)] \\
+ & \; [V_t^1(\overline{n})-V_t^1(\overline{n}-e_i)]\Bigg\{1-h\sum_{k=1}^{N}\lambda_k\mathbf{1}_{\{n_k<B_k\}}(\overline{n}) - h\sum_{k=1}^{N}\mu_k(m_k)p_{k0}\mathbf{1}_{\{n_k>0\}}(\overline{n}) \\
& \qquad -h\sum_{k=1}^{N}\sum_{l=1,l\neq k}^{N}\mu_k(m_k)p_{kl}\mathbf{1}_{\{n_l<B_l\}}(\overline{n})\mathbf{1}_{\{n_k>0\}}(\overline{n})\Bigg\} \\
- & \; h[V_t^1(\overline{n})-V_t^1(\overline{n}-e_i)][\mu_i(n_i)-\mu_i(n_i-1)]\left\{\sum_{l=1,l\neq i}^{N} p_{il}\mathbf{1}_{\{n_l<B_l\}}(\overline{n})+p_{i0}\right\}\mathbf{1}_{\{n_i>0\}}(\overline{n}) \\
+ & \; h[\mu_i(n_i)-\mu_i(n_i-1)]\left\{\sum_{l=1,l\neq i}^{N} p_{il}\mathbf{1}_{\{n_l<B_l\}}[V_t^1(\overline{n}-e_i+e_l)-V_t^1(\overline{n}-e_i)]\right\} \qquad (21)
\end{aligned}
$$

Notice that the coefficient of $V_t^1(\overline{n}) - V_t^1(\overline{n}-e_i)$ is equal to $p^1(\overline{n},\overline{n})$ and the sum of all other coefficients is $1-p^1(\overline{n},\overline{n})$. Therefore, if V_t^1 satisfies (19), then so does V_{t+1}^1. This proves (19). We here note that (18) can also be proved by using the lines of the above proof. □

Proof of Lemma 3.2. Let $M(S^2)$ be as in [7], the class of functions ϕ defined on S^2 such that

$$\phi(\overline{n}+e_i) \geq \phi(\overline{n}) \quad , \quad \forall i \in \{1,\ldots,N\}, \quad \forall \overline{n} \in S^2$$

We note that the fonction φ defined in (17) is in $M(S^2)$, since

$$\mathbf{1}_{\{n_i \geq B_i\}}(\overline{n}+e_j) - \mathbf{1}_{\{n_i \geq B_i\}}(\overline{n}) = \mathbf{1}_{\{n_i = B_i - 1\}}(\overline{n}).\delta_{ij}$$

where δ is the Kronecker symbol. Moreover, $M(S^2)$ is stable by T^3. In fact, similarly to (21) we have

$$\begin{aligned}
T^3\phi(\overline{n}+e_j) - T^3\phi(\overline{n}) = \\
& h\sum_{i=1}^{N}\lambda_i[\phi(\overline{n}+e_i+e_j) - \phi(\overline{n}+e_j)] \\
+ \ & h\sum_{i=1}^{N}\mu_i(n_i)\mathbf{1}_{\{n_i>0\}}(\overline{n})\sum_{l=0}^{N}p_{il}[\phi(\overline{n}-e_i+e_l+e_j) - \phi(\overline{n}-e_j+e_l)] \\
+ \ & h[\mu_j(n_j+1) - \mu_j(n_j)]\sum_{l=1}^{N}p_{jl}[\phi(\overline{n}+e_l) - \phi(\overline{n})] \\
+ \ & \left[1 - h\sum_{i=1}^{N}\lambda_i - h\sum_{i=1}^{N}\mu_i(n_i)\mathbf{1}_{\{n_i>0\}}(\overline{n}) - h[\mu_j(n_j+1) - \mu_j(n_j)]\right].[\phi(\overline{n}+e_j) - \phi(\overline{n})]
\end{aligned}$$

Noticing that, according to (1), the coefficient of $[\phi(\overline{n}+e_j) - \phi(\overline{n})]$ is $p^3(\overline{n}+e_j, \overline{n}+e_j)$ is always nonnegative, $T^3\phi(\overline{n}+e_j) - T^3\phi(\overline{n})$ appears to be a linear combination, with nonnegative coefficients, of terms like $\phi(\overline{n}+e_i) - \phi(\overline{n})$ which are supposed to be nonnegative. Similarly, Equation (9) shows that for all ϕ in $M(S^2)$,

$$\bar{T}^2\phi(\overline{n}) \leq T^3\phi(\overline{n}) \quad , \quad \forall \overline{n} \in S^2$$

The proof is then completed by induction on t, as follows. If $\bar{T}_t^2\phi(\overline{n}) \leq T_t^3\phi(\overline{n})$, then

$$\bar{T}_{t+1}^2\phi(\overline{n}) \leq \bar{T}^2[T_t^3\phi(\overline{n})] \leq T^3[T_t^3\phi(\overline{n})],$$

since $\overline{n} \to T_t^3\phi(\overline{n})$ is a function of $M(S^2)$. □

4 Numerical examples

We consider the following particular configuration :

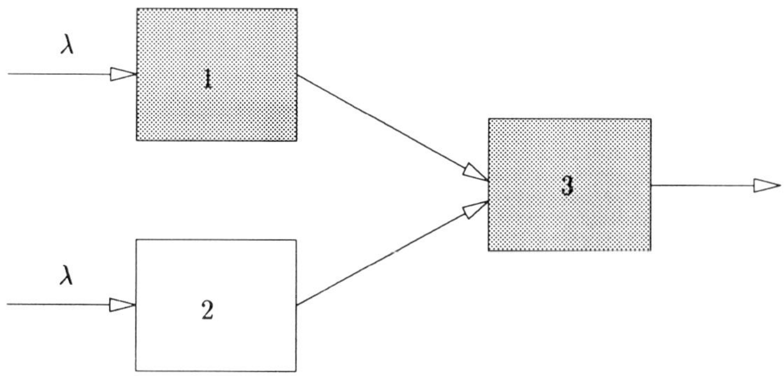

Figure 1: The configuration

where the buffer number 2 is infinite, whereas the buffers number 1 and 3 are supposed to be of size B. The input streams are independent Poisson processes of intensity λ and the services are all exponentially distributed with the same parameter μ. We study the relative bound, that is to say, the first bound in (2) divided by the throughput of the corresponding Network 3. From the model configuration, it can easily be seen that this bound reduces to $\rho_3^B(2/\rho_3 + 1/2^{B+1})$ where $\rho_i = \lambda_i/\mu$ is the traffic intensity at station i. We first determine the values of the pair (B, ρ_3) for which the overflow probability β is equal to a given value and then compute the corresponding relative bound. This procedure allows us to fix the boundary of the domain in which the relative throughput error is less than the bound computed according to the first inequality of (2)(see figures below) .

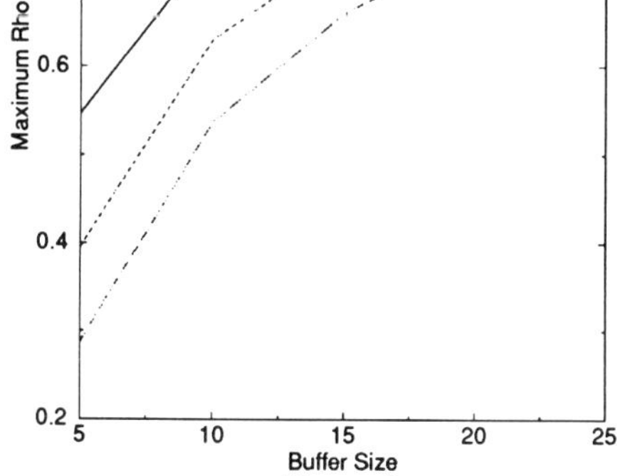

Figure 2

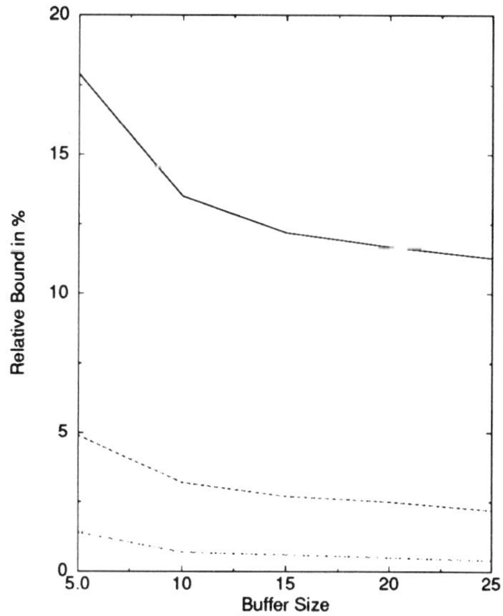

Figure 3

The solid (resp. dotted,dashed) curve is for $\beta = 5\%$ (resp. $\beta = 1\%$, $\beta = 0.2\%$). If (B, ρ_3) is in the domain below the curve on Figure 2, corresponding to the given overflow probability β, then the relative throughput error bound is below the value at B of the corresponding curve in Figure 3.

5 Concluding remarks

As real queueing networks do not usually have a simple analytic solution, simplified models are often suggested to approximate such networks. Recently an error bound technique has been developed, based on bounding the so-called bias-terms, to provide analytic error bounds on the accuracy of such approximations. This technique was successfully applied to queueing networks with blocking, provided blocked jobs were assumed to be lost. This paper extends the technique to queueing networks with a more realistic protocol consisting of sending jobs to only non blocked stations equivalent to the repetitve service random destination protocol. The extension has been made possible by introducing an intermediate network. This new approach is of interest in itself and seems promising for further investigations. Furthermore, the numerical results obtained for a particular configuration seem to justify *a posteriori* this sort of approximation.

References

[1] I.F. Akyildiz and H.G. Perros, Queueing networks with finite buffers. *Special Issue on Performance Evaluation*, 10(4):149–151, 1989.

[2] R. Onvural and H.G. Perros, On equivalencies of blocking mechanisms in queueing networks with blocking. *Operations Research Letters*, 5(6):293–298, 1986.

[3] R. Onvural, Survey of closed queuing networks with blocking. *ACM Computing Surveys*, 22(2):84–121, 1990.

[4] H.G. Perros, A bibliography of papers on queueing networks with finite buffer capacities. *Performance Evaluation*, 255–260, 1989.

[5] H.G. Perros, Open queueing networks with blocking. In Takagi H., editor, *Stochastic Analysis of Computer and Communications Systems*. Elsevier, North-Holland, 1989.

[6] N. Van Dijk, The importance of bias terms for error bounds and comparison results. In W.J. Stewart, editor, *Numerical Solutions of Markov Chains*, pages 617–642. Dekker, January 1990.

[7] N. Van Dijk and H. Korezlioglu, On product form approximations for communications networks with losses : error bounds. *Annals of Operations Research*, 35:69–94, 1992.

[8] N. Van Dijk ,P. Tsoucas, and J. Walrand, Simple bounds and monotonicity of the call congestion of finite multiserver delay systems. *Prob. Eng. Info. Sci.*, 2:129–138, 1988.

[9] N. Van Dijk and J. Van der Wal, Simple bounds and monotonocity results for finite multi-server exponential tandem queues. *Queueing Systems*, 4:1–16, 1989.

Queueing Networks with Finite Capacity
R.O. Onvural and I.F. Akyildiz (Editors)

A Decomposition Method for Throughput Analysis of Cyclic Queues with Production Blocking

Xiao-Gao Liu[a], Li Zhuang[a] and John A. Buzacott[b]

[a]Dept of Management Sciences, University of Waterloo, Waterloo, Ont., Canada N2L 3G1

[b]Faculty of Administrative Studies, York University, North York, Ont., Canada M3J 1P3

Abstract

We discuss a simple approximation method for throughput analysis of cyclic queues with production blocking. The method combines Norton's theorem for product form networks with a decomposition technique for exponential tandem queues with blocking.

1 Introduction

Consider a standard cyclic queue with M single server nodes and N customers. The nodes are arranged along a loop; the customers repeatedly visit the nodes to receive service, moving from nodes i to $i+1$ and then from M to 1 and so on. The service time of node i is exponentially distributed with rate μ_i, and in front of node i there is a buffer (queue space) of size B_i ($1 \leq B_i \leq \infty$), including the space at the server ($i = 1, \ldots, M$). For the case in which some $B_i < N$, blocking may occur in the system occasionally. Assume that a node can be blocked only when a customer upon finishing his service at the node cannot enter the next buffer because it is full. In other words, the blocking is the production type or type 1 [13]. For this system, we are interested in determining its throughput, TH, which is defined as the departure rate of node M. (By flow conservation, this rate is the same at any node in the system.)

In the past few years several authors have attempted to develop analytical approximation methods for this system. Suri and Diehl [20] considered the system with $B_1 = \infty$ and established a decomposition procedure for TH using a so-called variable buffer two-node model as the building block. This variable buffer model plays a role similar to that of Norton's theorem in the product form network [5]. The procedure was mainly tested for some three-node cases. Nevertheless, their idea of incorporating Norton's theorem into the solution of the closed queueing network with blocking is intuitively appealing. Perros et al. [16] applied this idea to some special closed queueing networks with blocking and found encouraging results. The system with $B_1 = \infty$ was also studied by Dallery and Frein [6]. They decomposed the system into a set of equivalent finite single queues by each buffer, assuming that the corresponding server of the single queue in the original system can be blocked only by its succeeding queue. These single queues are related to each other by imposing on them the conservation of flow, and a population constraint:

the summation of the mean queue lengths of these single queues must equal N. Their numerical tests showed an improvement of the method over Suri and Diehl [20]'s. But, since the method uses several different stopping criterions, the population constraint is often not satisfied at the stopping point. A further explanation for this behavior of the method is needed. (See [9] for a generalization of this method to the exponential cyclic queue with communication blocking where the assumption of $B_1 = \infty$ is relaxed.)

For the general exponential cyclic queue with production blocking, Onvural and Perros [14] established an empirical method for fitting the throughput curve against N from some known solutions (including one time-consuming numerical solution of the Markov chain of the system with blocking). They used an observation that the throughput of the cyclic system is a concave function of N (also see [3], [2]). The estimates of TH by their method is very accurate. In [1], Akyildiz and Liebeherr applied Norton's theorem to the cyclic queue as well as other more general cases. The main use of their approach is to study the effect of varying a selected node on the throughput of the system. The approach could be adapted to a decomposition procedure for throughput analysis of the system. But a difficulty is that the number of subsystems to be analyzed will likely explode due to the population effect. The reader is referred to [11] for a survey of other existing results on solving the cyclic queue and other closed queueing networks with blocking.

In this paper we present a new approximate decomposition method for throughput analysis of the cyclic queue with production blocking. This method is also based on Norton's theorem. But different from the previous authors, we try to combine Norton's theorem with a decomposition technique for the exponential tandem queue with finite buffers, a special case of the cyclic queue with blocking [15]. Our method thus satisfies the condition that it is identical to the Norton theorem decomposition procedure if the system becomes a product form network (a cyclic queue with infinite buffers); and it is identical to the decomposition procedure for the tandem queue if the system becomes such a queue. We believe that to satisfy such a condition should be the least requirement for a general approximation method for the cyclic queue. From extensive numerical tests the proposed method is found reasonably accurate and efficient (typically, the absolute relative error of the method $< 5\%$).

The organization of the paper is as follows. Section 2 provides some preliminary results of the cyclic queue. The new decomposition method is discussed in Section 3, along with numerical examples. Section 4 concludes the paper.

Throughout the rest of the paper, the decomposition of the cyclic queue will refer to decomposing the system, with respect to buffers $2, \ldots, M$, into $M-1$ ordered two-node cyclic queue subsystems, called two-node segments, i.e., the first two-node segment contains nodes 1 and 2, the second two-node segment contains nodes 2 and 3, and so on. In these two-node segments, the second node, $i+1$ say, has its original buffer with buffer size B_{i+1}; the buffer size of the first node, i, is specified as $B - B_{i+1}$ where $B = \sum_{i=1}^{M} B_i$.

2 Some Special Case Solutions

In this section we discuss some special case solutions and throughput properties of the cyclic queue with production blocking which are used in the proposed decomposition method for TH.

1. The full system: $N = B$ (the total buffer space). This is an unrealistic case where each space in the system is occupied by a customer. But, since we assume production blocking, the system can in theory be operated by moving the N customers synchronously after each customer occupying a server has finished his service. Hence, $TH = 1/E[max\{S_i\}]$ where S_i is the service time of node i, $i = 1, \ldots, M$ [14]. It can be shown that

$$\begin{aligned} E[max\{S_i\}] &= \sum_{i=1}^{M} \frac{1}{\mu_i} - \sum_{i<j} \frac{1}{\mu_i + \mu_j} + \\ &+ \sum_{i<j<k} \frac{1}{\mu_i + \mu_j + \mu_k} - \cdots + (-1)^{M-1} \frac{1}{\mu_1 + \mu_2 + \cdots + \mu_M}. \end{aligned} \tag{1}$$

Note that in this formula the number of terms in the summation involving c service rates is $\binom{M}{c}$. In fact, the subscripts of the c service rates constitute a combination of c numbers from $1, \ldots, M$. As we will see, although this case is not of much practical interest, it provides a useful check point in our solution procedure.

2. The two-node cyclic queue: $M = 2$. For the purpose of this paper we need a slightly more general result for this case: Node 1 has a state-dependent service rate $\mu_1(x)$ where x denotes the queue length at node 1. Following the standard result for birth-death queues [17], we have,

$$TH = \begin{cases} \frac{\mu_2 \phi(N, B_e)}{1 + \phi(N, B_e)} & \text{if } N \le B_1, \\ \frac{\mu_2 \eta(B_1)[1 + \phi(B_1, B_e)]}{1 + \eta(B_1)[1 + \phi(B_1, B_e)]} & \text{if } N > B_1, \end{cases} \tag{2}$$

where

$$\eta(n-k) = \prod_{x=n-k}^{n} \frac{\mu_1(x)}{\mu_2^{k+1}}, \quad 0 \le k < n, \tag{3}$$

$$\phi(n, K) = \sum_{k=0}^{K} \eta(n-k), \quad K < n, \text{ and} \tag{4}$$

$$B_e = \begin{cases} min[N, B_2] - max[0, N - B_1] - 1 & \text{if } N \le B_2, \\ min[N, B_2] - max[0, N - B_1] & \text{if } N > B_2. \end{cases} \tag{5}$$

This two-node model will be used as the building block in our decomposition procedure.

3. The product form network: $N \le B_{min} + 1$ ($B_{min} = min[B_i]$) [15]. By Norton's theorem for product form networks [5], this case can be solved by a decomposition procedure as follows:

step 1 Decompose the system;

step 2 Solve the first two-node segment for population $n = 1, \ldots, N$ by (2), denoting the corresponding throughput with n by $TH_2(n)$;

step 3 Replace μ_2, the rate of the first node in the second two-node segment, by $\mu_{u2}(x)$ ($= TH_2(x)$), a state-dependent rate, where x is the queue length at the first node of this segment; solve this modified two-node segment for population $n = 1, \ldots, N$ by (2), denoting the corresponding throughput with n by $TH_3(n)$;

step 4 Similarly, find $\mu_{u3}(x)$ and $TH_4(n)$, $\mu_{u4}(x)$ and $TH_5(n)$, etc., and finally, $\mu_{uM-1}(x)$ and $TH_M(N)$, which is the solution to TH of the system. As will be seen, it is helpful to express the state-dependent rate $\mu_{ui}(x)$ in the following equations, for $x = 1, \ldots, N$; $i = 2, \ldots, M-1$,

$$\mu_{di}(x) = \mu_i \tag{6}$$

$$\frac{1}{\mu_{ui}(x)} = \frac{1}{\mu_i} + \left(\frac{1}{TH_i(x)} - \frac{1}{\mu_{di}(x)}\right). \tag{7}$$

For the general exponential cyclic queue with finite buffers, the above procedure can be used to obtain an upper bound for TH by replacing all the buffers by infinite buffers in the original system [19]. Let TH_I be the throughput of the modified system. We have $TH \le TH_I$. *Intuitively,* TH_I *is a good estimate of* TH *for a relatively small* N.

4. The open tandem queueing network: $B - B_{max} < N \le B_{max}$ $(= max[Bi])$. In this case the cyclic queue is equivalent to the open exponential tandem queueing network with finite buffers [15]. Although the latter generally has no simple exact solution, it can be approximated well by various decomposition techniques (see, e.g., [8] for a list of references). Assume that $B_1 = B_{max}$, i.e., node 1 can never be starved and node M can never be blocked. We adopt the following simple decomposition procedure ([7], [10], [4]):

step 1 Decompose the system; (Note that here each two-node segment is a finite single queue, $M/M/1/B_{i+1}$, $i = 1, \ldots, M-1$.)

step 2 Replace μ_i and μ_{i+1} in each two-node segment, except for nodes 1 and M in the first and $(M-1)$st two-node segments, by equivalent rates, μ_{ui} and μ_{di+1}, respectively; let TH_{i+1} be the throughput of the segment with the equivalent rates; determine these rates by the following iteration equations,

$$\frac{1}{\mu_{di}^{(k)}} = \frac{1}{\mu_i} + \left(\frac{1}{TH_{i+1}^{(k-1)}} - \frac{1}{\mu_{ui}^{(k-1)}}\right), \tag{8}$$

$$\frac{1}{\mu_{ui}^{(k)}} = \frac{1}{\mu_i} + \left(\frac{1}{TH_i^{(k)}} - \frac{1}{\mu_{di}^{(k)}}\right), \tag{9}$$

$$\mu_{di}^{(1)} = \mu_i, \ \mu_{u1}^{(k)} = \mu_1, \text{ and } \mu_{dM}^{(k)} = \mu_M, \tag{10}$$

$$i = 2, \ldots, M-1, \ k = 1, 2, \ldots$$

stopping criterion:

$$| TH_M^{(k)} - TH_M^{(k-1)} | \le \varepsilon,$$

where $TH_i^{(k)}$, $\mu_{di}^{(k)}$ and $\mu_{ui}^{(k)}$ are the kth step solutions of TH_i, μ_{di} and μ_{ui}, respectively; $\varepsilon > 0$;

step 3 use $TH_M^{(k)}$ on convergence as the estimate of TH.

Dallery and Frien [7] showed that (8)-(10) converge to a unique solution. Using (2) with $\mu_1(x) = \mu_{ui-1}^{(k)}$, $\mu_2 = \mu_{di}^{(k)}$, $B_2 = B_i$, and $B_1 = B - B_i$, it can be shown that

$$\frac{1}{TH_i^{(k)}} = \frac{1}{\mu_{di}^{(k)}} + \frac{1}{\mu_{ui-1}^{(k)}}\delta(\rho_{i-1}^{(k)}, B_i) = \frac{1}{\mu_{ui-1}^{(k)}} + \frac{1}{\mu_{di}^{(k)}}\delta(1/\rho_{i-1}^{(k)}, B_i), \tag{11}$$

where $\rho_{i-1}^{(k)} = \mu_{ui-1}^{(k)}/\mu_{di}^{(k)}$, and function δ is defined as

$$\delta(\rho, K) = \frac{1}{\sum_{n=0}^{K} \rho^n}, \quad \rho \in (0, \infty), \ K \text{ integer}. \tag{12}$$

From our experience, the estimate of TH of the exponential tandem queueing network by (8)-(10) is in general acceptable, except for some cases where some $B_i = 1$. It is also interesting to note the similarity between (6)-(7) and (8)-(10). This is an indication that these equations may be combined.

5. The throughput curve and its pseudo-symmetry property. Using numerical tests, Onvural and Perros [14] observed that the throughput of the cyclic queue with production blocking is concave in N, i.e., as N increases, TH at first increases but then decreases upon reaching the peak point. This property of TH is intuitively true since when N is small the system can gain more efficiency by increasing work loads, but as N becomes larger blocking increases, which prevents the system from working more efficiently. For the two-node cyclic queue, it can be shown that the throughput curve is not only concave but also symmetrical. To see this we note that the two-node cyclic queue is equivalent to an $M/M/1/B_{2N}$ finite single queue where

$$B_{2N} = \begin{cases} N & \text{if } N \leq B_1 \text{ and } N \leq B_2, \\ min[B_1, B_2] + 1 & \text{if } N > min[B_1, B_2] \text{ and } N \leq max[B_1, B_2], \\ B_1 + B_2 - N + 2 & \text{if } N > B_1 \text{ and } N > B_2. \end{cases}$$

This suggests that for $M = 2$, $TH(2) = TH(B)$, $TH(3) = TH(B-1)$ and so on.

Unfortunately, in general the throughput curve of the cyclic queue with production blocking does not possess the symmetry property. However, it is interesting to note that although the throughput curve is not always perfectly symmetrical, the symmetry can in fact be used to find rather good estimates of TH along the curve. This property of the throughput curve, which we call *the pseudo-symmetry property of the cyclic queue with production blocking*, is supported by our extensive numerical tests. From our experience, the pseudo-symmetry can be simply located by finding

$$N_B = min\{N : TH(N) \geq TH(B)\}. \tag{13}$$

To use it, we estimate $TH(B-1)$ by $TH(N_B+1)$, $TH(B-2)$ by $TH(N_B+2)$, and so on. Table 1 gives one of our results comparing the estimates by the pseudo-symmetry property with the exact solution (see Columns TH_{Exact} and TH_{PS} in the table), where μ and b are respectively the service rate vector and the buffer size vector.

As can be seen, the pseudo-symmetry property is useful for reducing computational efforts in obtaining the throughput curve. More importantly, however, since very little is known about the complex interaction in the system when N is large, the pseudo-symmetry allows us to use the method for small N to estimate the part of the throughput curve corresponding to large N. The property also provides a way to estimate N^*, the N at which TH reaches the maximum.

6. The usefulness of the cyclic queue with $B_1 = \infty$. Clearly, if we replace one of the buffers in the cyclic queue by an infinite buffer, we can use the modified system to obtain an upper bound for TH [19]. It turns out, from our numerical tests for $M \leq 4$, that this upper bound is very tight for N up to N^*, if the modified buffer is properly chosen. This property, which we call *the infinite B_1 property of the cyclic queue with production blocking*, can be seen from Table 1 where the bound is denoted by $TH_{B1\infty}$.

This property suggests that one can use a method for the cyclic queue with $B_1 = \infty$ with the pseudo-symmetry property to obtain the estimate of the entire throughput curve of the system. Our proposed method is based on this hypothesis. We will discuss how to heuristically determine the buffer for modifying in the next section.

3 Approximation of the General Case

Let us first consider a simple three-node cyclic queue in which $B_1 = \infty$, i.e., node 3 can never be blocked. Following the discussion in [18] [12], we can in principle decompose this cyclic queue, with respect to each node, into three $M/M/1/K$ finite single queues with state-dependent arrival and service rates. The throughput of the cyclic queue can be solved using any one of these single queues. However, the difficulty is that the rates of these queues, in general, can only be obtained using the joint steady state probabilities of the original cyclic queue, which are not available from any one of the single queues. To analyze the cyclic queue by decomposition, we therefore must resort to an approximation method for these rates.

Note that since node 3 in the original cyclic queue is never blocked, the service rate of node 3 in the decomposed system is its original rate, μ_3. As a result, we only need to obtain the state-dependent arrival rate for this queue. A possible way to do this is to directly use Norton's theorem as suggested in [16]. That is, we decompose the original system, and use the throughput of the first two-node segment to estimate the state-dependent arrival rate for the queue with node 3 (the second two-node segment). But, we found that by this approach, if node 2 can be blocked in the original system, the state-dependent rate given by Norton's theorem seems always too small. As a result, this approach always underestimates the original throughput.

To explain why this could happen, we note that the effective interarrival time to the queue with node 3 must be the service time of node 2 plus some idle time experienced by the node in the original system. (The idle time is the time in which a node is empty.) We also note that the idle time of a node will increase if the node experiences less blocking, i.e., if a node can release customers faster, it will be empty more often. Now, the use of Norton's theorem to approximate the interarrival time is in fact to estimate the idle time of node 2 by ignoring blocking to the node. However, in reality, node 2 does experience blocking in the original system. As a result, the idle time of node 2 must have been overestimated. That is, the arrival rate must have been underestimated.

Clearly, we need to modify Norton's theorem for our purpose. From our experience with decomposition approaches for tandem systems with blocking, to modify Norton's theorem, the key is to add some blocking time to node 2 in the first two-node system so that the overestimate of the idle time can be reduced. This blocking time can be determined from the second two-node system. On the other hand, to solve the second two-node system we need to estimate the idle time from the first two-node system. We also note that the blocking experienced by node 2 will depend on the distribution of the customers in the system: if many customers are waiting at node 3, node 2 will more likely be blocked; otherwise, node 2 will less likely be blocked. This observation suggests that the blocking time added to node 2, or the effective rate of node 2, should depend on the population in the first two-node system.

A modified Norton theorem approximation. Now, let $\mu_{d2}(x)$ be the population-dependent rate of node 2 in the first two-node segment with a population x, and $\mu_{u2}(x)$ be the state-dependent (arrival) rate of node 2 in the second two-node segment when there are $N - x$ customers in buffer 3. We suggest the following iterative equations for the modified rates, (For a moment the subscript i of the modified rates in (14)-(17) below should be read as "2." But, as can be seen, the equations are readily applied to the system with an arbitrary M.)

$$\frac{1}{\mu_{di}^{(k)}(x)} = \begin{cases} \frac{1}{\mu_i} & \text{if } N \leq B_{i+1}+1, \\ \frac{1}{\mu_i} + \frac{1}{\mu_{di+1}^{(k-1)}(N)}\delta(1/\rho_i^{(k-1)}(x), B_{i+1}) & \text{if } N > B_{i+1}+1, \end{cases} \tag{14}$$

$$\frac{1}{\mu_{ui}^{(k)}(x)} = \frac{1}{\mu_i} + \left(\frac{1}{TH_i^{(k)}(x)} - \frac{1}{\mu_{di}^{(k)}(x)}\right), \tag{15}$$

$$\mu_{u1}^{(k)}(x) = \mu_1, \quad \mu_{di}^{(1)}(x) = \mu_i, \text{ and } \mu_{dM}^{(k-1)}(N) = \mu_M, \tag{16}$$

$$x = 1,\ldots,N, \quad i = 2,\ldots,M-1, \quad k \geq 1; \tag{17}$$

stopping criterion:

$$| TH_M^{(k)}(N) - TH_M^{(k-1)}(N) | \leq \varepsilon, \tag{18}$$

where

1. $\mu_{di}^{(k)}(x)$ and $\mu_{ui}^{(k)}(x)$ are the kth step solutions of $\mu_{di}(x)$ (the population-dependent service rate of two-node segment $i-1$) and $\mu_{ui}(x)$ (the state-dependent arrival rate of two-node segment i), respectively;
2. $TH_i^{(k)}(x)$ is the throughput of two-node segment $i-1$ calculated using (2) with $\mu_{ui-1}^{(k)}(\cdot)$, $\mu_{di}^{(k)}(x)$, and a population x;
3. $\rho_i^{(k-1)}(x) = \mu_{ui}^{(k)}(x)/\mu_{di+1}^{(k)}(N)$ and δ is as defined in (12);
4. $\varepsilon > 0$.

Observe two properties of (14)-(17): (i) *if $N \leq B_i+1$ for all i, these equations provide the Norton theorem decomposition for the product form network*; (ii) *if $B_1 \geq N > \sum_{i=2}^M B_i$, these equations provide the decomposition procedure for the tandem queueing network of the previous section.*

Property (i) is obvious. In fact, if $N \leq B_i+1$ for all i, we have, from (14), $\mu_{di}^{(k)}(x) = \mu_i$ for all x and k. Hence, from (15), $\mu_{ui}^{(k)}(x) = TH_i^{(k)}(x)$. These are exactly what is used in the Norton theorem decomposition for the product form network.

To see Property (ii), we note that the state-dependent arrival rates that are used in each two-node segment in order to obtain $TH_M^{(k)}(N)$ are identical if $B_1 \geq N > \sum_{i=2}^M B_i$. First, consider the case $M = 3$. In order to obtain $TH_3^{(1)}(N)$ we need $\mu_{u2}^{(1)}(x) = TH_2^{(1)}(x)$, $x = N, N-1, \ldots, N-B_3$ (see (15)). Since $B_1 \geq N - B_3 > B_2$, from (5), the B_e

used in (2) for $TH_2^{(1)}(N - B_3)$ should equal B_2. Also, it is easy to see that $B_e = B_2$ for $x = N, N-1, \ldots, N - B_3 + 1$. As a result, we must have that $TH_2^{(1)}(x)$ ($x = N, N-1, \ldots, N - B_3$) are all identical. Substituting $\mu_{u2}^{(1)}(x)$ in (14), we then have that $\mu_{d2}^{(2)}(x)$ ($x = N, N-1, \ldots, N - B_3$) are all identical. This in turn indicates that $\mu_{u2}^{(2)}(x)$ ($x = N, N-1, \ldots, N - B_3$) are all identical. This conclusion is clearly true for $k > 2$. Next, observe that in general in order to obtain $TH_M^{(1)}(N)$, tracing back to $i = 2$, we must find $\mu_{u2}^{(1)}(x) = TH_2^{(1)}(x)$, $x = N, N-1, \ldots, N - \sum_{i=3}^{M} B_i$. Since $B_1 > N - \sum_{i=3}^{M} B_i > B_2$, again by (5), we should use $B_e = B_2$ in (2) for $TH_2^{(1)}(x)$, $x = N, N-1, \ldots, N - \sum_{i=3}^{M} B_i$. The same argument used for $M = 3$ now also applies to the general case, which implies that $\mu_{ui}^{(k)}(x)$ ($x = N, N-1, \ldots, N - \sum_{j=i+1}^{M} B_j$) are identical for each i ($= 2, \ldots, M-1$) and $k \geq 1$. It is now clear that x is irrelevant to the calculation of $TH_M^{(k)}(N)$. However, if we drop the reference to x in (14)-(17), these equations will be exactly the same as (8)-(10) (also see (11)). That is, Property (ii) holds.

The above two properties of (14)-(17) imply that the modified Norton theorem approximation will "interpolate" the throughput estimate of a system from the solutions of the product form network and the tandem queueing network. Since we know very little about the behavior of the cyclic queue with blocking, this interpolation feature of the method is the least a general method should be expected. We also note that because of the above relationship between (14)-(17) and (8)-(10), by following the proof in Dallery and Frein [7], it can be shown that (14)-(17) converge to a unique solution.

A remark on using (14)-(17) for systems with $B_1 = \infty$ and $M > 3$. Clearly, These equations can be used for an arbitrary M. But, for $M > 3$ it seems that we could do better by strictly following the logic of Norton's theorem than by using (14)-(17). For example, for $M = 4$, to find the state-dependent arrival rate for the finite single queue with node 4, by the logic of Norton's theorem, we should use the throughput of the three-node cyclic queue with nodes 1, 2 and 3 over a range of populations, in which node 3 is modified. Since the three-node system can be solved using (14)-(17), the four-node system can also be solved. With the solution of the four-node system, we can then attempt to solve the five-node system, and so on. Conceptually, there is no difficulty solving the cyclic queue in this way. However, a major problem with such a method is that the number of subsystems to be solved in the procedure will explode as M and N increase. This is certainly undesirable for an approximation method. For this reason, we will still use (14)-(17) for $M > 3$. To explain why these equations are likely to work for $M > 3$, besides the above two properties, we note that if we decompose the system by the strict logic of Norton's theorem, we will eventually have to solve many identical two-node systems. By dealing only with selected two-node systems, (14)-(17) may indeed be able to cover most of the two-node systems necessary to be solved using strict Norton's theorem.

A decomposition method for the general case. As we pointed out in the previous section, the solution of the system with $B_1 = \infty$ may be used along with the pseudo-symmetry property of the throughput curve for estimating TH of the general system. To use this conjecture, however, we must replace one of the buffers in the general system by an infinite buffer. Since the main effect of replacing a finite buffer by an infinite buffer is the reduction of the blocking caused by that buffer, naturally, we should find the buffer that causes the least blocking in the system for modifying. Intuitively, if a node can work faster in the system, it will cause less blocking to its preceding node. Following this observation, we suggest the following heuristic method for determining the least blocking buffer:

1. find $B_{min} = min[B_i]$ and $B_{max} = max[B_i]$; set $x = B_{min} + 2$ if $B_{min} + 2 < B$; otherwise, set $x = 1$;
2. if $B_j = B_{max}$ and $x \leq B_j$, B_j will be modified; otherwise, solve all the two-node segment (including the two-node segment with nodes M and 1) with the original service rates of each node and a population x by (2); suppose two-node segment j have the largest throughput (ties, if any, are resolved arbitrarily); B_j will be modified.

This method is found working reasonably well within our framework. Note that since (14)-(17) assume $B_1 = \infty$, to use these equations we need to renumber the nodes in a general system after finding the B_j, i.e., set $j = 1, j+1 = 2$, (if $j+k > M$, set $1 = k$) and so on. Note also that since we will use the pseudo-symmetry property in our method, we only need to replace the buffer size of the first node in two-node segment i by $B - B_{i+1}$ for evaluating $TH_{i+1}^{(k)}(x)$ by (2) (see (14)-(17)). This treatment is consistent with our definition of the decomposition of the cyclic queue in Section 1, and it will allow our method to provide a correct solution for the special case $M = 2$.

Based on the above discussion, we establish the following decomposition method for the throughput curve of the cyclic queue with production blocking:

step 1 Decompose the system;
step 2 Calculate the full system throughput, $TH(B)$, using (1); Renumber all the nodes according to the above B_j;
step 3 For $N = 1, \ldots, B-1$, solve $TH_M(N)$ using (14)-(17); Also, calculate $TH_I(N)$ by modifying the system to a product form network;
step 4 Estimate the throughput of the cyclic queue by $TH(N) = TH_M(N)$ or by $TH(N) = TH_I(N)$ if $TH_M(N) > TH_I(N)$ (the upper bound from the product form network);
step 5 Finally, find N_B using $TH(B)$ with (13); Set $TH(B-1) = TH(N_B+1), TH(B-2) = TH(N_B+2)$, and so on, by the pseudo-symmetry property.

Numerical examples. The above decomposition method for TH of the cyclic queue with production blocking has been tested extensively with exact solutions and simulations. Due to space limitation, Tables 1-13 mainly present the results for comparing the new method with existing methods, where $\mu = (\mu_1, \ldots, \mu_M)$; $b = (B_1, \ldots, B_M)$; TH_{Exact} = exact solution; TH_{Sim} = simulation; TH_{New} = new solution; TH_{DF} = Dallery and Frein [6]; TH_{OP} = Onvural and Perros [14]. The exact results in Table 1 were obtained by solving Markov chains of the system with a Fortran program on an IBM 4381 mainframe. The new decomposition method is programmed in Turbo C with the emulation math mode on an IBM PC 386/33 compatible without co-processor. The stopping criterion (18) used is $\varepsilon = 10^{-4}$. In all the tables the CPU times are those for obtaining the entire throughput curves up to the largest N in the corresponding columns in each table. The TH_{Exact} and TH_{OP} in Tables 6-13 were obtained on a Vax 11/785 according to [14].

As can be seen, in all the cases the throughput estimates from our new method are acceptable and show good agreement with existing results. The worst cases of our test

N	4	5	6	7	8	9	10	11	CPU
TH_{Exact}	.890	.929	.942	.946	.945	.935	.881	.769	0.59 sec
TH_{PS}*	.890	.929	.942	.946	.942	.929	.890	.769	–
error(%)	0.0	0.0	0.0	0.0	-0.3	-0.6	1.0	0.0	
$TH_{B1\infty}$†	.890	.929	.943	.946	–	–	–	–	–
error(%)	0.0	0.0	0.1	0.0	–	–	–	–	
TH_{New}‡	.890	.928	.938	.941	.938	.928	.890	.769	0.95 sec
error(%)	0.0	-0.1	-0.4	-0.5	-0.7	-0.7	1.0	0.0	

* pseudo-symmetry; †infinite B_1 with $b = (3,2,7,3)$; ‡new method

Table 1: $M = 4$, $\mu = (2,4,2,1)$, $b = (3,2,3,3)$

N	5	8	11	14	17	20	23	26	CPU
TH_{Sim}*	.357	.471	.547	.596	.626	.637	.640	.640	N/A
TH_{DF}*	.338	.451	.531	.589	.626	.648	.660	.664	123.26 sec
error(%)	-5.3	-4.2	-2.9	-1.2	0.0	1.7	3.1	3.8	
TH_{New}	.357	.471	.550	.609	.622	.623	.623	.623	72.12 see
error(%)	0.0	0.0	0.5	2.2	0.6	2.2	2.7	2.7	

* Ex10 in [6]

Table 2: $M = 10$, $\mu_i = 1$, $b = (\infty,3,3,3,3,3,3,3,3,3)$

seem to occur when there are many $B_i = 1$ or the service rates are highly unbalanced in the system. The proposed method is rather efficient.

4 Conclusion

In this paper we developed a decomposition method for the cyclic queue with production blocking. The method is mainly based on combining the solutions of the product form network and open tandem queueing network with some available throughput properties of the cyclic queue such that the method satisfies all the special case solutions and the throughput properties. This way of developing an approximation method is justified because of the lack of general understanding of the behavior of closed queueing networks with blocking and the lack of a way to handle highly dependent random variables in the closed system. Our numerical tests showed that the method can provide acceptable

N	5	8	11	14	17	20	23	26	CPU
TH_{Sim}*	.354	.456	.517	.551	.567	.573	.573	.573	N/A
TH_{DF}*	.329	.425	.492	.540	.572	.593	.593	.593	108.97 sec
error(%)	-7.1	-6.8	-4.8	-2.0	0.9	3.5	3.5	3.5	
TH_{New}	.357	.471	.527	.543	.546	.547	.547	.547	89.2 sec
error(%)	0.8	3.3	1.9	-1.5	-3.7	-4.5	-4.5	-4.5	

* Ex11 in [6]

Table 3: $M = 10$, $\mu_i = 1$, $b = (\infty,5,1,5,1,5,1,5,1,3)$

N	5	8	11	14	17	20	23	26	CPU
TH_{Sim}*	.355	.448	.479	.483	.484	.484	.484	.484	N/A
TH_{DF}*	.339	.446	.506	.506	.506	.506	.506	.506	96.45 sec
error(%)	-4.5	-0.4	5.6	4.8	4.5	4.5	4.5	4.5	
TH_{New}	.357	.440	.445	.445	.445	.445	.445	.445	70.14 sec
error(%)	0.6	-1.8	-7.1	-7.9	-8.1	-8.1	-8.1	-8.1	

* Ex12 in [6]

Table 4: $M = 10$, $\mu_i = 1$, $b = (\infty, 5, 5, 5, 5, 1, 1, 1, 1, 3)$

N	5	8	11	14	17	20	23	26	CPU
TH_{Sim}*	.229	.294	.336	.362	.376	.381	.382	.382	N/A
TH_{DF}*	.218	.290	.342	.379	.403	.416	.418	.418	120.84 sec
error(%)	-4.8	-1.4	1.8	4.7	7.2	9.2	9.4	9.4	
TH_{New}	.229	.294	.337	.366	.386	.397	.397	.397	68.16 sec
error(%)	0.0	0.0	0.3	1.1	2.7	4.2	3.9	3.9	

* Ex13 in [6]

Table 5: $M = 10$, $\mu_i = (1, .5, 1, .5, 1, .5, 1, .5, 1, .5)$, $B_1 = \infty, B_i = 3, i > 1$

N	4	5	6	7	9	10	11	12	13	14	CPU
TH_{Exact}*	1.379	1.477	1.541	1.583	1.606	1.587	1.549	1.487	1.385	1.168	31.3 sec
TH_{OP}*	1.391	1.506	1.57	1.593	1.597	1.584	1.559	1.495	1.385	–	4.8 sec
error(%)	0.9	2.0	1.9	0.6	-0.6	-0.2	0.6	0.5	0.0	–	
TH_{New}	1.369	1.454	1.578	1.603	1.606	1.603	1.578	1.454	1.369	1.168	1.59 sec
error(%)	-0.7	-1.6	2.4	1.3	0.0	1.0	1.9	-2.2	-1.2	0.0	

* Table11 in [14]

Table 6: $M = 4$, $\mu = (3, 2, 4, 2)$, $b = (6, 2, 2, 4)$

N	4	5	6	7	9	10	11	12	13	14	CPU
TH_{Exact}*	.889	.935	.961	.973	.978	.974	.964	.936	.881	.769	39.3 sec
TH_{OP}*	.892	.94	.965	.975	.977	.972	.962	0.94	.899	–	5.9 sec
error(%)	0.3	0.5	0.4	0.2	-0.1	-0.2	-0.2	0.4	2.0	–	
TH_{New}	.889	.934	.963	.977	.979	.977	.963	.934	.889	.769	1.32 sec
error(%)	0.0	-0.1	0.2	0.4	0.1	0.3	-0.1	-0.2	0.9	0.0	

* Table12 in [14]

Table 7: $M = 4$, $\mu = (2, 1, 4, 2)$, $b = (3, 4, 5, 2)$

N	4	5	6	7	8	10	11	12	13	14	15	CPU
TH_{Exact}*	.849	.895	.917	.926	.930	.931	.929	.923	.909	.87	.751	36.3 sec
TH_{OP}*	.843	.888	.914	.925	.93	.931	.928	.924	.912	.892	–	23.4 sec
error(%)	-0.7	-0.8	-0.3	-0.1	0.0	0.0	-0.1	0.1	0.3	2.5	–	
TH_{New}	.854	.908	.931	.932	.933	.933	.932	.931	.908	.854	.751	2.2 sec
error(%)	0.6	1.5	1.5	0.6	0.3	0.2	0.3	0.9	-0.1	-1.8	0.0	

* Table13 in [14]

Table 8: $M = 5$, $\mu = (3, 2, 4, 2, 1)$, $b = (4, 3, 2, 4, 2)$

N	4	5	6	7	9	10	11	12	CPU
TH_{Exact}*	.590	.637	.660	.668	.668	.660	.629	.524	138.7 sec
TH_{OP}*	.583	.63	.656	.665	.667	.6597	.633	-	29.1 sec
error(%)	-1.2	-1.1	-0.6	-0.4	-0.1	0.0	0.6	–	
TH_{New}	.593	.645	.652	.654	.654	.652	.645	.524	1.87 sec
error(%)	0.5	1.3	-1.2	-2.1	-2.1	-1.2	2.5	0.0	

* Table14 in [14]

Table 9: $M = 6$, $\mu = (1, 1, 1, 3, 2, 3)$, $b = (2, 2, 2, 2, 2, 2)$

N	4	5	6	7	9	10	11	12	CPU
TH_{Exact}*	.699	.755	.793	.817	.804	.773	.722	.616	130.2 sec
TH_{OP}*	.704	.767	.803	.816	.815	.8	.738	-	30.3 sec
error(%)	0.7	1.6	1.3	-0.1	1.4	3.5	2.2	–	
TH_{New}	.700	.759	.801	.832	.832	.801	.759	.616	1.32 sec
error(%)	0.1	0.5	1.0	1.8	3.5	3.6	5.1	0.0	

* Table15 in [14]

Table 10: $M = 6$, $\mu = (2, 1, 4, 3, 1, 4)$, $b = (2, 2, 2, 2, 2, 2)$

N	4	5	6	7	8	9	11	12	13	14	CPU
TH_{Exact}*	.796	.860	.893	.913	.920	.925	.918	.902	.859	.734	983.2 sec
TH_{OP}*	.787	.847	.885	.907	.917	.922	.921	.913	.882	–	218.5 sec
error(%)	-1.1	-1.5	-0.9	-0.7	-0.3	-0.3	0.3	1.2	2.7	–	
TH_{New}	.797	.868	.910	.917	.918	.919	.917	.910	.868	.734	3.29 sec
error(%)	0.1	0.9	1.9	0.4	-0.2	-0.6	-0.1	0.9	1.0	0.0	

* Table16 in [14]

Table 11: $M = 7$, $\mu = (3, 2, 4, 5, 1, 2, 3)$, $b = (2, 2, 2, 2, 2, 2, 2)$

N	4	5	6	7	8	9	10	12	13	14	15	16	17	CPU
TH_{Exact}*	1.033	1.157	1.252	1.327	1.389	1.434	1.460	1.457	1.427	1.381	1.309	1.186	.968	N/A
TH_{OP}*	1.043	1.183	1.297	1.39	1.426	1.449	1.462	1.461	1.444	1.415	1.317	1.173	-	N/A
error(%)	1.0	2.2	3.6	4.7	2.7	1.0	0.1	0.3	1.2	2.5	0.6	-1.1	-	
TH_{New}	1.033	1.161	1.265	1.349	1.419	1.477	1.487	1.477	1.419	1.349	1.265	1.161	.968	3.35
error(%)	0.0	0.3	1.0	1.7	2.2	3.0	1.8	1.4	-0.6	-2.3	-3.4	-2.1	0.0	sec

* Table17 in [14]

Table 12: $M = 7$, $\mu = (4,2,2,3,5,2,3)$, $b = (3,2,3,3,2,2,2)$

N	4	5	6	7	8	9	10	12	13	14	15	16	17	CPU
TH_{Exact}*	.649	.712	.758	.790	.811	.823	.827	.826	.816	.795	.762	.704	.592	N/A
TH_{OP}*	.648	.713	.716	.795	.811	.821	.826	.825	.818	.805	.769	.714	-	N/A
error(%)	-0.2	0.1	-5.5	0.6	0.0	-0.2	-0.1	-0.1	0.2	1.3	0.9	1.4	-	
TH_{New}†	.649	.715	.765	.803	.831	.836	.837	.836	.831	.803	.765	.715	.592	3.14 sec
error(%)	0.0	0.4	0.9	1.6	2.5	1.6	1.2	1.2	1.8	1.0	0.4	1.6	0.0	

* Table18 in [14]

Table 13: $M = 7$, $\mu = (3,1,2,1,2,3,4)$, $b = (3,2,3,3,2,2,2)$

results for throughput analysis of the system, and is sufficiently efficient.

From our experience with the new method, it seems that the population effect plays a more important role in the closed system with blocking. This suggests that further study of such systems put more emphasis on modifying existing methods for product form networks. A good candidate in this direction is the mean value analysis technique. By adapting this technique to the closed system with blocking, we may also find an alternative way to estimate the mean queue length of the system.

References

[1] I.F. Akyildiz and J. Liebeherr, Application of Norton's Theorem on Queueing Networks with Finite Capacities. In the Proceedings of INFOCOM 89, 914-923.

[2] J.H. Bookbinder and T.R. Kotwa, Modeling an AGV Automobile Body-Framing System, Interfaces 17 (1987) 41-50.

[3] J. Buchan and E. Koenigsberg, *Scientific Inventory Management* (Prentice-Hall, Englewood Cliffs, New Jersey, 1963).

[4] J.A. Buzacott and J.G. Shanthikumar, *Stochastic Models of Manufacturing Systems* (Prentice-Hall, Englewood Cliffs, New Jersey, 1992).

[5] K.M. Chandy, U. Herzog and L. Woo, Parametric Analysis of Queueing Networks, IBM J. Res. Develop. 19 (1975) 36-42.

[6] Y. Dallery and Y. Frein, A Decomposition Method for the Approximate Analysis of Closed Queueing Networks with Blocking, in: *Queueing Networks with Blocking*, H.G. Perros and T. Altiok, eds. (Elsevier Science Publishers, North Holland, 1989) 193-215.

[7] Y. Dallery and Y. Frein, On Decomposition Methods for Tandem Queueing Networks with Blocking, Technical Report MASI, Universite Pierre et Marie Curie, Paris, July 1989. To appear in Operations Research.

[8] Y. Dallery and S.B. Gershwin, Manufacturing Flow Line Systems: A Review of Models and Analytical Results, to appear in Queueing Theory.

[9] Y. Frein and Y. Dallery, Analysis of Cyclic Queueing Networks with Finite Buffers and Blocking Before Service, Performance Evaluation 10 (1989) 197-210.

[10] X-G. Liu, Toward Modeling Assembly Systems: Applications of Queueing Networks with Blocking, Ph.D. Thesis, University of Waterloo, Waterloo, Ontario (May 1990).

[11] R.O. Onvural, A Survey of Closed Queueing Networks with Blocking, SCM Computing Surveys 22 (1990) 83-121.

[12] R.O. Onvural, On the Exact Decomposition of Deadlock Free Closed Queueing Networks with Blocking, in: *Queueing Networks with Blocking*, H.G. Perros and T. Altiok, eds. (Elsevier Science Publishers, North Holland, 1989) 73-83.

[13] R.O. Onvural and H.G. Perros, On Equivalencies of Blocking Mechanisms in Queueing Networks with Blocking, Operations Research Letters 5-6 (1986) 293-298.

[14] R.O. Onvural and H.G. Perros, Approximate Throughput Analysis of Cyclic Queueing Networks with Finite Buffers, IEEE Transactions on Software Engineering 15 (1989) 800-808.

[15] R.O. Onvural and H.G. Perros, Some Equivalencies between Closed Queueing Networks with Blocking, Performance Evaluation 9 (1989) 111-118.

[16] H.G. Perros, A.A. Nilsson and Y.C. Liu, Approximate Analysis of Product Form Type Queueing Networks with Blocking and Deadlock, Performance Evaluation 8 (1988) 19-40.

[17] S.M. Ross, *Stochastic Process* (John Wiley & Sons, New York, 1983).

[18] P.J. Schweitzer, Aggregation Methods for Large Markov Chains, in: *Mathematical Computer Performance and Reliability*, G. Iazeolla et al., eds. (North Holland, 1984) 275-286.

[19] J.G. Shanthikumar and D.D. Yao, Monotonicity and Concavity Properties in Cyclic Queueing Networks with Finite Buffers, in: *Queueing Networks with Blocking*, H.G. Perros and T. Altiok, eds. (Elsevier Science Publishers, North Holland, 1989) 325-344.

[20] R. Suri and G.W. Diehl, A Variable Buffer-Size Model and Its Use in Analyzing Closed Queueing Networks with Blocking, Management Science 32 (1986) 206-224.

Queueing Networks with Finite Capacity
R.O. Onvural and I.F. Akyildiz (Editors)

On the Tail Approximation of the Blocking Probability of Single Server Queues with Finite Buffer Capacity

C. Bisdikian, J. S. Lew, and A. N. Tantawi

IBM Research Division, Thomas J. Watson Research Center, P.O.Box 704, Yorktown Heights, NY 10598, USA (e-mail: bisdik, lew, or tantawi @watson.ibm.com)

Abstract

Due to analytical complexity, blocking probabilities in queues with finite buffer capacity are often approximated by the tail of the distribution of the queue size of the corresponding infinite buffer queues. Integrated services networks have a quality of service measure which is expressed by the blocking (or loss) probability of information cells. Hence, in such a network, buffer dimensioning of queueing elements in the network is frequently influenced by a desirable blocking probability. In this paper, we investigate the error introduced by the tail of the queue distribution approximation of the blocking probability for single server queues and its effects on buffer sizing.

1. INTRODUCTION

Broadband Integrated Services Digital Networks (B-ISDNs) with the plethora of diverse types of traffic that they carry have usually very stringent specifications for information loss probabilities. The quality of service (QOS) of B-ISDN, especially under the Asynchronous Transfer Mode (ATM) transport mechanism [1], is quite often measured with such loss probabilities (referred to in ATM networks as the *cell loss ratio* (CLR)). Acceptable values of the CLR range from 10^{-4} to 10^{-11} or even less [2], depending on the particular applications transported through the network. Hence, often a decision on the buffer sizes of the various statistical multiplexers, switches, etc., at the nodes of a B-ISDN is based (or at least strongly influenced) by the sufficiency of the (suggested) buffer sizes to achieve the desirable blocking probabilities.

Queues with finite buffer capacity are excellent models for the queueing elements in a network (e.g., statistical multiplexers) with limited storing capabilities. But in general, the analysis of finite buffer capacity queueing systems are usually more numerically demanding than is that of their infinite capacity counterparts. In many cases, it is altogether impossible to obtain any form of a solution for a finite buffer queue while for its infinite buffer counterpart a (form of a) solution is possible.

A performance measure of great importance for queues with finite buffer capacity is the blocking probability of a customer, i.e., the probability that a customer upon its arrival to the queue finds the queue filled to its capacity and the customer is lost. Due to numerical complexity, the blocking probability is often approximated by the tail of the queue distribution of the corresponding infinite buffer queue [3-9]. Assuming a stable infinite buffer queue, the latter approximation is asymptotically (i.e., as the buffer size of the finite buffer queue increases to ∞) exact. Yet, since very small probabilities of error ($\approx 10^{-9}$) are commonplace in B-ISDN specifications, not only the absolute value of the approximation is important but also the relative error of the approximation needs to be quantified[1]. The latter is the topic of this paper.

This paper is organized as follows. In Section 2, we present the queueing model and we define its parameters. In Section 3, we investigate the $M/M/1$ queue, while in Section 4 we discuss the $M/G/1$ queue. In Section 5, we investigate a discrete time $Geo^{[X]}/D/1$ queue with batch arrivals. In Section 6, we present some numerical results and we conclude with a summary of our results in Section 7.

2. MODEL DESCRIPTION AND DEFINITIONS

Initially, we will consider a continuous time queueing model. In Section 5 we will present a slightly modified discrete time model but we will maintain the terminology introduced in this section.

We consider a single server, finite buffer queue. The arrival times of customers to the queue constitute a stationary and ergodic process with interarrival time (first order) distribution A and a finite arrival rate λ. The service requirements of customers also form a stationary and ergodic process with (first order) distribution S and a finite mean value $\bar{S}$. For the analysis in the following sections, the arrival and service processes belong to very special cases, but, for the presentation in this section no particular restrictions are necessary. The *offered load* ρ to the queue is given by

$$\rho = \lambda \bar{S}. \tag{1}$$

The service discipline of the server is of no particular interest as long as it is work conserving (i.e., the server never idles as long as there is work to be performed), nonpreemptive, and no sharing of the server by several customers is allowed. The queue is assumed to have a buffer of size $K < \infty$. A corresponding infinite buffer queue is defined in such a way that it has exactly the same arrival and service processes (sample path copies of the original processes) and the same service discipline but with an infinite buffer size, (or $K = \infty$).

Let Q_t, $t \geq 0$, denote the queue size at time t^-, including the customer in service (if any). If at time t a customer arrives at the queue, the customer is *not* counted in Q_t. Whenever an infinite buffer queue is considered in this paper, $\rho < 1$ is necessary and sufficient for the queue to be stable and to possess a steady state, i.e., $Q_t \to Q$ as $t \to \infty$, where Q is a proper random variable independent of any initial conditions [10]. Since the queue size in a finite buffer queue is impossible to increase to infinity, it is always stable no matter what the value of ρ is. The existence of a steady state for the finite queue will be assumed in this paper (but it can formally be shown to exist for the particular queues studied in the following sections).

In the sequel, it will always be assumed that $\rho < 1$ holds true. In addition, whenever we refer to a finite buffer queue, we will include in our notations the buffer size K, e.g., $Q_t(K)$. For the infinite buffer queues, i.e., $K = \infty$, we will drop the K from our notation.

If upon the arrival instant t_a of a customer to the finite buffer queue $Q_{t_a}(K) = K$, the customer is assumed to be blocked and never accepted in the queue, i.e., the customer is considered lost. With each arriving customer, we associate an indicator random variable (r.v.) $B(n; K)$, $n = 1, 2, \ldots$, defined as follows:

---

[1] Note that the approximation of 10^{-9} with a 0 is accurate in the first 8 decimal digits. Nevertheless, any buffer dimensioning based on such an approximation is totally meaningless.

$$B(n; K) = \begin{cases} 1, \text{ the } n\text{-th customer to arrive to the queue is blocked, and} \\ 0, \text{ otherwise.} \end{cases} \tag{2}$$

We assume that the limit

$$\lim_{N \to \infty} \frac{1}{N} \sum_{n=1}^{N} B(n; K) \tag{3}$$

exists (either with probability 1 or in probability) and that it is a constant (i.e., non-random) quantity. We will call the "customer average" limit in (3) the *blocking probability* and denoted by $\tau_b(\rho; K)$ [2].

Due to the increased complexity in analyzing finite buffer queues, the blocking probability $\tau_b(\rho; K)$ has frequently been approximated by the probability $\hat{\tau}_b(\rho; K)$ that the (steady state) size Q of the infinite buffer queue is in excess of K, i.e.,

$$\hat{\tau}_b(\rho; K) = Pr(Q > K) = Pr(Q \geq K + 1), \tag{4}$$

[3-9]. We will call this approximation the *tail approximation* because it can be calculated from the tail of the distribution of the queue size of the infinite buffer queue. Although different tail approximations may be defined, see Section 7, in this paper we concentrate in the tail approximation as defined in (4) due to its repeated appearance in the literature. Notice that, since the random variable Q is a limit as $t \to \infty$, then strictly speaking, a tail approximation attempts to approximate the following "time average" limit (assuming that it exists)

$$\lim_{t \to \infty} \frac{\hat{B}(t; K)}{t}, \tag{5}$$

where $\hat{B}(t; K)$ is a step function counting the number of blocked customers in the interval $[0, t]$. For arrival processes with the *ASTA* (Arrivals See Time Averages) property [11], e.g., a Poisson process, the later limit coincides with that in (3).

As $K \to \infty$ both $\tau_b(\rho; K)$ and $\hat{\tau}_b(\rho; K)$ converge to 0, and, hence, the tail approximation is asymptotically exact, i.e., $|\hat{\tau}_b(K; \rho) - \tau_b(K; \rho)| \to 0$ as $K \to \infty$. For the study of the relative accuracy of the tail approximation two "quality" measures will be used: the *relative magnitude* $\gamma(\rho; K)$ of the tail approximation, which is defined as the ratio

$$\gamma(\rho, K) = \frac{\hat{\tau}_b(\rho; K)}{\tau_b(\rho; K)}, \tag{6}$$

and the *relative error* $\varepsilon(\rho; K)$ of the tail approximation, which is given by

$$\varepsilon(\rho; K) = \frac{\hat{\tau}_b(\rho; K) - \tau_b(\rho; K)}{\tau_b(\rho; K)} = \gamma(\rho; K) - 1. \tag{7}$$

The limits of $\gamma(\rho; K)$ and $\varepsilon(\rho; K)$ as $K \to \infty$ (if they exist) will be denoted by $\bar{\gamma}(\rho)$ and $\bar{\varepsilon}(\rho)$, respectively.

Next in Section 3, we consider the $M/M/1(/K)$ queue, while in Section 4, we consider the $M/G/1(/K)$ queue and investigate the accuracy of the tail approximation in both cases. Most of the results of Section 3 can be derived from those of Section 4, but an $M/M/1(/K)$ queue

---

[2] The probability $\tau_b(\rho; K)$ is generally dependent on the whole arrival and service processes as well as on the service discipline. For notation brevity such dependences will be assumed only implicitly in this paper.

provides (as always) additional insight and, hence, deserves a separate section. The material in the next two sections may be complemented (and vice versa) by the work in [12] where the asymptotic form of $\tau_b(\rho; K)$ (as $K \to \infty$) for the $M/G/1/K$ and $G/M/1/K$ queues has been studied, including the case that $\rho \geq 1$.

3. THE BLOCKING PROBABILITY IN THE $M/M/1/K$ QUEUE

Under the assumption that the queue under investigation can be modelled as an $M/M/1/K$ queue, it follows that [13]

$$\tau_b(\rho; K) = (1-\rho)\frac{\rho^K}{1-\rho^{K+1}}. \tag{8}$$

From the corresponding infinite buffer $M/M/1$ queue, we obtain

$$\hat{\tau}_b(\rho; K) = (1-\rho)\sum_{i=K+1}^{\infty} \rho^i = (1-\rho)\frac{\rho^{K+1}}{1-\rho} = \rho^{K+1}. \tag{9}$$

It follows from (8) that if a blocking probability of magnitude τ_b at an offered load ρ is desired, then a buffer of size[3]

$$K \geq \frac{\ln\left(\dfrac{\tau_b}{1-(1-\tau_b)\rho}\right)}{\ln(\rho)} = \frac{\ln(\tau_b) - \ln(1-(1-\tau_b)\rho)}{\ln(\rho)} \stackrel{\text{def}}{=} K^* \tag{10}$$

should be selected. Under the same conditions, (9) results in a buffer of size

$$K \geq \frac{\ln\left(\dfrac{\tau_b}{\rho}\right)}{\ln(\rho)} = \frac{\ln(\tau_b)}{\ln(\rho)} - 1 \stackrel{\text{def}}{=} \hat{K}^*. \tag{11}$$

From (10) and (11) it follows that if

$$\begin{cases} (2-\tau_b)\rho \leq 1, & \text{then, } K^* \geq \hat{K}^*, \text{ and if} \\ (2-\tau_b)\rho \geq 1, & \text{then, } K^* \leq \hat{K}^*. \end{cases} \tag{12}$$

Notice that the conclusions in (12) hold true for any τ_b if $\rho \leq 0.50$. If $\rho > 0.50$, the conclusions in (12) hold true for sufficiently small τ_b.

Next, we investigate the accuracy of the tail approximation. Dividing (9) by (8) yields

$$\gamma(\rho; K) = \frac{\hat{\tau}_b(\rho; K)}{\tau_b(\rho; K)} = \frac{\rho^{K+1}}{(1-\rho)\dfrac{\rho^K}{1-\rho^{K+1}}} = \frac{\rho}{1-\rho}(1-\rho^{K+1}). \tag{13}$$

The relative magnitude $\gamma(\rho; K)$ is an increasing function of K and it converges to

$$\bar{\gamma}(\rho) \stackrel{\text{def}}{=} \lim_{K \to \infty} \gamma(\rho; K) = \frac{\rho}{1-\rho}. \tag{14}$$

---

[3] $\stackrel{\text{def}}{=}$ stands for "equal by definition".

Denoting the *golden ratio* irrational number by $\phi = (1+\sqrt{5})/2 \approx 1.618$, we conclude from (13) that if

$$\left\{\begin{array}{l} 0 \le \rho \le 0.50, \text{ then } \gamma(\rho; K) \le 1, \forall K, \text{ (i.e., underestimation of } \tau_b(\rho; K)); \quad \text{or, if} \\ 0.50 < \rho \le \phi^{-1}, \text{ then } \left\{\begin{array}{l} \gamma(\rho; K) \le 1, \text{ for } K \le [\ln(2\rho - 1)/\ln(\rho)] - 2, \\ \gamma(\rho; K) > 1, \text{ otherwise; or, if} \end{array}\right. \\ \rho > \phi^{-1}, \text{ then } \gamma(\rho; K) > 1, \forall K, \end{array}\right. \tag{15}$$

where it follows from the definition of ϕ that $\phi^{-1} = \phi - 1$. Also, for $\rho > 10/11 \approx 0.909$, $\hat{\tau}_b(\rho; K)$ overestimates $\tau_b(\rho; K)$ (for a sufficiently large K) by at least an order of magnitude.

The relative error $\varepsilon(\rho; K)$ of the tail approximation is given by

$$\varepsilon(\rho; K) = \gamma(\rho; K) - 1 = \rho\frac{1-\rho^{K+1}}{1-\rho} - 1 \underset{K\to\infty}{\to} \frac{2\rho - 1}{1-\rho} \overset{\text{def}}{=} \bar{\varepsilon}(\rho). \tag{16}$$

4. THE BLOCKING PROBABILITY IN THE *M/G/1/K* QUEUE

It can be shown, e.g., see [14], that the blocking probability $\tau_b(\rho; K)$ can be obtained from the steady state probability mass function of the corresponding infinite buffer queue, in particular

$$\tau_b(\rho; K) = \frac{1 - \rho + (\rho - 1)\sum_{i=0}^{K-1} Pr(Q = i)}{1 - \rho + \rho\sum_{i=0}^{K-1} Pr(Q = i)} = (1-\rho)\frac{Pr(Q \ge K)}{1 - \rho Pr(Q \ge K)}. \tag{17}$$

Thus, the relative magnitude $\gamma(\rho; K)$ is given by

$$\begin{aligned} \gamma(\rho; K) &= \frac{\hat{\tau}_b(\rho; K)}{\tau_b(\rho; K)} = \frac{1}{1-\rho}\,\frac{[1 - \rho Pr(Q \ge K)]Pr(Q \ge K+1)}{Pr(Q \ge K)} \\ &= \frac{1}{1-\rho}\,\frac{1 - \rho Pr(Q \ge K)}{\dfrac{Pr(Q > K)}{Pr(Q \ge K+1)}} = \frac{1}{1-\rho}\,\frac{1 - \rho Pr(Q \ge K)}{1 + \dfrac{Pr(Q - K)}{Pr(Q \ge K+1)}}. \end{aligned} \tag{18}$$

From (18) one may notice once again that as $K \to \infty$, $\gamma(\rho; K) - B(\rho)/(1-\rho)$, where $B(\rho)$ is the inverse of the limit (assuming that it exists) of the denominator of (18). In general, nothing additional can be stated about $B(\rho)$ except that $0 \le B(\rho) \le 1$. In practice, quite often the tail of the queue size distribution of Q decays exponentially. The latter condition necessitates the finiteness of all the moments of Q and, hence, of the service time S. In such a case, $Pr(Q = m)$ can be satisfactorily approximated for a sufficiently large m by

$$\begin{aligned} &Pr(Q = m) \approx C(\rho)[z_0(\rho)]^{-m}, \\ &\Rightarrow Pr(Q \ge m) \approx \frac{C(\rho)z_0(\rho)}{z_0(\rho) - 1}[z_0(\rho)]^{-m}, \end{aligned} \tag{19}$$

where $C(\rho)$ is a constant (with respect to m) independent of $z_0(\rho)$. The above approximation is asymptotically (as $m \to \infty$) exact. Notice, that for an $M/M/1$ queue $C(\rho) = (1-\rho)$ and

$z_0(\rho) = \rho^{-1}$ and (19) (which, in this case, is exact for every m) reduces to (9). $z_0(\rho)$ and $C(\rho)$ can easily be obtained with the help of an inverse FFT of the z-transform $Q^*(z)$ of the queue size [15], see also the Appendix for a uniqueness property of $z_0(\rho)$. $Q^*(z)$, $|z| \le 1$, is given from the Pollaczeck-Khinchine formula:

$$Q^*(z) = (1-\rho)\frac{(1-z)S^*(\lambda(1-z))}{S^*(\lambda(1-z)) - z}, \tag{20}$$

where $S^*(\cdot)$ is the Laplace-Stieltjes transform (LST) of the service time S. When $S^*(\cdot)$ is a rational function, $z_0(\rho)$ corresponds to the dominant pole of the $Q^*(z)$. For sufficiently large K, substituting (19) in (18) yields

$$\gamma(\rho; K) \approx \frac{z_0(\rho) - 1 - \rho C(\rho)[z_0(\rho)]^{-K+1}}{(1-\rho)(z_0(\rho)-1)z_0(\rho)} \underset{K\to\infty}{\to} \frac{[z_0(\rho)]^{-1}}{1-\rho} \overset{\text{def}}{=} \bar{\gamma}(\rho), \tag{21}$$

where the limit as $K \to \infty$ is the exact limit of $\gamma(\rho; K)$. The relative error $\varepsilon(\rho; K)$ is given by

$$\begin{aligned}\varepsilon(\rho; K) = \gamma(\rho; K) - 1 &\approx \frac{(z_0(\rho)-1)[1-(1-\rho)z_0(\rho)] - \rho C(\rho)[z_0(\rho)]^{-K+1}}{(1-\rho)(z_0(\rho)-1)z_0(\rho)} \\ &\underset{K\to\infty}{\to} \frac{[z_0(\rho)]^{-1} + \rho - 1}{1-\rho} \overset{\text{def}}{=} \bar{\varepsilon}(\rho).\end{aligned} \tag{22}$$

5. THE BLOCKING PROBABILITY IN THE $Geo^{[X]}/D/1/K$ QUEUE

In this section, we digress somewhat from the models of the previous sections. In particular, we consider a discrete (or slotted) time queue with batch arrivals and deterministic service time. Such a queue frequently arises in the modelling and analysis of ATM based networks, where the unit of time is taken to be the fixed size of a cell (or slot).

We consider a single server queue that at each slot boundary receives a batch of customers. The batches are assumed to be i.i.d. and have a mean λ, in accordance to the arrival rate λ of the previous sections. We will assume that there is a nonzero probability that at least two customers may arrive at the same slot boundary. Customers from the same batch are "selected" in a random order. Each customer requires a fixed amount of service equal to one slot. It follows from the later assumption that the offered load ρ to the queue equals the arrival rate λ. Under the previous assumptions, the infinite buffer queue is stable and possesses a steady state if and only if $\rho < 1$, which will be assumed to hold true in the sequel. Although $\rho = \lambda$, we will continue to write ρ in our notations in accordance to the previous sections. The (steady state) queue size $Q(K)$ is defined as the queue occupancy (including the customer (if any) in service) just following the leading boundary of a slot. Let $\tau_b(\rho; K)$ denote the blocking probability that a (randomly selected) customer arriving to the queue finds K customers already in queue and it is thus blocked (and forever lost) from entering the queue. Let $\lambda'(\rho; K)$ denote the rate of customers entering (i.e., accepted by) queue, then, $\lambda'(\rho; K) = 1 - Pr(Q(K) = 0)$. A simple flow conservation argument results in

$$\lambda'(\rho; K) = \lambda[1 - \tau_b(\rho; K)] \Rightarrow \tau_b(\lambda; K) = 1 - \frac{\lambda'(\rho; K)}{\lambda}. \tag{23}$$

It has been shown in [16] that

$$Pr(Q(K)=m)=\frac{Pr(Q=m)}{\sum_{i=0}^{K} Pr(Q=i)}, \tag{24}$$

where the probabilities in the RHS of (24) are obtained from the corresponding infinite buffer queue. Combining (23) and (24) yields

$$\tau_b(\rho;K)=\frac{1-\rho}{\rho}\,\frac{1-\sum_{i=0}^{K} Pr(Q=i)}{\sum_{i=0}^{K} Pr(Q=i)}=\frac{1-\rho}{\rho}\,\frac{Pr(Q\geq K+1)}{1-Pr(Q\geq K+1)}. \tag{25}$$

Dividing both sides of the above expression by $\hat{\tau}_b(\rho;K)=Pr(Q\geq K+1)$ and inverting yields

$$\gamma(\rho;K)=\frac{\hat{\tau}_b(\rho;K)}{\tau_b(\rho;K)}=\frac{\rho}{1-\rho}\left[1-Pr(Q\geq K+1)\right]\underset{K\to\infty}{\to}\frac{\rho}{1-\rho}\overset{\text{def}}{=}\bar{\gamma}(\rho). \tag{26}$$

Thus, the relative error $\varepsilon(\rho;K)$ is given by

$$\varepsilon(\rho;K)=\gamma(\rho;K)-1=\frac{\rho\left[2-Pr(Q\geq K+1)\right]-1}{1-\rho}\underset{K\to\infty}{\to}\frac{2\rho-1}{1-\rho}=\bar{\varepsilon}(\rho). \tag{27}$$

Notice that the limits as $K\to\infty$ depend on the customer batch size only through its mean. If, once again, the assumption of the exponential decay of the queue size is made, an assumption that necessitates the finiteness of all the moments of a customer batch size, then (26) yields

$$\gamma(\rho;K)\approx\frac{\rho}{1-\rho}\left(1-\frac{C(\rho)}{z_0(\rho)-1}\left[z_0(\rho)\right]^{-K}\right). \tag{28}$$

Similarly, the relative error $\varepsilon(\rho;K)$ is given by

$$\varepsilon(\rho;K)=\gamma(\rho;K)-1\approx\frac{(2\rho-1)(z_0(\rho)-1)-\rho C(\rho)\left[z_0(\rho)\right]^{-K}}{(1-\rho)(z_0(K)-1)}. \tag{29}$$

6. NUMERICAL RESULTS

In this section, we will present some numerical results related to the $M/M/1(/K)$ queue and the $M/G/1(/K)$ queue. For further discussion on the tail approximation for the $Geo^{[X]}/D/1(/K)$ queue see [16].

Figure 1 depicts the relative error

$$\delta(\tau_b;\rho)=\frac{\hat{K}^*-K^*}{K^*} \tag{30}$$

in estimating the buffer size of an $M/M/1/K$ queue in order to achieve a given blocking probability τ_b, see (10) and (11). $\delta(\tau_b;\rho)$ is an increasing function of ρ and "explodes" to

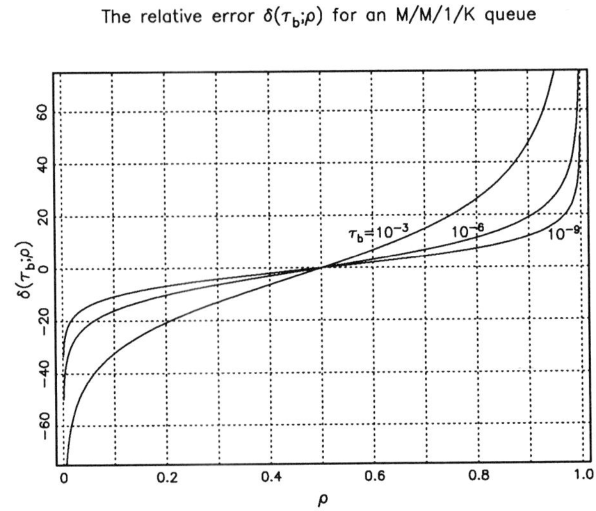

Figure 1. The relative error $\delta(\tau_b;\rho)$ for an $M/M/1/K$ queue as a function of ρ and various blocking probabilities τ_b

infinity as $\rho \to 1$. $\delta(\tau_b;\rho)$ is initially negative implying that $\hat{K}^*$ underestimates K^*. As ρ increases, $\delta(\tau_b;\rho)$ eventually becomes positive. The crossover point for the small blocking probabilities considered here is located at (about) $\rho = 0.50$. For larger blocking probabilities,

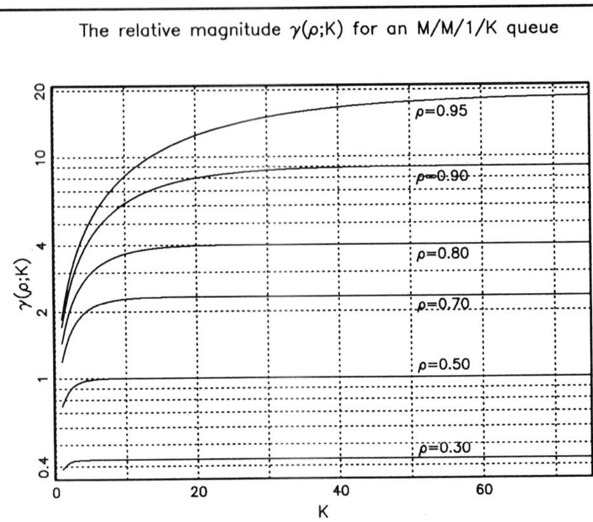

Figure 2. The relative magnitude $\gamma(\rho;K)$ for an $M/M/1/K$ queue as a function of the buffer size K and for various ρ

the crossover point is offset to the right as described by (12). Finally, the absolute value of $\delta(\tau_b; \rho)$ decreases with decreasing τ_b, showing that, as $\tau_b \to 0$, $\hat{K}^*$ becomes a more and more reliable estimate of K^*. Figure 2 depicts the relative magnitude $\gamma(\rho; K)$ for an $M/M/1/K$ queue as a function of K for various levels of offered load ρ, see (13). $\gamma(\rho; K)$ is increasing on both ρ and K. But, $\gamma(\rho; K)$ saturates to its limit $\overline{\gamma}(\rho)$ (relatively) fast with an increasing K, see (14). For example, when $K \geq 50$ $\gamma(0.95; K)$ is very close to $\overline{\gamma}(0.95) = 19$. As discussed in Section 3, when $\rho > 0.50$ then $\hat{\tau}_b(\rho; K)$ (eventually) overestimates the true blocking probability $\tau_b(\rho; K)$, i.e., $\gamma(\rho; K) > 1$. When $K \geq 15$, $\gamma(0.95; K)$ is larger than 10, which implies that $\hat{\tau}_b(0.95; K)$ overestimates $\tau_b(0.95; K)$ by at least an order of magnitude.

Next, we consider more general service times S but all with the same mean $\overline{S} = 1$. Due to its flexibility, the Gamma distribution with a parameter α has been selected. In particular, the LST $\Gamma^*_\alpha(\sigma)$, $\mathrm{Re}(\sigma) \geq 0$, of the service time distribution is given by

$$\Gamma^*_\alpha(\sigma) = E\left(e^{-\sigma S}\right) = \left[\frac{\alpha}{\alpha+\sigma}\right]^\alpha, \quad \mathrm{Re}(\sigma) \geq 0, \ \alpha > 0. \tag{31}$$

The service time distribution so selected has mean value $\overline{S} = 1$ and squared coefficient of variation $c_v^2 = 1/\alpha$. Notice that when $\alpha = 1$ the Gamma distribution becomes the exponential and as $\alpha \to \infty$ it converges to the constant service time. Hence, the Gamma distribution can represent a wide range of distributions for which the first two moments are being specified. Figure 3 depicts the dominant pole $z_0(\rho)$, of the Pollaczeck-Khinchine formula in (20), as a function of the offered load ρ. For a given value of ρ, the dominant pole gets smaller (and closer to 1) as α decreases, i.e., the service time becomes more variable, which implies a heavier tail for the distribution of the infinite buffer queue size. In the Appendix, we prove that for a class of service distributions, which includes the Gamma distribution, there is a unique dominant real pole. Figure 4 depicts the corresponding coefficient $C(\rho)$. The latter two quantities can be used in the approximations of $\gamma(\rho, K)$ and $\varepsilon(\rho, K)$ given in (21) and (22). For various values of α, Figure 5 depicts the relative magnitude of the tail approxi-

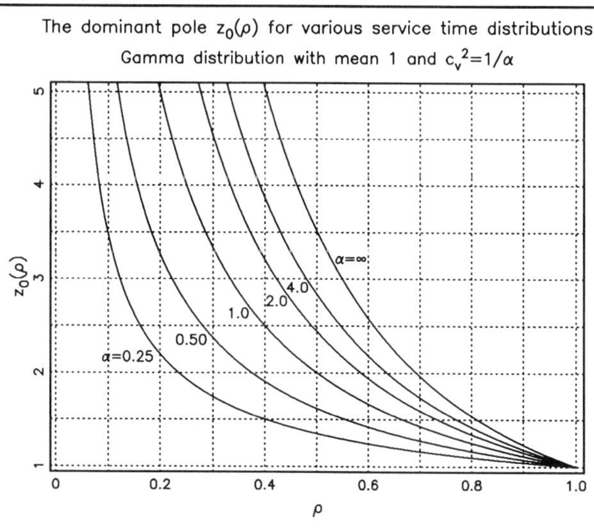

Figure 3. The dominant pole $z_0(\rho)$ as a function of ρ for various α

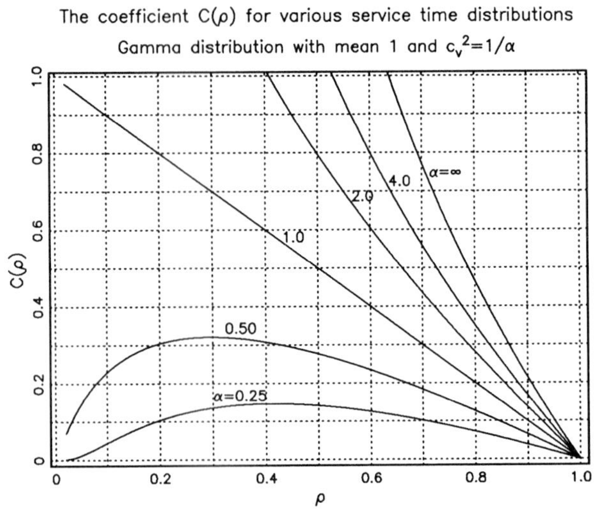

Figure 4. The coefficient $C(\rho)$ as a function of ρ for various α

mation $\gamma(\rho; K)$ as a function of the buffer size K for $\rho = 0.90$. The curves have been obtained with the help of an inverse FFT of the Pollaczeck-Khinchine formula, but the approximation in (21) provides practically the same curves when $K > 5$. As $K \to \infty$, $\gamma(0.90; K)$ increases to its upper bound given by $\bar{\gamma}(0.90)$ in (21). As expected, in the limit, the tail approximation becomes less reliable (manifested by a larger $\bar{\gamma}(0.90)$) as the variability of the service time

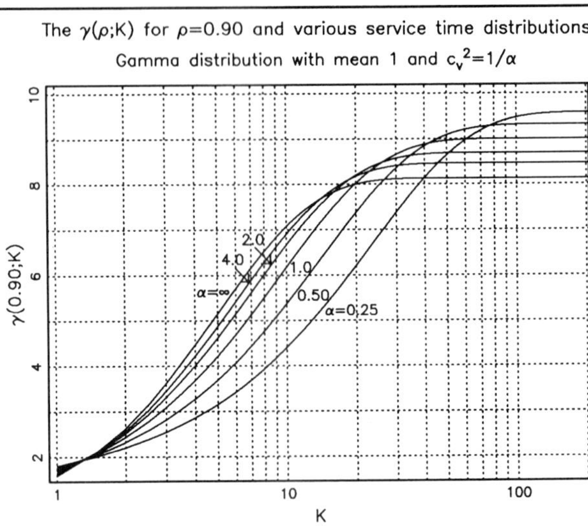

Figure 5. The relative magnitude $\gamma(0.90;K)$ as a function of the buffer size and for various α

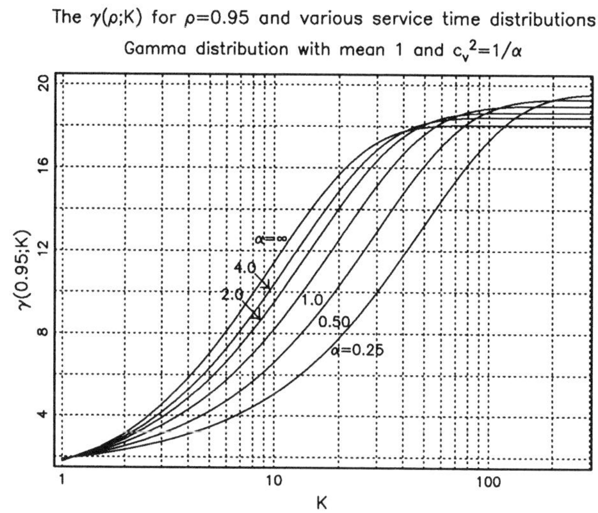

Figure 6. The relative magnitude $\gamma(0.95;K)$ as a function of the buffer size and for various α

increases (i.e., as α decreases). Surprisingly, the latter expected result does not hold true for all K. No solid explanation can be provided for this behavior, which is more prominent in Figure 6 which depicts $\gamma(\rho; K)$ for $\rho = 0.95$. From Figure 6, it is also apparent that the limits $\bar{\gamma}(0.95)$ are very close to each other, as it would have been expected from Figure 3 where $z_0(\rho) \approx \rho^{-1} \approx 1$ as $\rho \to 1$. Now, since

$$\tau_b(\rho; K) = \frac{\hat{\tau}_b(\rho; K)}{\gamma(\rho; K)}, \tag{32}$$

when both K and ρ are expected to be (sufficiently) large (as it would be the case within the ATM framework), we may improve on the estimate of the blocking probability provided by the tail approximation by multiplying the latter by $\hat{\tau}_b(\rho; K)$ by $(1 - \rho)/\rho$.

7. CONCLUSIONS

In this paper, we have investigated the accuracy of the frequently used tail approximation $\hat{\tau}_b(\rho; K) = Pr(Q > K) = Pr(Q \geq K + 1)$ of the blocking probability $\tau_b(\rho; K)$ for a number of widely used single server queueing models.

A number of very interesting results have been obtained. Firstly, the tail approximation $\hat{\tau}_b(\rho; K)$ does not always provide an upper bound on the true blocking probability. For the cases investigated, this is true only for moderately or heavily loaded queues ($\rho > 0.50$). Hence, any bounding property claim of the tail approximation must be very carefully stated. Secondly, the relative magnitude $\gamma(\rho; K)$ and error $\varepsilon(\rho; K)$ of the tail approximation are proportional to $(1 - \rho)^{-1}$ and, thus, they both explode to infinity as $\rho \to 1$. This implies that the tail approximation considerably overestimates the buffer size necessary to achieve a prespecified blocking probability. Substantial savings in buffer size may be achieved if a more accurate estimate of the true blocking probability can be obtained. Our final result addresses the latter point. In all the three cases we have investigated, the limit $\bar{\gamma}(\rho)$ of the relative magnitude as $K \to \infty$ has the simple form $B/(1-\rho)$, where $B \to 1$ (from below) as $\rho \to 1$. For both the $M/M/1(/K)$ and $Geo^{[X]}/D/1(/K)$, B is equal to ρ, while for the $M/G/1(/K)$ queue $B \approx \rho$ as $\rho \to 1$. Hence, under heavy load, e.g., $\rho > 0.90$, for these widely used queueing models a very simple correction adjustment to the tail approximation can be provided, in the form of

$$\tau_b(\rho; K) \approx \frac{1-\rho}{\rho} \hat{\tau}_b(\rho; K). \tag{33}$$

The extent of the validity of the above simple adjustment to more general arrival processes beyond the memoryless Poisson and i.i.d. batches for the continuous and discrete time models, respectively, is still an open issue. Nevertheless, it is our belief that such a simple adjustment to the tail approximation may still be valid for more general arrival and service processes.

In closing, the work presented here can be verbatim repeated for a tail approximation defined as $Pr(Q \geq K)$ [17]. For the queues investigated in this paper, it is trivial to show that $\tau_b(\rho; K) \leq Pr(Q \geq K)$ *for every* $0 < \rho < 1$ and $K \geq 1$. Moreover, the asymptotic relative measures of the latter tail approximation still behave like $(1-\rho)^{-1}$, signifying once again a substantial overestimation of the blocking probability as $\rho \to 1$. The latter observation restricts the range of validity of the claim in [18] that a tail approximation should provide a *tight* upper bound on the blocking probability. Since $Pr(Q \geq K+1) < Pr(Q \geq K)$, the tail approximation studied in this paper may provide a tighter upper bound than the latter tail approximation, a fact that holds true for at least the queueing models investigated in this paper for (eventually) any $\rho > 0.50$.

Acknowledgments

The authors are thankful to Dr. I. Iliadis for his constructive observations and comments on an earlier version of the paper.

APPENDIX: On the Uniqueness of the Dominant Pole

Let S be a nonnegative random variable that is assumed to possess all of its moments. It is assumed that $Pr(S > 0) > 0$. Let $s(\cdot)$, $\bar{S}$, and $S^*(\cdot)$ denote the density, the mean value, and the LST of S, respectively. The latter function is given by

$$S^*(\sigma) = E(e^{-\sigma S}) = \int_0^\infty e^{-\sigma t} s(t) \mathrm{d}t, \tag{34}$$

where $\sigma = x + iy$ is a complex variable with real and imaginary parts to be denoted by x and y, respectively. Since all the moments of S are assumed to exist, $S^*(\cdot)$ is analytic and absolutely convergent when $x_{abs} \leq x < 0$, where x_{abs} is a strictly negative real number defined by

$$x_{abs} = \inf \{ x : S^*(x) \text{ is convergent } \}, \tag{35}$$

where $x_{abs} = -\infty$ is allowed. Notice that for $x > x_{abs}$

$$| S^*(x+iy) | \le S^*(x) < \infty. \tag{36}$$

Let us now assume that $S^*(x)$ increases to $+\infty$ as x decreases towards x_{abs}. Once again, due to the existence of all moments, it can be shown that $S^*(x)$ is a convex function which combined with the previous assumption results in $S^*(x)$ being monotonically decreasing for $x_{abs} < x$. It can be easily shown that for $x_{abs} < x < 0$

$$S^*(x) \ge e^{-t_0 x}\left[\int_{t_0}^{\infty} s(t)\mathrm{d}t\right] = [\text{a positive constant}]e^{-t_0 x}, \tag{37}$$

where t_0 is any strictly positive real number. A large class $\mathcal{A}$ of densities satisfy the above assumptions, including the Gamma densities and all the densities with finite support (i.e., bounded random variables).

We will now consider an $M/G/1$ queue with service time S with a density belonging to the class $\mathcal{A}$. We are interested in finding the dominant pole of the Pollaczeck-Khinchine formula in (20), i.e., the root of

$$S^*(\lambda(1-\sigma)) - \sigma = 0, \tag{38}$$

with the smaller magnitude larger than 1. Here, λ denotes the intensity of the Poisson arrivals. For the stability of the queue, $\lambda < 1/\overline{S}$ is required to hold true. Let $w = W(\sigma) = 1 - (\sigma/\lambda)$, then, all the roots of (38) have the form $w = W(\sigma)$ where $S^*(\sigma) = W(\sigma)$.

<u>*Lemma*</u>

If the random variable S has a density from the class $\mathcal{A}$, then the equation $S^*(\sigma) = W(\sigma)$ has a unique real root x^* such that $x_{abs} < x^* < 0$.

<u>Proof:</u> By Taylor's theorem for a negative x sufficiently close to 0, $S^*(x) = 1 - \overline{S}x + \cdots < 1 - x/\lambda = W(x)$. From the defining assumptions of class $\mathcal{A}$, as x decreases to x_{abs}, then, eventually $W(x) < S^*(x)$ holds true. Thus, $S^*(x) - W(x)$ has a root x^* in $(x_{abs}, 0)$. Since $S^*(x)$ is convex, so is $D(x) = S^*(x) - W(x)$ and given the roots of $D(x)$ at 0 and x^*, $D(x)$ cannot have any other roots in $(x_{abs}, 0)$.

QED

<u>*Theorem*</u>

If $S^*(\sigma) = W(\sigma)$ then either $\sigma = 0$ or $\sigma = x^*$ or $\mathrm{Re}(\sigma) < x^*$.

<u>Proof:</u> Since the z-transform of a random variable (in this case, the queue size) is analytic for $|\sigma| \le 1$, it can be easily shown that $S^*(\sigma) - W(\sigma)$ has no roots on the half plane $\mathrm{Re}(\sigma) \ge 0$, except at $\sigma = 0$.

Let us now consider the left half plane where $\mathrm{Re}(\sigma) < 0$. If $x^* < x < 0$, then from the previous Lemma $0 < S^*(x) < W(x)$ and of course $0 < S^*(x^*) = W(x^*)$. Increasing $|y|$ increases $|W(x+iy)|$ but from (36) $|S^*(x+iy)| \le S^*(x)$. Hence, the strip where $x^* \le Re(\sigma) < 0$, contains no root except $\sigma = x^*$.

QED

<u>*Corollary*</u>

$w^* = 1 - (x^*/\lambda)$ is the dominant pole (i.e., the smallest (in magnitude) pole with real part in $(1, \infty)$) of the Pollaczeck-Khinchine formula, it is real, and it is unique.

REFERENCES

[1] M. de Prycker, *Asynchronous Transfer Mode: Solution for Broadband ISDN*, New York: Ellis Horwood, 1991.

[2] M. E. Anagnostou, M. E. Theologou, K. M. Vlakos, D. Tournis, and E. N. Protonotarios, "Quality of Service Requirements in ATM-based B-ISDNs," *Computer Communications*, vol. 14, no. 4, pp. 197-204, May 1991.

[3] M. Schwartz, *Computer-Communication Network Design and Analysis*, Englewood Cliffs, N.J.: Prentice-Hall, Inc., chap. 7, 1977.

[4] J. Y. Hui and E. Arthurs, "A Broadband Packet Switch for Integrated Transport," *IEEE J. Select. Areas Commun.*, vol. 5, no. 8, pp. 1264-1273, Oct. 1987.

[5] Y.-S. Yeh, M. G. Hluchyj, and A. S. Acampora, "The Knockout Switch: A Simple, Modular Architecture for High-Performance Packet Switch," *IEEE J. Select. Areas Commun.*, vol. 5, no. 8, pp. 1274-1283, Oct. 1987.

[6] J. S.-C. Chen and R. Guérin, "Input Queueing of an Internally Non-Blocking Packet Switch with Two Priority Classes," *Proc. of IEEE INFOCOM'89*, pp. 529-535, Ottawa, Ont., Canada, Apr. 23-24, 1989.

[7] E. Desmet, B. Steyaert, H. Bruneel, and G. H. Petit, "Tail Distributions of Queue Length and Delay in Discrete-Time Multiserver Queueing Models, Applicable in ATM Networks," *in Proc. 13th Int'l Teleltr. Congress*, (J. W. Cohen and C. D. Pack (ed.)), pp. 1-6, Copenhagen, Denmark, June 19-26, 1991.

[8] Y. Xiong and H. Bruneel, "Performance of Statistical Multiplexers with Finite Number of Inputs and Train Arrivals," *in Proc. IEEE INFOCOM'92*, pp. 2036-2044, Florence, Italy, May 6-8, 1992.

[9] I. Widjaja and A. Leon-Garcia, "Design and Performance Analysis of a Flexible Output-Buffered ATM Switch," *in Proc. 1st Int'l Conf. Comp. Commun. & Netw. IC^3N*, (E. K. Park and K. Chae (ed.)), pp. 178-182, San Diego, CA, USA, June 8-10, 1992.

[10] A. A. Borovkov, *Stochastic Processes in Queueing Theory*, New York: Springer-Verlag, chap. 1.3, 1976.

[11] B. Melamed and W. Whitt, "On Arrivals that See Time Averages," *Operations Research*, vol. 38, no. 1, pp. 156-172, Jan.-Feb. 1990.

[12] A. Baiocchi, "Asymptotic Behaviour of the Loss Probability of the M/G/1/K and G/M/1/K Queues," *Queueing Systems*, vol. 10, no. 3, pp. 235-248, Jan. 1992.

[13] L. Kleinrock, *Queueing Systems, Vol.1: Theory*, New York: John Wiley & Sons, chap. 3.6, 1975.

[14] R. B. Cooper, *Introduction to Queueing Theory*, Washington, D.C.: CEEPress Books, chap. 5.9, 1990.

[15] J. N. Daigle, Y. Lee, and M. N. Magalhães, "Discrete Time Queues with Phase Dependent Arrivals," *Proc. of IEEE INFOCOM'90*, pp. 728-732, San Francisco, CA, USA, June 5-7, 1990.

[16] C. Bisdikian, "On the Theory of Output Queueing Space-Division Packet Switches with Renewal Arrivals", IBM Research Report, RC 17482, Dec. 1991.

[17] M. G. Hluchyj and M. J. Karol, "Queueing in Space-Division Packet Switching," *Proc. of IEEE INFOCOM'88*, pp. 334-343, New Orleans, LA, USA, Mar. 29-31, 1988.

[18] T.-C. Hou and D. M. Lucantoni, "Buffer Sizing for Synchronous Self-Routeing Broadband Packet Switches with Bursty Traffic," *Int'l J. of Digital and Analog Cabled Syst.*, vol. 2, pp. 253-260, 1989.

Queueing Networks with Finite Capacity
R.O. Onvural and I.F. Akyildiz (Editors)

A NOTE ON THE ARRIVAL THEOREM FOR CLOSED QUEUEING NETWORKS WITH BLOCKING

Nico M. van Dijk
University of Amsterdam, The Netherlands

Abstract: The arrival theorem is investigated for closed form queueing networks with different types of blocking. Using two and three station networks, it is shown that the arrival theorem holds only under additional constraints for both the delay and recycle protocols. For the jump over protocol, on the other hand, it is proven to generally apply for two station networks with arbitrary randomized blocking provided visits during jump over are taken into account as arrivals. The results illustrate that special care is necessary in using the arrival theorem for product form networks that involve blocking.

1. Introduction

The "arrival theorem" is a well known result for closed Jacksonian queueing networks as a consequence of the product form expressions for the steady state queue length distributions, cf. [8], [12], [15]. Roughly speaking, the arrival theorem applies at a station if *an arriving job at station i sees the system as in steady state at an arbitrary epoch for the system without that job.* The same definition applies to the "departure theorem" if an "arriving job" is replaced by a "departing job". By virtue of workload preservation, the departure theorem iis directly related to the arrival theorem.

These intuitively appealing results have proved most useful for both computational and theoretical purposes. Most notably, the arrival theorem is a fundamental pilar of the MVA algorithm to efficiently compute throughputs and related performance measures for large product form networks (e.g. [1-7], [9-10], [13], [16]).

Considering both exact and approximate MVA-methods, the general perception seems to have grown that the arrival theorem is an absolute consequence of product form results. For example, appealing though intuitive arguments based on "quasi reversibility" of the individual stations are sometimes given to argue the arrival theorem (e.g. [15]), but such arguments can easily be misinterpreted as being valid when station or job interferences are involved. More precisely, all references above only deal with product form networks which donot allow blocking and no explicit results to support this perception for networks with blocking seem to be available. In contrast, general results on Palm distributions (e.g. [11]) indicate that further conditions might be needed to this end.

In this paper, we provide some further insight in the validity of the arrival theorem for product form networks with blocking. The intention is not so much a characterization of the arrival theorem in general terms of transition rates, as this will not give practical insights, but more to bring over to practitioners soem awareness of the consequences of blocking by means of simple examples.

In particular, using two-station examples, we illustrate the following results:

1. The arrival theorem does not generally apply for product form queueing networks with blocking. It fails under both the delay and recycle protocols

   The two exceptions to this result happens when

   -recycled arrivals are taken into account, and

   -the job could not be blocked by the station it come from

2. The arrival theorem appears to apply under the jump over protocol provided station visits during jumping over are included as arrivals

These results indicate that one has to be most careful to conclude or employ the arrival theorem, such as when using MVA, for queueing networks with blocking. Similar complications or specialized results can be expected for product form networks with other types of station or job interdependencies.

2. Delay Blocking

In this section, we will give a simple example to illustrate that a so-called delay blocking will lead to failure of the arrival theorem when blocking delay factors are involved.

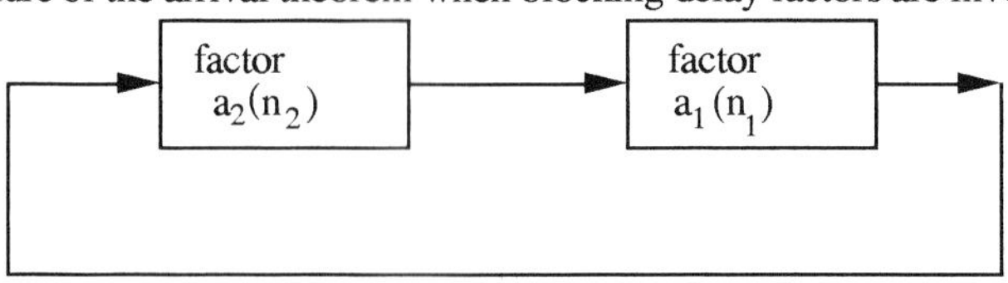

Figure 1: A Two-station cyclic network

Consider a closed two station cyclic queueing network with M jobs and a single service unit at each station. A job requires an exponential amount of service with parameter μ_i at station i. With (n_1,n_2) denoting the number of jobs n_i at station i, i=1,2, let $a_1(n_1)$ and $a_2(n_2)$ be the service speeds at station 2 and 1, respectively, which can be interpreted as *service delay factors* depending on the loads at the next station. For example, if $a_2(n_2)=0$ then service at station 1 is completely stopped. In particular,

$$N_1=\min\{k \mid a_1(k)=0\} \text{ and } N_2=\min\{k \mid a_2(k)=0\}$$

can be interpreted as finite capacity constraints at stations 1 and 2. Let $C_M=\{(n_1,n_2) \mid 0\le n_1\le N_1,\ 0\le n_2\le N_2,\ n_1+n_2=M\}$ denote the set of admissible states. With $\pi_M(n_1,n_2)$ denoting the steady state distribution at C_M, the global balance equations are given as:

$$\pi_M(n_1,n_2)[\mu_1 a_2(n_2)+\mu_2 a_1(n_1)]=\pi_M(n_1-1,n_2+1)\ \mu_2 a_1(n_1-1)\ 1_{n_2<N_2}\ 1_{n_1>0}+$$
$$\pi_M(n_1+1,n_2-1)\ \mu_1 a_2(n_2-1)\ 1_{n_1<N_1}\ 1_{n_2>0},\quad (n_1,n_2)\ \varepsilon\ C_M \qquad (2.1)$$

With $a_1(N_1)=a_2(N_2)=0$, and with c_M being the normalizing constant at C_M, these equations are directly verified by substituting:

$$\pi_M(n_1,n_2)=c_M\,(1/\mu_1)^{n_1}\,(1/\mu_2)^{n_2}\prod_{k=1}^{n_1}a_1(k-1)\prod_{k=1}^{n_2}a_2(k-1),\qquad (n_1,n_2)\,\varepsilon\,C_M \tag{2.2}$$

Let $\pi^2(k_1,k_2)$ represent the steady state probability that a job which arrives at station 2 leaves k_1 jobs behind at station 1 and sees k_2 jobs at station 2. Then, with $N_2<M$, one directly argues that

$$\pi^2(k_1,N_2)=0 \text{ and } \pi_{M-1}(k_1,N_2)>0 \tag{2.3}$$

which corresponds to the case ($n_2=N_2$) that servicing at station 1 is stopped and that there can be no arrivals at station 2, whereas there is a positive probability that $n_2=N_2$ also in the system with one less job, provided $N_1+N_2<M$. Then, we immediately conclude that:

Under Delay Blocking $\pi^2(.,.)\neq\pi_{M-1}(.,.)$, *i.e. the arrival theorem fails when* $N_2<M$.

More generally, by standard arguments and by substituting (2.2), the arrival distribution at station 2 is computed by:

$$\pi^2(k_1,k_2)=\frac{\pi_M(k_1+1,k_2)\,\mu_1 a_2(k_2)}{\sum_{(m_1;m_2)}\pi_M(m_1+1,m_2)\,\mu_1 a_2(m_2)}$$

$$=\frac{(1/\mu_1)^{k_1}\,(1/\mu_2)^{k_2}\,a_1(k_1)\,a_2(k_2)\prod_{k=1}^{k_1}a_1(k-1)\prod_{k=1}^{k_2}a_2(k-1)}{\sum_{(m_1;m_2)}(1/\mu_1)^{m_1}\,(1/\mu_2)^{m_2}\,a_1(m_1)\,a_2(m_2)\prod_{k=1}^{m_1}a_1(k-1)\prod_{k=1}^{m_2}a_2(k-1)} \tag{2.4}$$

From this expression, we conclude that:

Under delay blocking $\pi^2(.,.)\neq\pi_{M-1}(.,.)$, *i.e. the arrival theorem fails at station 2 when* $a_2(.)=1$, and $\pi^2(k_1,k_2)=0$ and $\pi_{M-1}(k_1,k_2)$*i.e., the arrival theorem applies only if there is no blocking at station 2 at all.*

3. Recycle Blocking

In this section, we consider the recycle protocol. Under this protocol, a blocked job always recycled to the station it came from, regardless of whether or not this station itself allows blocking. It will turn out that the arrival theorem requires a special interpretation of what jobs should be seen as arrivals.

3.1. Recycled jobs are not included, deterministic routing

Let us now reconsider the two station model given in figure 1 after the state dependent delay factors $a_i(n_i)$ reflecting blocking are replaced by recycling probabilities upon blocking. More precisely, let $1-a_i(n_i)$ denote the probability that an arriving job at station i is blocked when there are n_i jobs at the station. Furthermore, let us assume that a blocked job returns to the station it came from to undergo a new service at that station.

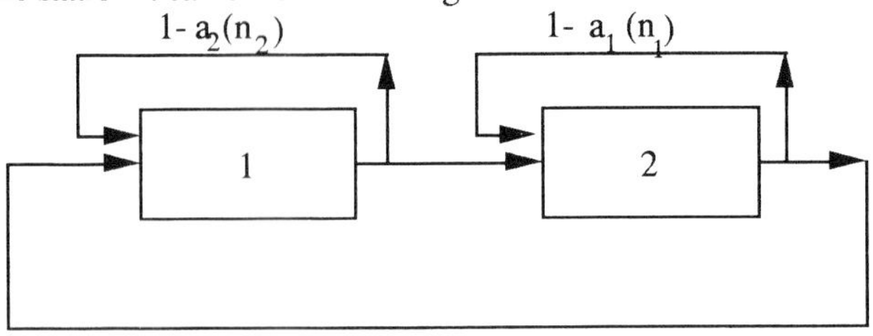

Figure 2: A Two-station cyclic network with recycle blocking

As blocked jobs effectively leave the state unchanged, blocked transitions can be left out from both the out and in-rate of global balance equations. As a consequence, the global balance equations and its solution (2.2) still apply. Similar to (2.4), the arrival distribution at station 2 can be calculated as follows:

$$\pi^2(k_1,k_2)=\frac{\pi_M(k_1+1,k_2)\mu_1}{\sum_{(m_1;m_2)}\pi_M(m_1+1,m_2)\mu_1}$$

$$=\frac{(1/\mu_1)^{k_1}(1/\mu_2)^{k_2}a_1(k_1)\prod_{k=1}^{k_1}a_1(k-1)\prod_{k=1}^{k_2}a_2(k-1)}{\sum_{(m_1;m_2)}(1/\mu_1)^{m_1}(1/\mu_2)^{m_2}a_1(m_1)\prod_{k=1}^{m_1}a_1(k-1)\prod_{k=1}^{m_2}a_2(k-1)} \tag{3.1}$$

First, let us assume that $a_1(k)=1$ if $k<N_1$ and 0 otherwise, and, $a_2(.)$ be any arbitrary function. Furthermore, note that $(k_1+1,k_2)\ \varepsilon\ C_M$ implies that $k_1<N_1$, so that for all values of k_1 and m_1 in (3.1), we have $a_1(k_1)=1$and $a_1(m_1)=2$. As a consequence, expression (3.1) in this case leads to the arrival theorem result:

Under recycle blocking $\pi^2(k_1,k_2)=\pi_{M-1}(k_1,k_2)$, $(k_1,k_2)\ \varepsilon\ C_{M-1}$ (3.2)

However, with randomized blocking at station 1, i.e. $0< a_1(k)<1$ for some $k<N_1$, the reduction presented at section 2 is not possible as these $a_1(.)$ values are differently involved in both the denominator and numerator. We note that this also applies when station 2 has no blocking, i.e. $a_2(.)=1$. Hence, we have:

Under recycle and randomized blocking at station 1, the arrival theorem fails at station 2

3.2. Recycled jobs are not included: random routing

Consider the network in figure 3 with M jobs and a single server at each station. The service requirements at each station is exponential with rate μ_i. Upon completion at station

3, a job routes to station i with probability p_i, i=1,2, and $p_1+p_2=1$. An arrival at station i is blocked with probability $1-a_i(n_i)$ when n_i jobs are already present at the station. When blocked, a job has to return to the station that it just came from to undergo a new service, thus without facing blocking. As before, let $N_i=\min\{k \mid a_i(k) > 0\}$.

Let $\pi_M(n_1,n_2,n_3)$ be the steady state distribution and $C_M=\{(n_1,n_2,n_3) \mid 0\le n_i \le N_i,\ i=1,2,3,$ and $n_1+n_2+n_3=M\}$ denote the set of admissible states.

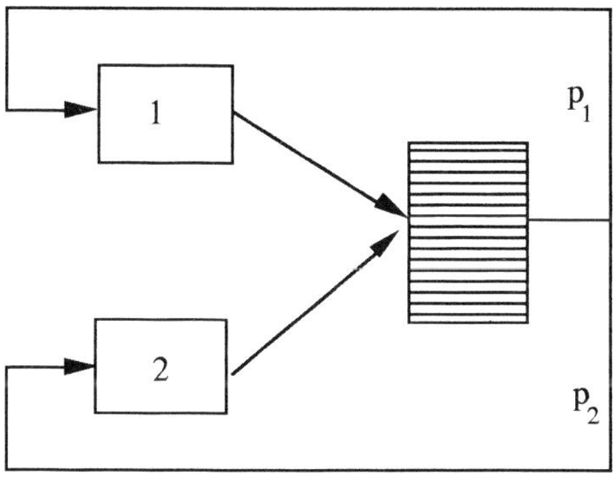

Figure 3: A three station network

The gobal balance equations are then given as follows:

$$\pi_M(n_1,n_2,n_3)\ [\mu_1 a_3(n_3)+\mu_2\ a_3(n_3)+\mu_3\, p_1\ a_1(n_1)+\mu_3\ p_2\ a_2(n_2)]= \tag{3.3}$$

$$\pi_M(n_1-1,n_2,n_3-1)\ \mu_1 a_3(n_3-1)\ 1_{n_1<N_1}\ 1_{n_3>0} + \pi_M(n_1,n_2+1,n_3-1)\ \mu_2 a_3(n_3-1)\ 1_{n_2<N_2}\ 1_{n_3>0} +$$

$$\pi_M(n_1-1,n_2,n_3+1)p_1\ \mu_3 a_1(n_1-1) 1_{n_3<N_3}\ 1_{n_1>0} + \pi_M(n_1,n_2-1,n_3+1)\ p_2\ \mu_3 a_2(n_2-1)\ 1_{n_3<N_3} 1_{n_2>0}$$

With c_M being the normalizing constant at C_M, these equations are directly verified by substituting:

$$\pi_M(n_1,n_2,n_3)=c_M\ (1/\mu_1)^{n_1}\ (1/\mu_2)^{n_2}\ (1/\mu_3)^{n_3} \prod_{k=1}^{n_1} a_1(k-1) \prod_{k=1}^{n_2} a_2(k-1) \prod_{k=1}^{n_3} a_3(k-1), \tag{3.4}$$

Let $\pi^3(k_1,k_2,k_3)$

$$\pi^2(k_1,k_2)= \frac{\pi_M(k_1+1,k_2,k_3)\ \mu_1+\pi_M(k_1,k_2+1,k_3)\ \mu_2}{\sum_{(m_1;m_2;m_3)} \pi_M(m_1+1,m_2,m_3)\ \mu_1+\pi_M(m_1,m_2+1,m_3)\ \mu_2} = \tag{3.5}$$

$$\frac{(1/\mu_1)^{k_1}\ (1/\mu_2)^{k_2}\ (1/\mu_3)^{k_3}\ \{p_1\ a_1(k_1)+p_2\ a_2(k_2)\} \prod_{k=1}^{k_1} a_1(k-1) \prod_{k=1}^{k_2} a_2(k-1) \prod_{k=1}^{k_3} a_3(k-1)}{\sum_{(m_1;m_2;m_3)} (1/\mu_1)^{m_1}\ (1/\mu_2)^{m_2}\ (1/\mu_3)^{m_3}\ \{p_1\ a_1(m_1)+p_2\ a_2(m_2)\} \prod_{k=1}^{m_1} a_1(k-1) \prod_{k=1}^{m_2} a_2(k-1) \prod_{k=1}^{m_3} a_3(k-1)}$$

Hence,

with $a_3(.)$ arbitrary and $a_1(.)=a_2(.)=1$, the arrival theorem applies at station 3 (3.6)

However, in contrast with section 2, stations 1 and 2 may lead to blocking and not necessarily randomized as in section 3.1 (note that as soon as either one of the functions $a_1(.)$ or $a_2(.)$ has a value less than 1, arrival theorem fails at station 3). For example, suppose that $a_1(.)=a_2(.)=1$ but that $a_2(k)=1$ if $k<N_2$ and 0 otherwise and consider a state (k_1,N_2,k_3). Then, $\{a_1(k_1)p_1+ a_2(k_2)p_2\}=p_1$ while $\{a_1(m_1)p_1+ a_2(m_2)p_2\}=1$ if $m_2<N_2\}$ and p_1 otherwise. Seeing that these terms donot cancel at the denominator and the numerator pf (3.7), we have:

Under the recycle protocol, and even without blocking at station 3 while blocking can arise at station 1 or 2, the arrival theorem fails at station 3 for jobs completing service at station 1 or 2.

Remark: Note the difference with (3.4) which allows strict blocking due to a finite capacity constraint at station 1 while here no blocking ,ay take place. This is due to the fact that also arrivals at station 3 need to be considered which comes from station 2 while station 1 is saturated. Such situations can not arise in the two station model.

3.3. Distribution with recycled arrivals

Reconsider the two station model of figure 2 but now also include recycled departures from station 2 as arriving jobs at station 2. Similar to the above cases, we have:

$$\pi^2(k_1,k_2)=\frac{\pi_M(k_1+1,k_2)\mu_1+\pi_M(k_1,k_2+1)\mu_2\,\{1-a_1(k_1)\}}{\sum\limits_{(m_1;m_2)}\pi_M(m_1+1,m_2)\mu_1+\pi_M(m_1,m_2+1)\mu_2\,\{1-a_1(m_1)\}} = \tag{3.7}$$

$$\frac{(1/\mu_1)^{k_1}(1/\mu_2)^{k_2}a_1(k_1)\prod\limits_{k=1}^{k_1}a_1(k-1)\prod\limits_{k=1}^{k_2}a_2(k-1)\{a_1(k_1)+a_2(k_2)[1-a_1(k_1)]\}}{\sum\limits_{(m_1;m_2)}(1/\mu_1)^{m_1}(1/\mu_2)^{m_2}a_1(m_1)\prod\limits_{k=1}^{m_1}a_1(k-1)\prod\limits_{k=1}^{m_2}a_2(k-1)\{a_1(m_1)+a_2(m_2)[1-a_1(m_1)]\}}$$

This expression reduces to $\pi_{M-1}(k_1,k_2)$ under either one of the following conditions: $a_1(.)=1$, $a_1(.)=1$, or in no state both stations can block. Otherwise this reduction is not possible. Hence,

Under the recycle protocol with recycled jobs at station 2 included as arrivals, the arrival theorem at station 2 applies only when one station allows blocking. Otherwise, when both station allow blocking at the same time, then it fails.

A similar and even more restrictive statement can be given for the randomized model from section 3.2. In summary, also under recycling blocking, the arrival theorem seemd to apply only under restrictive conditions and with inclusion of recycled job arrivals.

4. Jump-over Blocking

In this section, it will be shown that the arrival theorem does apply for two station models under the so called "jump-over" protocol provided that visits at a station during jumping over are taken into account as arrivals.

Under jump-over blocking, a job which is blocked to enter a station, jumps over or flows through at infinite speed and attempts to enter a proceeding station as if it was served normally. Repeated jump-overs can take place. For example, with $a_i(n_i)$ denoting the probability of acceptance at station i per visit, $[1-a_1(n_1)][1-a_2(n_2)][1-a_1(n_1)]a_2(n_2)$ is the probability that an arriving job at station 1 will jump over three times before it gets accepted at station 2.

We first note that the recycle and jump over blocking mechanisms are effectively the same if blocking can arise at only one station as in that case a blocked job will always end up at the other station. Hence, we have:

With blocking at only one station, the "jump-over" and recycle protocols are equivalent and the arrival theorem applies at both stations provided arrivals due to jumping over or recycling are included (4.1)

Now assume that with probability $1-a_i(n_i)$ a job attempting to enter station i when n_i jobs are present jumps over station i to the next station, where again it has to attempt to enter, independent of the unsuccessful attempts already experienced.

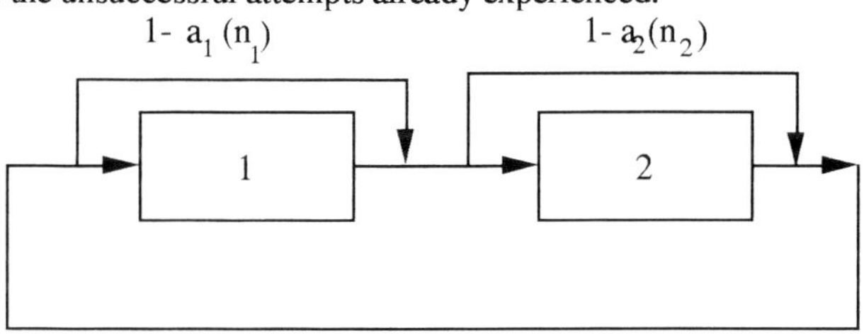

Figure 4: A Two station cyclic network

First, let us investigate whether a product form still applies. Only for the case of nonrandomized blocking due to finite capacity constraints at stations, this is intuitively argued by interpreting jump overs as jobs serviced at infinite speeds and formally proven in [14] for arbitrary Jackson networks by using absorbing Markov chain analysis. For randomized blocking, i.e. with $a_i(k) < 1$ for k-values smaller than N_i, no such result seems to be available. In that case, the global balance equations (2.1), after allowing transitions from a state to itself, become:

$\pi_M(n_1,n_2)\ [\mu_1+\mu_2]=$

$\pi_M(n_1-1,n_2+1)\ \mu_2\ 1_{n_2<N_2}\ 1_{n_1>0}\ a_1(n_1-1)\ \sum_{k=0}^{\infty}[1-a_1(n_1-1)]^k\ [1-a_2(n_2-1)]^k\ +$

$\pi_M(n_1,n_2)\ \mu_1\ a_1(n_1-1)[1-a_2(n_2-1)]\ \sum_{k=0}^{\infty}[1-a_2(n_2)]^k\ [1-a_1(n_1-1)]^k\ +$

$\pi_M(n_1+1,n_2-1)\ \mu_1\ 1_{n_1<N_1}\ 1_{n_2>0}\ a_2(n_2-1)\ \sum_{k=0}^{\infty}[1-a_1(n_1-1)]^k\ [1-a_2(n_2-1)]\ +$

$\pi_M(n_1,n_2)\ \mu_2\ a_2(n_2-1)[1-a_1(n_1-1)]\ \sum_{k=0}^{\infty}[1-a_2(n_2-1)]\ [1-a_1(n_1-1)]^k$

We aim to show that the product form given earlier still applies. To this end, by virtue of symmetry in stations 1 and 2, it sufficies to verify that the outrate due to station 1 (i.e. $\pi_M(n_1,n_2)\ \mu_1$ is balanced by the inrate due to station 1. Indeed, substituting (2.2) and noting that $a_2(N_2)=0$ and $a_1(n_1-1)>0$ for $n_1\leq N_1$, we have:

$\pi_M(n_1,n_2)\mu_1\{a_2(n_2)+a_1(n_1-1)[1-a_2(n_2-1)]\}\{1-[1-a_1(n_1-1)][1-a_2(n_2-1)]\}^{-1} = \pi_M(n_1,n_2)\ \mu_1.$

Recalling the symmetry in the description for station 1 and 2, the product form (2.2) is thus proven. Now, let us investigate the applicability of the arrival theorem for jobs arriving at station. Here, an arrival should include any visit made by a job at station 2, andthus, any visit at station 2 during the process of instantenous "jumping over" until a job gets accepted at one of two stations. The arrival distribution is then given by:

$\pi_M^2(k_1,k_2)=$

$$\frac{\text{mean \# of arrivals/time at station 2 when the system is in state } (k_1,k_2)}{\text{mean \# arrivals/time at station 2 when the system is in an arbitrary state } (m_1,m_2)}$$

$=$ {$\pi_M(k_1+1,k_2)\ \mu_1$ Expected # of visits at station 2 by a job leaving station 1 when the system is in state (k_1,k_2) $+\pi_M(k_1,k_2+1)\ \mu_2$ Expected # of visits at station 2 by a job leaving station 1 when the system is in state (k_1,k_2)}

[$\sum_{(m_1;m_2)}$ {$\pi_M(m_1+1,k_2)\ \mu_1$ Expected # of visits at station 2 by a job leaving station 1 when the system is in state (k_1,k_2) $+\pi_M(k_1,k_2+1)\ \mu_2$ Expected # of visits at station 2 by a job leaving station 1 when the system is in state (k_1,k_2)}]

$$= \pi_M(k_1+1,k_2)\,\mu_1\{a_2(k_2)+a_1(k_1)[1-a_2(k_2)]\}\sum_{k=0}^{\infty}[1-a_1(k_1)]^k\,[1-a_2(k_2)]^k(k+1)\;+$$

$$\pi_M(k_1,k_2+1)\,\mu_2(1-a_1(k_1))\{a_2(k_2)+a_1(k_1)[1-a_2(k_2)]\}\sum_{k=0}^{\infty}[1-a_1(k_1)]^k\,[1-a_2(k_2)]^k(k+1)$$

$$\overline{\sum_{(m_1;m_2)}\pi_M(m_1+1,m_2)\,\mu_1\{a_2(m_2)+a_1(m_1)[1-a_2(m_2)]\}\sum_{k=0}^{\infty}[1-a_1(m_1)]^k\,[1-a_2(m_2)]^k(k+1)\;+}$$

$$\pi_M(m_1,m_2+1)\,\mu_2(1-a_1(m_1))\{a_2(m_2)+a_1(m_1)[1-a_2(m_2)]\}\sum_{k=0}^{\infty}[1-a_1(m_1)]^k\,[1-a_2(m_2)]^k(k+1)$$

Now, noting the following equality for arbitrary a_1 and a_2:

$$\{a_2+[1-a_2]\,a_1\}\sum_{k=0}^{\infty}[1-a_1]^k\,[1-a_2]^k(k+1)=[a_1+a_2-a_1a_2]^{-1}$$

and substituting the product form (2.2) again, this equation reduces to:

$$\frac{(1/\mu_1)^{k_1}(1/\mu_2)^{k_2}\prod_{k=1}^{k_1}a_1(k-1)\prod_{k=1}^{k_2}a_2(k-1)\{a_1(k_1)+a_2(k_2)[1-a_1(k_1)]\}[a_1(k_1)+a_2(k_2)-a_1(k_1)a_2(k_2)]^{-1}}{\sum_{(m_1;m_2)}(1/\mu_1)^{m_1}(1/\mu_2)^{m_2}\prod_{k=1}^{m_1}a_1(k-1)\prod_{k=1}^{m_2}a_2(k-1)\{a_1(m_1)+a_2(m_2)[1-a_1(m_1)]\}[a_1(m_1)+a_2(m_2)-a_1(m_1)a_2(m_2)]^{-1}}$$

which is equal to $\pi_{M-1}(k_1,k_2)$. Hence, we have:

Under jump-over blocking and for arbitrary, possibly randomized, blocking at both stations 1 and 2, the arrival theorem applies at both stations 1 and 2 provided all visits of jobs due to jumping over are incvluded.

Remark: The validity of both a product form result and the arrival theorem under the jump over protocol can also be expected for arbitrary Jacksonian network structures with randomized blocking. However, the technicalities will become substantially more complicated (c.f. [14] for an algebraic proof of the product form result in the case of nonrandomized blocking). This section merely aimed to illustrate, in contrast with the delay and recycle cases, that only under the special jump over protocol and under appropriate arrival interpretation, the arrival theorem appears to be justified.

Acknowledgement

The author is grateful to Dick Muntz for stimulating discussions on this topic.

References

[1] Akyildiz, I.F. (1988), On the exact and approximate throughput analysis of closed queueing networks with blocking, IEEE Transactions Software Eng., SE-14, 62-71

[2] Akyildiz, I.F. and von Brand, H. (1989), Exact Solutions for Open, Closed, and Mixed Queueing Networks with Rejection Blocking, Theoret. Comp. Sci., 64, 203-219

[3] S.C. Bruell and G. Balbo, Computational Algorithms for Closed Queueing Networks, North Holland, Amsterdam, 1980

[4] S.C. Bruell, G. Balbo and P.V. Afshari, (1984) Mean Value Analysis of Mixed Multiple Class BCMP networks with Load Dependent Service Stations, Performance Evaluation, 4-4, 241-260

[5] A.E. Conway, E. de Souza e Silva, and S.S. Lavenberg, (1989) Mean Value Analysis by Chain of Product Form Queueing Networks, IEEE Trans. Computers, 38, 432-442

[6] P.J.B. King, Computer and Communication System Modeling, Prentice Hall, 1990

[7] S.S. Lam, (1983) A Simple Derivation of the MVA and LBLANC algorithms from the Convolution Algorithm, IEEE Trans. Computers, 32, 1062-1064

[8] S.S. Lavenberg and M. Reiser, (1980), Stationary State Probabilities at Arrival Instants for Closed Queueing Networks with Multiple Types of Customers, J. Applied Probability, 17-4, 1048-1061

[9] P.J. Schweitzer (1979), Approximate Analysis of Multiclass Closed Networks of Queues, in Proc. Int. Conf. on Stochastic Control and Optimization, Amsterdam, 1979

[10] P.J. Schweitzer (1990), A Survey of Mean Value Analysis, its generalizations, and applications for Networks of Queues, Research Rep., W.E. Simon Graduate School of Business Administration, University of Rochester

[11] R. Serfozo, (1989), Markovian Network Processes: Congestion Dependent Routing and Processing, Queueing Systems, 5-36

[12] K.C. Sevcik and I. Mitrani (1981), The Distribution of Queueing Network States at Input and Output Instants, J. ACM, 28-2, 358-371

[13] K.S. Trivedi, Probability and Statistics with Reliability,: Queueing and Computer Science Applications, Prentice Hall, 1983

[14] N.M. van Dijk (1988), On Jackson's Product Form with Jump Over Blocking, Oper. Res. Letters, 7, 233-235

[15] J. Walrand, An Introduction to Queueing Networks, Prentice Hall, 1988

[16] J. Zahorjan (1984), The Distribution of Network States During Residence Times in Product Form Queueing Networks, Performance Evaluation, 4, 99-104

Queueing Networks with Finite Capacity
R.O. Onvural and I.F. Akyildiz (Editors)

Variance of Output of a Tandem Production System

Stanley B. Gershwin
Laboratory for Manufacturing and Productivity
Massachusetts Institute of Technology
Cambridge, Massachusetts 02139

Abstract

This paper develops a method for calculating the variance of the number of parts produced by a transfer line during a fixed time interval, *e.g.*, one week. It provides an exact formula for the production variance of a single machine. It uses decomposition for an approximate extension to two-machine and longer lines. Comparison with simulation indicates that the method is effective. Some relationships between production variance and line parameters are explored.

1 Introduction

There is a long literature on the prediction of performance of transfer or flow lines (Dallery and Gershwin, 1991) but the *variance* of the production during a time period has been almost entirely neglected. Nearly all the papers in this literature deal with steady-state average production rates and average buffer levels. Prediction of production variability is no less important than that of the prediction of the mean. The buyer of a transfer line cares about more than its mean production rate; the firm must be able to deliver a certain amount of finished product each week with a certain probability.

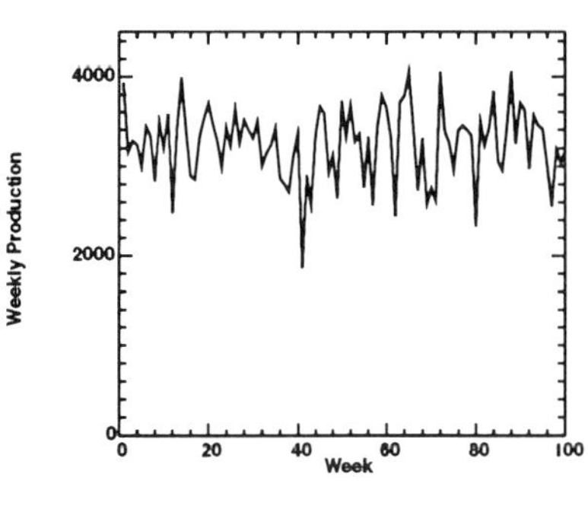

Figure 1: Production Variability

To illustrate the importance of transfer line production variability, we examine a representative system (not a model of an actual line). Consider a deterministic processing time (Buzacott model) transfer line consisting of 20 machines and 19 buffers. Assume that all the machines and all the buffers are identical. The system does operations with a one-minute cycle. It is run for two shifts (16 hours) a day, five days a week, or 4800 minutes per week. Each machine has a mean time to fail (MTTF) of 3880 minutes, or just under 65 hours. It has a mean time to repair (MTTR) of 120 minutes.

(In terms of quantities that are defined in Section 2, each machine has $r = .00833$ and $p = .000258$.) The buffers are all assumed to have a capacity of 10 parts. Using the decomposition technique of Gershwin (1987) and Dallery, David, and Xie (1988), we find that the average production rate of the line is .6764 parts per minute, or 3247 parts per week.

A simulation model of this line was run for 100 weeks. Its average production rate during the whole interval was .6769 parts per minute, about 2 parts per week greater than the decomposition technique. However, there was tremendous variability from week to week. Figure 1 is a graph of weekly production.

It is apparent that there are wide variations in weekly performance. In fact, the production of the worst week during this 100-week period was 1863 parts, or 57% of average, and the best week's production was 4061 parts, or 125% of average. The standard deviation was 403 parts, or 12.4% of the weekly average. The standard deviation predicted by the method of Section 3 is 408.24.

Conclusions We can draw the following conclusions from this example.

1. First, randomness can be extremely important on a day-to-day and week-to-week basis in a manufacturing system, and not just on a minute-to-minute basis. That is, there can be large, nerve-wracking variations in performance which can affect customer satisfaction. Life with this system consists of periods when the line easily exceeds requirements, alternating with periods of high pressure, anxiety, and complaints. Furthermore, although the predicted performance will be observed in the long run, we have to live in the short run. One hundred weeks may be too long to wait to get the required production rate; very few manufacturing processes and products last that long without a major change.

2. Second, mathematical and simulation models can be accurate. Some engineers, who live with the enormous variability that these systems exhibit, reasonably conclude that simple mathematical or simulation models cannot capture the diversity of the disruptions that occur on these systems, and therefore that such models cannot predict performance. However, simulation experiments with standard models demonstrate the same kind of variability that real systems exhibit.

Literature Lavenberg (1975) calculates the mean and variance of the steady-state queueing time in a $M/G/1$ queue with a finite buffer using Laplace transform methods. Miltenburg (1987) studies the variance of production in a variety of models of small transfer lines. His objective is essentially the same as ours. In fact, we obtain very similar results for one of his problems (Section 5). His techniques are more general and possibly more accurate for two-machine lines, because he analyzes the underlying Markov chains exactly. For example, in addition to the model that is described here, he also analyzes a two-machine line with limited repair personnel, and a two-machine line with exponentially distributed repair, failure, and processing times. These papers only treat two-machine lines, and obtain results that are difficult to use and to understand intuitively. It is not clear how to extend the results of these papers either exactly or approximately, although Miltenburg's two-machine method might replace the line augmentation technique described in Section 3.

Gnedenko, Belyayev, and Solovyev (1969) study the distribution of *total operating time*, the total time that a single unreliable machine works during an interval, for a production process with somewhat different assumptions. They obtain a formula for the asymptotic variance of total operating time. Glukhov (1973) extends this for a more general system of machines with more general failure and repair processes.

Duenyas and Hopp (1990) calculate an estimate of the variance of the output of a cyclic (single closed loop) queuing system consisting of a set of single-server exponential queues. Duenyas, Hopp, and Spearman (1992) extend this to deterministic processing time queues with exponentially distributed repairs and failures.

Outline The variability of a single machine is analyzed in Section 2 and in the Appendix. A procedure for longer lines is proposed in Section 3. The results of simulation experiments are described in Section 4. Numerical experiments are reported on in Section 5. Conclusions and directions for new research appear in Section 6.

2 Variance of Production of a Single Machine

In this section and the Appendix, we develop the formula for the variance of production of a single machine. We use this formula in Section 3 as part of a procedure to calculate the production variance of a long line.

Model We assume a Buzacott-type model of a machine. That is, we assume a synchronous, or discrete-time, system. During a time unit when the machine is operational, it may either perform an operation (with probability $1-p$) or it may fail without completing an operation (with probability p). During a time unit when the machine is down, it may either stay down (with probability $1-r$), or its repair may be completed (with probability r). Two performance measures are of interest for such a machine: the mean and the variance of the number of parts produced during a specified time interval.

Define $\pi(n,\alpha,t)$ = the probability that the machine is in state α at time t and n parts have been produced in $[0,t]$ Then $\pi(n,\alpha,t)$ satisfies

$$\pi(n,\alpha,t) = 0, t \geq 0, n > t \tag{1}$$

$$\pi(t,0,t) = 0, t \geq 1 \tag{2}$$

$$\pi(0,1,t) = 0, t \geq 1 \tag{3}$$

$$\pi(n,0,t+1) = (1-r)\pi(n,0,t) + p\pi(n,1,t), \quad t \geq 1, 0 \leq n \leq t \tag{4}$$

$$\pi(n,1,t+1) = r\pi(n-1,0,t) + (1-p)\pi(n-1,1,t), \quad t \geq 1, 0 \leq n \leq t \tag{5}$$

In addition, we assume that the system starts in steady state. Therefore,

$$\pi(0,1,0) = \frac{r}{r+p} \tag{6}$$

$$\pi(0,0,0) = \frac{p}{r+p} \tag{7}$$

The mean and variance of the amount of material produced in t time steps are:

$$\bar{n}(t) = \sum_{n=0}^{t} n\left(\pi(n,0,t) + \pi(n,1,t)\right)$$

$$V(t) = E(n(t)^2) - (\bar{n}(t))^2$$

$$E(n(t)^2) = \sum_{n=0}^{t} n^2\left(\pi(n,0,t) + \pi(n,1,t)\right)$$

Formulas for the mean and variance of production are derived in the Appendix using the difference equations for $\pi(n,\alpha,t)$. They are

$$\bar{n}(t) = \frac{r}{r+p}t \tag{8}$$

$$V(t) = \frac{rp}{(r+p)^2}\left(\frac{2}{r+p} - 1\right)t + \frac{2rp}{(r+p)^4}\left((1-r-p)^t - 1\right)(1-r-p) \tag{9}$$

Note that for large t,

$$\frac{V(t)}{t} \approx \frac{rp}{(r+p)^2}\left(\frac{2}{r+p} - 1\right). \tag{10}$$

3 Variance of Production of a Long Line

We propose a procedure in this section for calculating the variance of a long transfer line.

Model The system is a transfer line consisting of k machines as described in Section 2 with parameters r_i and p_i $(i = 1, ..., k)$ and $k-1$ buffers with sizes N_i $(i = 1, ..., k-1)$. The production rate and average buffer levels of this system have been analyzed by decomposition (Gershwin, 1987; Dallery, David, and Xie, 1988).

Review of Decomposition Method In the decomposition method, a k-machine transfer line is represented by $k-1$ two-machine transfer lines. The size of the buffer of Line i is N_i. The parameters of its upstream machine are $r_u(i)$ and $p_u(i)$, and the parameters of its downstream machine are $r_d(i)$ and $p_d(i)$, where $r_u(i), p_u(i), r_d(i)$, and $p_d(i)$ are chosen to satisfy a set of equations derived in Gershwin (1987). The equations are such that upstream parameters $r_u(i)$ and $p_u(i)$ approximately describe the random process by which parts enter Buffer i. That is, the lengths of the intervals during which parts enter the buffer without interruption (except possibly by the buffer filling up) are approximately distributed geometrically, with mean $r_u(i)^{-1}$, and the lengths of the intervals during which no parts enter the buffer (other than those intervals in which the buffer is full) are approximately distributed geometrically, with mean $p_u(i)^{-1}$. The downstream parameters have similar meanings. An efficient algorithm for solving those equations is developed in Dallery, David, and Xie (1988).

Augmented System In one of the examples of Gershwin (1987), the last machine of a transfer line is extremely reliable; that is, p_k is close to 0 and r_k is relatively large, so that its isolated efficiency $r_k/(r_k + p_k)$ is close to 1, and much closer to 1 than that of the other machines. The production rate of the line is approximately the same as that of the line without the last machine and the last buffer[1].

In this case we can interpret $r_u(k-1)$ and $p_u(k-1)$ as representing the output behavior of the line consisting of Machines 1, ..., $k-1$ and Buffers 1, ..., $k-2$. In other words, the machine with parameters $r_u(k-1)$ and $p_u(k-1)$ has an output process which is approximately the same as that of the line. Therefore, the variance of the output of the line is approximately that of a machine with parameters $r_u(k-1)$ and $p_u(k-1)$.

Procedure This suggests the following procedure to determine the variance of the output of a transfer line consisting of Machines 1, ..., k with parameters r_i and p_i ($i = 1, ..., k$) and Buffers 1, ..., $k-1$ with sizes N_i ($i = 1, ..., k-1$):

1. Form the augmented line with one more buffer and one more machine. The sizes of Buffers 1, ..., k are N_i ($i = 1, ..., k-1$) and N_k. The value of N_k does not matter, as long as it is not too small. The machine parameters are r_i and p_i ($i = 1, ..., k$) and $r_{k+1} \approx 1$ and $p_{k+1} \approx 0$.

2. Analyze the augmented line by the Dallery, David, and Xie (1988) algorithm.

3. The variance is given by equation (9) with $r = r_u(k)$ and $p = p_u(k)$.

4 Simulation Results

A set of simulations are described here that were performed to test the results described in Sections 2 and 3. In each case, an initial transient period is ignored in tabulating results, and M samples T_i of throughput from intervals of length t are observed. The sample variance is calculated according to

$$\bar{V} = \frac{\sum_{i=1}^{M} \left(T_i - \bar{T}\right)^2}{M-1} \tag{11}$$

[1]and the average level of the last buffer is equal to the production rate, the average number of parts produced in a time unit

where

$$\bar{T} = \frac{\sum_{i=1}^{M} T_i}{M}.$$

The analytical estimate of variance is calculated from (9). Note that $\bar{E}$, the average production rate, is given by

$$\bar{E} = \frac{\bar{T}}{t}$$

which should be close to $e = r/(r+p)$. The variance error is calculated according to

$$\text{Error} = \frac{\text{Sample Variance} - \text{Analytical Variance}}{\text{Analytical Variance}}$$

First, single-machine simulation results are described that tend to confirm equation (9). Examples 1 is the single-machine approximation of the longer lines of Example 2. That is, in Example 1, r and p are $r_u(k-1)$ and $p_u(k-1)$ of Example 2. They indicate the magnitude of the discrepancy between analytic and simulated values of the variance. Such the other source of error is the approximation. Other longer lines are treated in Examples 3-5.

Example 1 In this example, $r = 0.1$ and $p = 0.151429$. Table 1 shows the results of a set of simulations in which a set of intervals of different lengths t are used. The initial transient period is 100 time steps. The sample variance is clearly close to the analytic. The error seems to be decreasing as t increases, but it is always negative. Here, $e = .398$.

t	Analytical	Example 1			Example 2		
		E	Sample Variance	Error	E	Sample Variance	Error
50	77.622	0.402	69.822	-.100	0.381	63.240	- 0.1853
100	160.916	0.400	137.845	-.143	0.388	126.132	- 0.2162
150	244.211	0.395	220.003	-.0991	0.397	191.932	- 0.2141
200	327.506	0.391	315.506	-.0366	0.389	277.561	- 0.1525
300	494.095	0.394	465.807	-.0573	0.389	374.850	- 0.2413
400	660.685	0.400	625.481	-.0533	0.394	528.978	- 0.1993
500	827.274	0.398	795.653	-.0382	0.393	705.937	-0.1467

Table 1: Example 1 and 2 Results

Confidence intervals: Additional runs were performed to determine confidence intervals[2]. In each run, 10,000 periods of length t were used to calculate one sample of variance, and ten samples of variance were used in the calculation of the confidence interval. For $t = 100$, the 95% confidence interval was [149.9, 150.9]; this should be compared with the estimate of variance, from equation (9), which is 160.916 (from Table 1). For $t = 1000$, the 95% confidence interval was [1526.6,1556.8]; the estimate from (9) is 1660.217. In both cases, the error is about 7% from the center of a fairly narrow interval.

[2] Details on these and other confidence interval estimates appear in Maselli (1992). The Student-t method described in Lavenberg (1983) was used.

Example 2 In this two-machine example, $r_1 = 0.1$, $p_1 = 0.1$, $r_2 = 0.1$, $p_2 = 0.1$, and $N = 10$. When the procedure of Section 3 is applied to this system, $r = r_u(2) = 0.1$ and $p = p_u(2) = 0.151429$. These are exactly the parameters of Example 1. The analytical values of the variance are identical, and both sets of sample variances are close to the analytical values. Table 1 shows the results of a set of simulations in which a set of intervals of different lengths t are used. The initial transient period is 100 time steps.

Example 3 In this five-machine example, $r_i = 0.09$, and $p_i = 0.01$ for $i = 1, ..., 5$, and $N_i = 50$ for $i = 1, ..., 4$. When the procedure of Section 3 is applied to this system, $r = r_u(5) = 0.09$ and $p = p_u(5) = 0.0157532$. In each run, 10,000 periods of length t were used to calculate one sample of variance, and ten samples of variance were used in the calculation of the confidence interval. For $t = 100$, the 95% confidence interval was [142.5, 147.5]; this should be compared with the estimate of variance, from equation (9), which is 206.801. For $t - 1000$, the 95% confidence interval was [1531.1,1556.5]; the estimate from (9) is 2250.465. For $t = 2500$, the 95% confidence interval was [3527.7,3596.8]; the estimate from (9) is 5656.575. All the intervals are narrow, and the errors from the centers of the intervals are 30%, 31%, and 37%, respectively.

Example 4 In this ten-machine example, $r_i = 0.09$, and $p_i = 0.01$ for $i = 1, ..., 10$, and $N_i = 50$ for $i = 1, ..., 9$. When the procedure of Section 3 is applied to this system, $r = r_u(10) = 0.09$ and $p = p_u(10) = 0.0168958$. In each run, 10,000 periods of length t were used to calculate one sample of variance, and ten samples of variance were used in the calculation of the confidence interval. For $t = 100$, the 95% confidence interval was [144.0, 148.1]; this should be compared with the estimate of variance, from equation (9), which is 214.873. For $t = 1000$, the 95% confidence interval was [1520.1,1568.3]; the estimate from (9) is 2335.947. For $t = 2500$, the 95% confidence interval was [3538.9,3608.5]; the estimate from (9) is 5871.077. All the intervals are narrow, and the errors from the centers of the intervals are 32%, 34%, and 39%, respectively.

Example 5 In this five-machine example, the machines and buffers are all different[3]. When the procedure of Section 3 is applied to this system, $r = r_u(5) = 0.0902075$ and $p = p_u(5) = 0.0145211$. In each run, 10,000 periods of length t were used to calculate one sample of variance, and ten samples of variance were used in the calculation of the confidence interval. For $t - 100$, the 95% confidence interval was [120.4, 125.7]; this should be compared with the estimate of variance, from equation (9), which is 196.635. For $t = 1000$, the 95% confidence interval was [1367.7,1393.3]; the estimate from (9) is 2141.814. For $t = 2500$, the 95% confidence interval was [3286.9,3363.1]; the estimate from (9) is 5383.786. All the intervals are narrow, and the errors from the centers of the intervals are 37%, 36%, and 38%, respectively.

Comments on accuracy The largest error in *variance* that we observed was about 40%. This is equivalent to an error in *standard deviation* of about 17%. This error

[3] $(r_1, p_1, N_1) = (0.090, 0.010, 50)$; $(r_2, p_2, N_2) = (0.080, 0.006, 40)$; $(r_3, p_3, N_3) = (0.100, 0.012, 60)$; $(r_4, p_4, N_4) = (0.095, 0.010, 50)$; $(r_5, p_5) = (0.085, 0.007)$.

is comparable to that of Duenyas and Hopp (1990), and somewhat larger than that of Duenyas, Hopp, and Spearman (1992).

5 Numerical Results

Figures 2 – 4 are graphs of $\pi(n,1,t)$ and $\pi(n,0,t)$ for $r = 0.09$ and $p = 0.01$, and $t = 100, 500$, and 1000. In all cases, the mean $\bar{n}(t)$ of the distribution is $0.9t$. As t increases, the distributions increasingly resemble normal distributions.

This suggests that production lines can be designed in ways similar to that of the items produced. For example, if it is required that a given production volume n^* is met in at least 99.9% of all weeks, then the parameters of the machines and the buffers of the line must be such that

$$n^* = \bar{n}(t) + 3V(t)^{\frac{1}{2}}$$

where t is the number of machine cycles in a week. The coefficient of the standard deviation is determined from the normal distribution.

Note that in all cases, when the distributions are not close to normal, the value of $\pi(t,1,t)$ is much higher than $\pi(n,1,t), n < t$. As the distribution approaches normal, $\pi(t,1,t)$ decreases and comes close to $\pi(t-1,1,t)$.

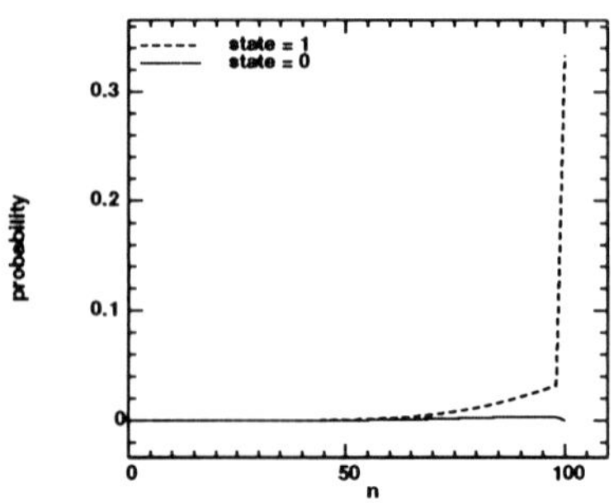

Figure 2: $\pi(n,1,t)$ and $\pi(n,0,t)$ for $r = 0.09$, $p = 0.01$, and $t = 100$

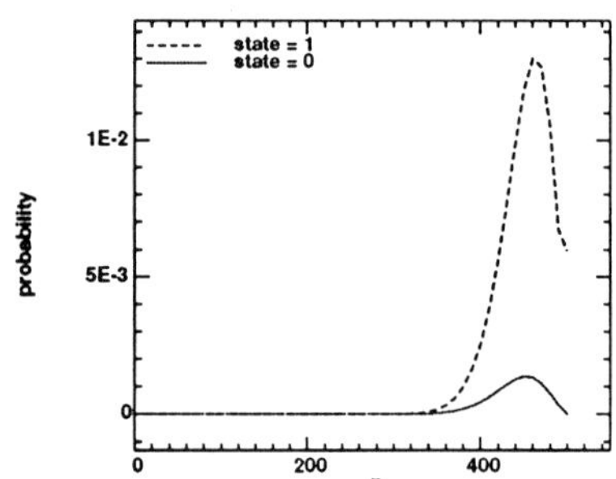

Figure 3: $\pi(n,1,t)$ and $\pi(n,0,t)$ for $r = 0.09$, $p = 0.01$, and $t = 500$

Miltenburg (1987) calculates the variance of production of a Buzacott model of two-machine transfer line. The parameters are $r_1 = r_2 = 0.3$, $p_1 = p_2 = 0.2$, and $N = 3$. For $t = 800$, he finds that $E = 0.50$ and $V = 401.6$. We obtain $E = 0.47$ and $V = 426.7$. The difference in the results may be due to slight differences in the models; for example, Miltenburg follows Buzacott's early practice of neglecting products of probabilities in the Markov chain formulation. There may also be differences in the blocking model.

Figure 5 is a graph of $V(t)$, and Figure 6 is a graph of the coefficient of variation

$$\frac{V(t)^{\frac{1}{2}}}{\bar{n}(t)}$$

as a function of t for $r = 0.0009, p = 0.0001$. The variance approaches the asymptote (10), and the limiting coefficient of variation is proportional to $t^{-\frac{1}{2}}$.

Figure 7 is a graph of $V(t)$ as a function of r for $p = 0.0001, t = 100$. The coefficient of variation has a similar shape. The most striking feature of this graph is the maximum at an intermediate value of r. The graph of $V(t)$ as a function of p is similar.

Figure 8 is a graph of $V(t)$ as a function of K, where $r = r_0 K$ and $p = p_0 K$ ($r_0 = 0.0009, p_0 = 0.0001$) and $t = 100$. The coefficient of variation looks similar.

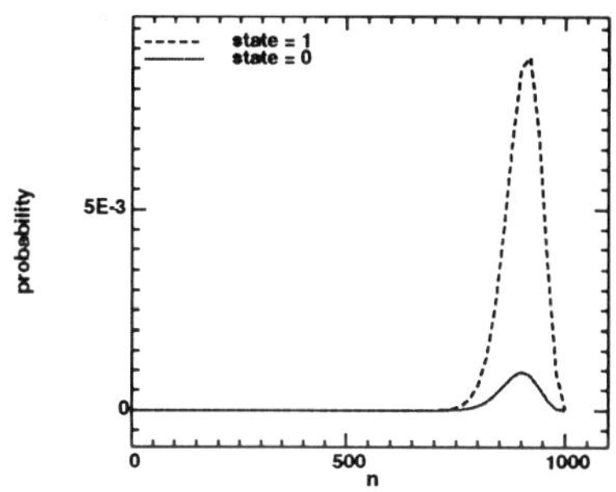

Figure 4: $\pi(n,1,t)$ and $\pi(n,0,t)$ for $r = 0.09$, $p = 0.01$, and $t = 1000$

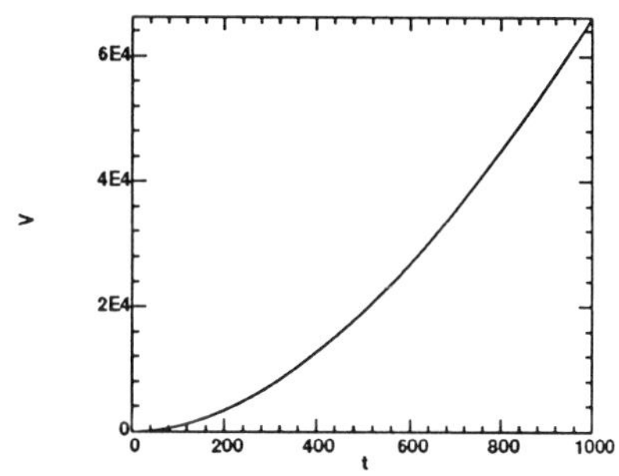

Figure 5: $V(t)$ for $r = 0.0009, p = 0.0001$

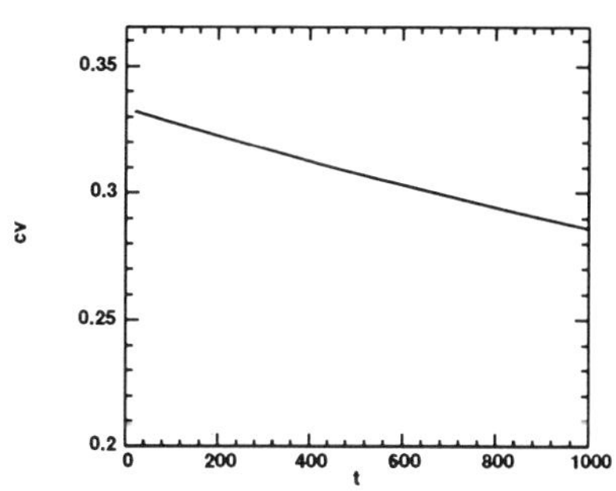

Figure 6: Coefficient of Variation *vs.* t for $r = 0.0009, p = 0.0001$

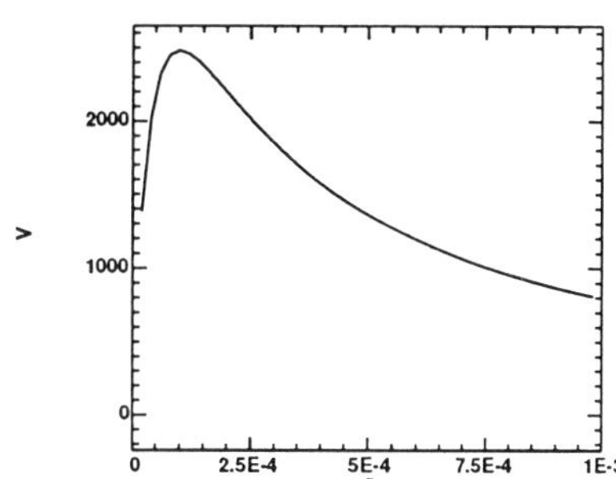

Figure 7: $V(t)$ *vs.* r for $p = 0.0001, t = 100$

Figure 9 is a graph of the variance as a function of N_3 in a four-machine, three buffer transfer line, in which $r_1 = r_2 = r_3 = 0.08$, $p_1 = p_2 = p_3 = 0.02$, $r_4 = p_4 = 0.9$, $N_1 = N_2 = 3$, and $t = 500$. This example can be thought of as representing a three-machine transfer line with a nearly constant demand process represented by the last machine. The third buffer then represents a finished goods inventory. This example shows that a large inventory can decouple the effects of machine randomness from the service provided to the customer.

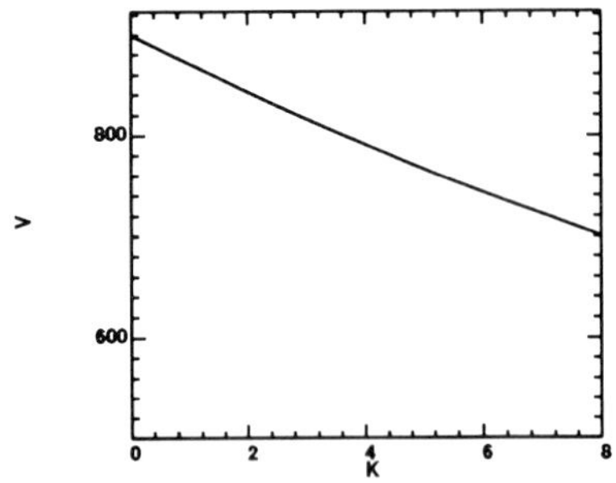

Figure 8: $V(t)$ *vs.* K, where $r = r_0 K$ and $p = p_0 K$ ($r_0 = 0.0009, p_0 = 0.0001$) and $t = 100$

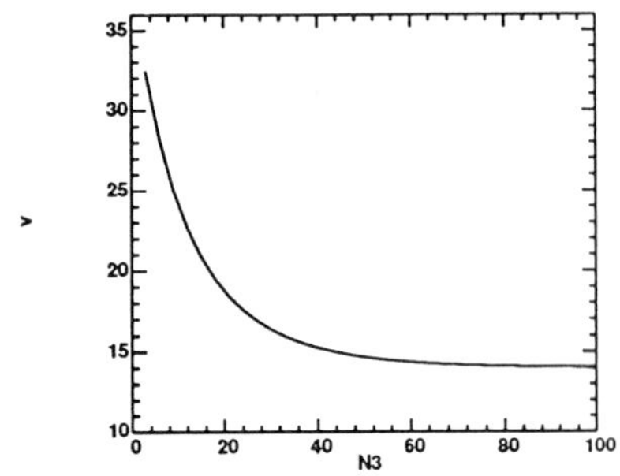

Figure 9: $V(t)$ *vs.* N_3

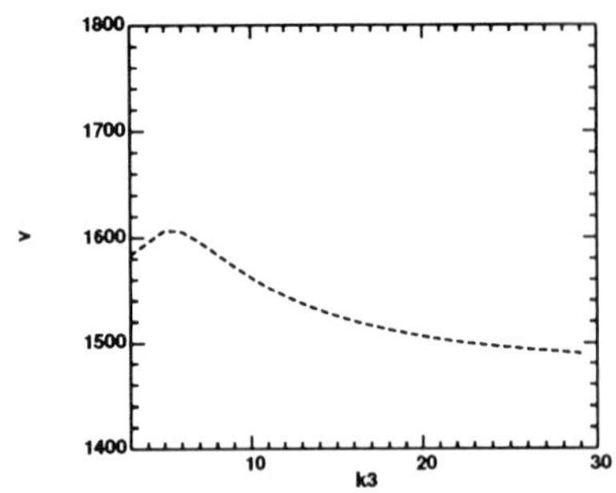

Figure 10: $V(t)$ *vs.* k

Figure 10 is a graph of variance as a function of line length k, where each machine in the line has $r = 0.085$ and $p = 0.015$, and each buffer has capacity 10. The time t is 500. Note the surprising local maximum, and that the variance is almost independent of k.

6 Conclusion

This paper has emphasized the importance of production variance as a measure of performance of a manufacturing system. It develops a formula for the production variance of a single discrete-time machine, and it proposes a procedure for calculating the production variance of a synchronous transfer line. This procedure is a small extension to an existing decomposition method which is fast and reliable.

Comparison with earlier work and with simulation suggests that the method provides a good first approximation for the variance, but more effort is warranted to reduce the error. The error seems to increase with increasing line length (k), duration (t), and line imbalance. The increase due to line imbalance may reflect error due to the decomposition method.

There is much research to do. In addition to reducing the error, the alternative approaches described in Section 3 should be investigated. The variance of a wide variety of different models of lines should be investigated.

The variance of the number of parts in buffers, or in the whole line, is also important. The time spent by a part in a line is highly influenced by the number of parts currently in the line. Since manufacturers must respond quickly and predictably to customers, a variable response time is just as undesirable as a variable production rate. In addition, the variance of the length of time required to produce a given number of parts is of interest (Dallery, 1991). See Gnedenko, Belyayev, and Solovyev (1969).

Finally, design procedures for transfer lines, which take into account the variance issues and procedure discussed here, should be developed.

Acknowledgments I am grateful to the MIT Leaders for Manufacturing Program, which brought this important issue to my attention. Important preliminary work was performed by Cathy Strosser at the General Motors Quad Four Engine Plant at Lansing, Michigan. The simulation program used here was written by Marc Maselli. He did many runs and provided the confidence interval work. I appreciate the helpful comments of Yves Dallery, N. Srivatsan, and Mitchell Burman. John Buzacott suggested some prior literature. I am indebted to Raif Onvural and Ian Akyildiz for making it possible for me to participate in the Second International Workshop on Queuing Networks with Finite Capacity. Raif has been generous with his patience and advice. I am grateful to participants of the workshop and to two anonymous reviewers for their helpful comments. This work was partly supported by the National Science Foundation under Grant DDM-8914277.

References

Dallery, Y. (1991), private communication.

Dallery, Y., David., R., and Xie, X.-L. (1988), "An Efficient Algorithm for Analysis of Transfer Lines with Unreliable Machines and Finite Buffers," *IIE Transactions*, Vol. 20, pp. 280-283.

Dallery, Y., and Gershwin, S. B. (1991), "Manufacturing Flow Line Systems: A Review of Models and Analytical Results," MIT Laboratory for Manufacturing and Productivity Report LMP 91-002, February, 1991.

Duenyas, I. and Hopp, W. J. (1990), "Estimating Variances of Output from Cyclical Exponential Queuing Systems," *Queuing Systems* Vol. 7, pp. 337-354.

Duenyas, I., Hopp, W. J., and Spearman, M. L. (1992) "Characterizing the Output Process of a CONWIP Line with Deterministic Processing and Random Outages," Technical Report 91-11, Department of Industrial Engineering and Management Sciences, Northwestern University, January, 1992.

Gershwin, S. B. (1987), "An Efficient Decomposition Algorithm for the Approximate Evaluation of Tandem Queues with Finite Storage Space and Blocking," *Operations Research*, Vol. 35, pp. 291-305.

Gnedenko, B. V., Belyayev, Y. K., and Solovyev, A. D. (1969), *Mathematical Methods of Reliability Theory*, Academic Press.

Glukhov, V. N. (1973), "Execution Time and Other Reliability Characteristics," *Automation and Remote Control*, March 15, 1973, pp. 1726-1733, Consultants Bureau, Plenum Publishing Company; English translation from *Avtomakika i Telemekhanika*, Vol. 33, No. 10, pp. 184-192, October, 1972.

Lavenberg, S. S. (1975), "The Steady-State Queueing Time Distribution for the $M/G/1$ Finite Capacity Queue," *Management Science*, Vol. 21, No. 5, pp. 501-506, January, 1975.

Lavenberg, S. S. *Computer Performance Modeling Handbook*, Academic Press, 1983.

Maselli, M. (1992) "Transfer Line Simulation as Applied to Evaluating Production Variance," MIT Department of Mechanical Engineering B. S. Thesis.

Miltenburg, G. J. (1987), "Variance of the Number of Units Produced on a Transfer Line with Buffer Inventories During a Period of Length T," *Naval Research Logistics Quarterly*, Vol. 34, pp. 811-822.

A Derivation of Variance Equation

We derive an equation for the variance of production by a single Buzacott-like machine during a time interval $[0, t]$. We first derive a set of intermediate results.

Lemma 1

$$U(t) \equiv \sum_{n=0}^{t} \pi(n,1,t) = \frac{r}{r+p} \tag{12}$$

$$D(t) \equiv \sum_{n=0}^{t} \pi(n,0,t) = \frac{p}{r+p} \tag{13}$$

Proof: It is easy to show that

$$D(t+1) = (1-r)D(t) + pU(t) \tag{14}$$

$$U(t+1) = rD(t) + (1-p)U(t) \tag{15}$$

Then,

$$rD(t+1) - pU(t+1) = (1-r-p)\left(rD(t) - pU(t)\right).$$

Since (from (6) and(7)), $rD(0) - pU(0) = 0$, we have $rD(t) - pU(t) = 0$. Also, from (14) and (15),

$$U(t+1) + D(t+1) = U(t) + D(t).$$

Then

$$U(t) + D(t) = 1$$

since $U(0) + D(0) = 1$ by assumption. The lemma follows.

Lemma 2 *Let*

$$E_\alpha(t) = \sum_{n=0}^{t} n\pi(n, \alpha, t) \tag{16}$$

Then,

$$E_0(t) = \frac{rp}{(r+p)^2}t + \frac{rp}{(r+p)^3}\left((1-r-p)^t - 1)\right)$$
$$E_1(t) = \frac{r^2}{(r+p)^2}t - \frac{rp}{(r+p)^3}\left((1-r-p)^t - 1)\right)$$

Proof: It is easy to show that

$$E_0(t+1) = (1-r)E_0(t) + pE_1(t) \tag{17}$$
$$E_1(t+1) = rE_0(t) + (1-p)E_1(t) + \frac{r}{r+p} \tag{18}$$

Difference equations (17) and (18) have a general solution:

$$E_\alpha(t) = A_\alpha + B_\alpha t + C_\alpha X^t$$

where A_α, B_α, C_α ($\alpha = 0, 1$), and X are constants that must satisfy

$$A_0 + B_0 = (1-r)A_0 + pA_1 \tag{19}$$
$$B_0 = (1-r)B_0 + pB_1 \tag{20}$$
$$C_0X = (1-r)C_0 + pC_1 \tag{21}$$
$$A_1 + B_1 = rA_0 + (1-p)A_1 + \frac{r}{r+p} \tag{22}$$
$$B_1 = rB_0 + (1-p)B_1 \tag{23}$$
$$C_1X = rC_0 + (1-p)C_1 \tag{24}$$
$$\tag{25}$$

Then

$$B_0 = \frac{rp}{(r+p)^2}; \qquad B_1 = \frac{r^2}{(r+p)^2}$$
$$\frac{rp}{(r+p)^2} = -rA_0 + pA_1. \tag{26}$$

Eliminating X from (21) and (24) leads to

$$p\left(\frac{C_1}{C_0}\right)^2 + (p-r)\left(\frac{C_1}{C_0}\right) - r = 0.$$

One solution of this equation, with the corresponding value of X, is:

$$\frac{C_1}{C_0} = -1, X = 1 - r - p$$

The second solution is constant in t, since the corresponding $X = 1$. It is thus already included in the A_0 and A_1 terms, and need not be treated separately.

By adding E_0 and E_1 and comparing with (16), we have

$$E_0(t) + E_1(t) = A_0 + A_1 + \frac{r}{r+p}t = \bar{n}(t) = \frac{r}{r+p}t \tag{27}$$

As a consequence, $A_0 + A_1 = 0$, so, from (26)

$$A_1 = -A_0 = \frac{rp}{(r+p)^3}$$

Finally,

$$\begin{aligned} E_0(0) &= 0 = A_0 + C_0 \\ E_1(0) &= 0 = A_1 - C_0, \end{aligned}$$

so

$$C_0 = \frac{rp}{(r+p)^3}$$

and the lemma is proved. We can now prove the theorem.

Theorem 1 *The variance of the number of parts produced during* $[0, t]$ *is given by*

$$V(t) = \frac{rp}{(r+p)^2}\left(\frac{2}{r+p} - 1\right)t + \frac{2rp}{(r+p)^4}(1-r-p)\left((1-r-p)^t - 1\right). \tag{28}$$

Proof: The variance is given by

$$V(t) = E(n(t)^2) - (\bar{n}(t))^2$$

The second term is known. Define

$$S(t) = E(n(t)^2) = \sum_{n=0}^{t} n^2\left(\pi(n,0,t) + \pi(n,1,t)\right)$$

Then it is straightforward, but tedious, to show that

$$S(t+1) = S(t) + 2rE_0(t) + 2(1-p)E_1(t) + \frac{r}{r+p}$$

$$= S(t) + 2\frac{r^2}{(r+p)^2}t - 2\frac{rp}{(r+p)^3}(1-r-p)\left((1-r-p)^t - 1\right) + \frac{r}{r+p}. \tag{29}$$

Equation (29) is a first order difference equation with initial condition $S(0) = 0$. The solution is

$$\begin{aligned} S(t) &= \left(\frac{r}{r+p}t\right)^2 + \frac{rp}{(r+p)^2}\left(\frac{2}{r+p} - 1\right)t \\ &\quad + \frac{2rp}{(r+p)^4}(1-r-p)\left((1-r-p)^t - 1\right). \end{aligned}$$

and the theorem is proved.